Ordered and Turbulent Patterns
in Taylor–Couette Flow

NATO ASI Series

Advanced Science Institutes Series

A series presenting the results of activities sponsored by the NATO Science Committee, which aims at the dissemination of advanced scientific and technological knowledge, with a view to strengthening links between scientific communities.

The series is published by an international board of publishers in conjunction with the NATO Scientific Affairs Division

A	Life Sciences	Plenum Publishing Corporation
B	Physics	New York and London
C	Mathematical and Physical Sciences	Kluwer Academic Publishers
D	Behavioral and Social Sciences	Dordrecht, Boston, and London
E	Applied Sciences	
F	Computer and Systems Sciences	Springer-Verlag
G	Ecological Sciences	Berlin, Heidelberg, New York, London,
H	Cell Biology	Paris, Tokyo, Hong Kong, and Barcelona
I	Global Environmental Change	

Recent Volumes in this Series

Volume 290—Phase Transitions in Liquid Crystals
        edited by S. Martellucci and A. N. Chester

Volume 291—Proton Transfer in Hydrogen-Bonded Systems
        edited by T. Bountis

Volume 292—Microscopic Simulations of Complex Hydrodynamic Phenomena
        edited by Michel Mareschal and Brad Lee Holian

Volume 293—Methods in Computational Molecular Physics
        edited by Stephen Wilson and Geerd H. F. Diercksen

Volume 294—Single Charge Tunneling: Coulomb Blockade Phenomena in Nanostructures
        edited by Hermann Grabert and Michel H. Devoret

Volume 295—New Symmetry Principles in Quantum Field Theory
        edited by J. Fröhlich, G.'t Hooft, A. Jaffe, G. Mack, P. K. Mitter, and R. Stora

Volume 296—Recombination of Atomic Ions
        edited by W. G. Graham, W. Fritsch, Y. Hahn, and J. A. Tanis

Volume 297—Ordered and Turbulent Patterns in Taylor–Couette Flow
        edited by C. David Andereck and F. Hayot

Series B: Physics

Ordered and Turbulent Patterns in Taylor-Couette Flow

Edited by
C. David Andereck and
F. Hayot

The Ohio State University
Columbus, Ohio

Plenum Press
New York and London
Published in cooperation with NATO Scientific Affairs Division

Proceedings of a NATO Advanced Research Workshop on
Ordered and Turbulent Patterns in Taylor–Couette Flow,
held May 22–24, 1991,
in Columbus, Ohio

NATO-PCO-DATA BASE

The electronic index to the NATO ASI Series provides full bibliographical references (with keywords and/or abstracts) to more than 30,000 contributions from international scientists published in all sections of the NATO ASI Series. Access to the NATO-PCO-DATA BASE is possible in two ways:

—via online FILE 128 (NATO-PCO-DATA BASE) hosted by ESRIN, Via Galileo Galilei, I-00044 Frascati, Italy.

—via CD-ROM "NATO-PCO-DATA BASE" with user-friendly retrieval software in English, French, and German (© WTV GmbH and DATAWARE Technologies, Inc. 1989)

The CD-ROM can be ordered through any member of the Board of Publishers or through NATO-PCO, Overijse, Belgium.

Library of Congress Cataloging-in-Publication Data

Ordered and turbulent patterns in Taylor-Couette flow / edited by C.
  David Andereck and F. Hayot.
      p.   cm. -- (NATO ASI series. Series B, Physics ; v. 297)
    "Proceedings of a NATO Advanced Research Workshop on Ordered and
  Turbulent Patterns in Taylor-Couette Flow, held May 22-24, 1991, in
  Columbus, Ohio"--Verso t.p.
    "Published in cooperation with NATO Scientific Affairs Division."
    Includes bibliographical references and index.
    ISBN 0-306-44238-8
    1. Vortex-motion--Congresses. 2. Fluid dynamics--Congresses.
  I. Andereck, C. David. II. Hayot, F. III. North Atlantic Treaty
  Organization. Scientific Affairs Division. IV. NATO Advanced
  Research Workshop on Ordered and Turbulent Patterns in Taylor
  -Couette Flow (1991 : Columbus, Ohio) V. Title: Taylor-Couette
  flow. VI. Series.
  QA925.073 1992
  532'.0595--dc20   92-18440
                                                                    CIP

ISBN 0-306-44238-8

© 1992 Plenum Press, New York
A Division of Plenum Publishing Corporation
233 Spring Street, New York, N.Y. 10013

All rights reserved

No part of this book may be reproduced, stored in a retrieval system, or transmitted in any form or by any means, electronic, mechanical, photocopying, microfilming, recording, or otherwise, without written permission from the Publisher

Printed in the United States of America

PREFACE

Seldom does a physical system, particularly one as apparently simple as the flow of a Newtonian fluid between concentric rotating cylinders, retain the interest of scientists, applied mathematicians and engineers for very long. Yet, as this volume goes to press it has been nearly 70 years since G. I. Taylor's outstanding experimental and theoretical study of the linear stability of this flow was published, and a century since the first experiments were performed on rotating cylinder viscometers. Since then, the study of this system has progressed enormously, but new features of the flow patterns are still being uncovered. Interesting variations on the basic system abound. Connections with open flows are being made. More complex fluids are used in some experiments. The vigor of the research going on in this particular example of nonequilibrium systems was very apparent at the NATO Advanced Research Workshop on "Ordered and Turbulent Patterns in Taylor-Couette Flow," held in Columbus, Ohio, USA May 22-24, 1991. A primary goal of this ARW was to bring together those interested in pattern formation in the classic Taylor-Couette problem with those looking at variations on the basic system and with those interested in related systems, in order to better define the interesting areas for the future, the open questions, and the features common (and not common) to closed and open systems. This volume contains many of the contributions presented during the workshop.

The organization of the contents is similar to that of the workshop itself. It begins with a detailed historical survey of the experimental aspects of the study of the Taylor-Couette system. This is followed by several contributions devoted to the classical Taylor-Couette problem, including numerical and experimental studies of Taylor vortices, wavy vortices, and spirals, and the phase dynamics approach to flows both near and far from onset. These papers highlight the fact that even the simplest form of the problem continues to provide a challenge to theory and experiment. The next section concerns Taylor-Couette flow, but with broken rotational symmetry. It has long been recognized that the basic system is very special in that it combines rotational symmetry with translational invariance. Recently several experiments have been undertaken to understand the effects on the flow patterns when the rotational symmetry is broken, either by adding a Coriolis force or by running the system horizontally with only a partially filled gap. In either case wholly new flow patterns have emerged and new routes to turbulence are seen.

Turbulence in Taylor-Couette flow and in plane Couette flow are covered in the next section, the first instance in this volume in which direct connections between flows in open and closed systems are discussed. In the following section, connections with more general theoretical modeling are made in papers on the Eckhaus and Benjamin-Feir instabilities, and on turbulence in the one-dimensional complex Ginzburg-Landau equation.

Several papers discussing variations on the basic system were presented. Throughflows have been imposed, both radially and axially, thus producing an effectively open system.

In contrast to most open systems, however, the modified Taylor-Couette system offers the experimentalist a very highly controlled environment for study. Other interesting variations described here include different geometries, such as rotating cones and cylinders, and magnetohydrodynamic instabilities with radial temperature gradients. Another paper in this group discusses the flow of a radically different fluid, superfluid helium, in an otherwise simple Taylor apparatus. The Taylor-Couette system is then actually serving as a testing ground not just for the solution of the governing equations, but for the very form of these equations.

The final group of papers deals with related open flow systems. Of particular importance is the interest in vortex formation and transition in flows through curved channels or over curved surfaces. The centrifugal instabilities in these cases are naturally related to the Taylor-Couette problem, although the spatial development is typically quite different in the open flows.

During a summary panel discussion a number of points were brought out regarding the status and future of work on the Taylor-Couette system and related pattern-forming systems. Motivation for further work remains strong:

- The Taylor-Couette system is a paradigm of pattern-forming systems
- It serves as a test bed for mathematical modeling of such systems, including symmetries, the identification of the correct normal form, the importance of nonlocal terms, the treatment of fluxes of various sorts, and the amplitude equations for various circumstances
- It is a problem in which detailed quantitative comparisons between theory and experiment are possible, partly because remarkably good laboratory control is possible
- It has long been a test bed for novel experimental techniques
- It is an appropriate test bed for computational fluid dynamics modeling
- It provides a suitable environment for new mathematical models of physical behavior such as that exhibited by complex fluids or superfluids.

There were several broadly defined open questions:

- Is fully developed turbulence universal, or is it special in this system?
- What can we learn from the Taylor-Couette system about the evolution of complexity in general systems?
- What can we learn about convective and absolute instabilities using variations of the basic system? What is the general relationship of the Taylor-Couette system to open flows?
- Are there new flow visualization or other diagnostic methods deserving of use?

It was clear that even though great progress has been made in understanding some of the basic instabilities and resulting flow patterns in the Taylor-Couette system, there remains a great deal of work to be done. We hope that this volume will serve to stimulate this line of research. To assist in this process we have included as an appendix a bibliography, prepared by Randall Tagg, that contains a very complete listing of references on the Taylor-Couette system and related problems.

We wish to thank the other members of the organizing committee, Russell J. Donnelly, Yves Pomeau and Daniel Walgraef, for their ideas and suggestions in making the workshop a success. We much appreciate the key speakers, including Donnelly, Phil Hall, Lorenz Kramer, Pomeau and Harry Swinney, who gave overview talks that broadened considerably the scope of the workshop and provided a context for our discussions. We also thank the North Atlantic Treaty Organization Scientific Affairs Division, the Ohio State University

Office of Research and Graduate Studies, the College of Mathematical and Physical Sciences, the Department of Physics and the Ohio State Fluids Research Institute for their generous financial support. Finally we thank Debra Dunson, Lynn McGraner and Natalie Novak for their dedication to the details of organizing the workshop and producing this volume.

Columbus, December 1991   C. David Andereck
                                                                 F. Hayot

CONTENTS

TAYLOR-COUETTE FLOW, EXPERIMENT AND THEORY

Evolution of Instrumentation for Taylor-Couette Flow 1
  R.J. Donnelly

Mode Competition and Coexistence in Taylor-Couette Flow 29
  J. Brindley and F.R. Mobbs

Low-Dimensional Spectral Truncations for Taylor-Couette Flow 43
  K.T. Coughlin and P.S. Marcus

The Couette-Taylor Problem in the Small Gap Approximation 51
  Y. Demay, G. Iooss, and P. Laure

Structure of Taylor Vortex Flow and the Influence of Spatial Amplitude Variations
  on Phase Dynamics 59
  D. Roth, M. Lücke, M. Kamps, and R. Schmitz

Chaotic Phase Diffusion Through the Interaction of Phase Slip Processes 67
  H. Riecke and H.-G. Paap

Phase Dynamics in the Taylor-Couette System 75
  M. Wu and C.D. Andereck

Spiral Vortices in Finite Cylinders 83
  E. Knobloch and R. Pierce

TAYLOR-COUETTE FLOW SYSTEMS WITH BROKEN ROTATIONAL SYMMETRY

A Model of the Disappearance of Time-Dependence in the Flow Pattern
  in the Taylor-Dean System 91
  L. Fourtune, I. Mutabazi, and C.D. Andereck

End Circulation in Non-Axisymmetrical Flows 99
  C. Normand, I. Mutabazi, and J.E. Wesfreid

Bifurcation Phenomena in Taylor-Couette Flow Subject to a Coriolis Force 107
  P.W. Hammer, R.J. Wiener, and R.J. Donnelly

Instability of Taylor-Couette Flow Subjected to a Coriolis Force 121
  R.J. Wiener, P.W. Hammer, and R. Tagg

Bifurcations to Dynamic States in Taylor-Couette Flow with External Rotation 131
  L. Ning, G. Ahlers, and D.S. Cannell

On the Stability of Taylor-Couette Flow Subjected to External Rotation 141
  M. Tveitereid, L. Ning, G. Ahlers, and D.S. Cannell

TURBULENCE IN TAYLOR-COUETTE AND PLANE COUETTE FLOW

Numerical Simulation of Turbulent Taylor-Couette Flow 149
  S. Hirschberg

Intermittent Turbulence in Plane and Circular Couette Flow 159
  J. Hegseth, F. Daviaud, and P. Bergé

INSTABILITIES, PATTERN FORMATION AND TURBULENCE IN MODEL EQUATIONS

On the Eckhaus and the Benjamin-Feir Instability in the Vicinity
  of a Tricritical Point 167
  H.R. Brand

Phase vs. Defect Turbulence in the 1-D Complex Ginzburg-Landau Equation 173
  A. Pumir, B.I. Shraiman, W. van Saarloos, P.C. Hohenberg, H. Chaté, and M. Holen

Double Eigenvalues and the Formation of Flow Patterns 179
  R. Meyer-Spasche

EXTENSIONS OF TAYLOR-COUETTE FLOW

The Effect of Throughflow on Rayleigh Benard Convective Rolls 187
  H.W. Müller, M. Lücke, and M. Kamps

Taylor Vortex Flow with Superimposed Radial Mass Flux 197
  K. Bühler

Vortex Patterns Between Cones and Cylinders 205
  M. Wimmer

Instability of Taylor-Couette Flow of Helium II 213
  C.J. Swanson and R. J. Donnelly

Effects of Radial Temperature Gradient on MHD Stability of Couette Flow
  Between Conducting Cylinders - A Wide Gap Problem 221
  H.S. Takhar, M.A. Ali, and V.M. Soundalgekar

OPEN FLOWS

Structure and Perturbation in Görtler Vortex Flow 245
  P. Petitjeans and J.E. Wesfreid

Transition to Turbulence in Görtler Flow 253
W. Liu and J.A. Domaradzki

Effect of Curvature Plane Orientation on Vortex Distortion in Curved
    Channel Flow 263
H. Peerhossaini and Y. Le Guer

Splitting, Merging and Wavelength Selection of Vortex Pairs in Curved and/or
    Rotating Channels 273
W.H. Finlay and Y. Guo

Transient, Oscillatory and Steady Characteristics of Dean Vortex Pairs in a
    Curved Rectangular Channel 281
P.M. Ligrani

On the Subharmonic Instability of Finite-Amplitude Longitudinal Vortex Rolls
    in Inclined Free Convection Boundary Layers 289
C.C. Chen, A. Labhabi, H.-C. Chang, and R.E. Kelly

Centrifugal Instabilities in Rotating Frame about Flow Axis 297
I. Mutabazi, C. Normand, M. Martin, and J.E. Wesfreid

APPENDIX

A Guide to Literature Related to the Taylor-Couette Problem 303
R. Tagg

INDEX 355

EVOLUTION OF INSTRUMENTATION FOR TAYLOR-COUETTE FLOW

Russell J. Donnelly

Department of Physics
University of Oregon
Eugene, Oregon 97403

1. INTRODUCTION

The organizers of this meeting discussed with me some time ago just what they would like to hear about. I recognized the wish to provide some long term perspective on our field, and I ventured to suggest an overview of experimental techniques might be of some interest. I would like to make the point that the study of Taylor-Couette flow has almost inevitably brought forth the highest level of experimental technique from the very beginning. I will try to illustrate that fact by showing you examples of a wide variety of instrumentation from the last 100 years of our subject.

My introduction to stability problems came as a graduate student at Yale. Onsager was interested in anything us low temperature students would do because of his invention of quantized vorticity in helium II. He gave me an article to look at on helium II in which he remarked that stability considerations, generally, depend on the kinematic viscosity of the fluid. I asked Lars what "stability" was, and he remarked that it is a field carried on by a "rare crew": C.C. Lin, G.I. Taylor and S. Chandrasekhar. He then proceeded to arrange for me to meet with them all. G.I. Taylor was first: he was coming to New York for a meeting, and Onsager and I went together to meet him. This began a friendship which lasted until Taylor's death many years later. C.C. Lin was at the same meeting, and I came to know Chandrasekhar as a faculty colleague at the University of Chicago over the period 1956 to 1966.

In my younger days participation in this field was not much larger than the groups attached to Taylor, Chandrasekhar and Lin. By the 1980's the field had grown to the point where scores of people attend a meeting such as this. It is interesting to ask what is the nature and significance of a field such as Taylor-Couette flow which has attracted the attention of giants in the past and continues to attract some of the best and brightest young investigators? First of all, the flow is simple in conception and a working apparatus can be put together in a day by almost anyone. Visualization of the flow produces striking and fascinating patterns which vary in complicated ways with changes in the rotation rates of the cylinders. Improvements to the basic apparatus are readily conceived. We shall see that early investigators Mallock and Couette built state-of-the-art viscometers, and investigators today bring the entire artillery of modern computer - controlled diagnostic techniques to bear on the problem at hand. The instabilities and flow patterns which emerge are challenging the most ingenious theorists to explain them.

2. THE EARLY DAYS

Our subject begins with Sir Isaac Newton (1642-1727) who in 1687 considered the circular motion of fluids in Book II of the *Principia* Section IX.[1] He begins with the definition of what is now called a Newtonian fluid with the following *Hypothesis*:

"The resistance, arising from the want of lubricity in the parts of a fluid, is *caeteris paribus*, proportional to the velocity with which the parts of the fluid are separated from each other." Today we would say the viscous stresses are proportional to the velocity gradient for a Newtonian fluid.

In proposition 51, Newton says, "If a solid cylinder infinitely long, in an uniform and infinite fluid, revolve with an uniform motion about an axis given in position, and the fluid be forced round by only this impulse of the cylinder, and every part of the fluid continues uniformly in its motion, I say that the periodic times of the parts of the fluid are as their distances from the axis of the cylinder." (Figure 1).

Corollary 2. "If a fluid be contained in a cylindric vessel of an infinite length, and contain another cylinder within, and both the cylinders revolve about one common axis, and the times of their revolutions be as their semidiameters, and every part of the fluid continues in its motion, the periodic times of the several parts will be as the distances from the axis of the cylinders."

The flow in Figure 1 is about a centrally rotating cylinder. The flow imagined in Corollary 2 is that which results if the flow is bounded by a second concentric outer cylinder. This reference must be one of the earliest discussing flow in the annulus between rotating cylinders.

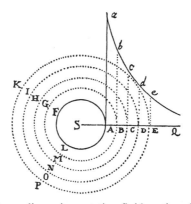

Figure 1. Newton's figure discussing rotating fluid motion about a cylinder AFL.

Sir George Gabriel Stokes (1819-1903), one of the great pioneers of theoretical fluid dynamics and Newton's successor in the Lucasian chair at Cambridge, writing 161 years later in the *Transactions of the Cambridge Philosophical Society* in 1848 says in part, "Let us now consider the motion of a mass of uniform inelastic fluid comprised between two cylinders having a common axis, the cylinders revolving uniformly about their axis and the fluid being supposed to have altered its permanent state of motion...."[2] He then solves for the velocity of the fluid and continues:

"The case of motion considered in this article may perhaps admit of being compared with experiment, without knowing the conditions which must be satisfied at the surface of a solid. A hollow, and a solid cylinder might be so mounted as to admit of being turned with different uniform angular velocities round their common axis, which is supposed to be vertical. If both cylinders are turned, they ought to be turned in opposite directions, if only one, it ought to be the outer one; for if the inner one were made to revolve too fast, the fluid near it would have a tendency to fly outwards in consequence of the centrifugal force, and eddies would be produced. As long as the angular velocities are not great, so that the surface of the liquid is very nearly plane, it is not of much importance that the fluid is there terminated; for the conditions which must be satisfied at a free surface are satisfied for any section of the fluid made by a horizontal plane, so long as the motion about that section is supposed to be the same as it would be if the cylinders were infinite. The principal difficulty would probably be to measure accurately the time of revolution, and distance from the axis, of the different annuli. This would probably be best done by observing motes [dust particles] in the fluid. It might be possible also to discover in this way the conditions to be satisfied at the surface of the cylinders; or at least a law might be suggested, which could be afterwards compared more accurately with experiment by means of the discharge of pipes and canals."

Several points in the discussion are worth noting. First, we see that Stokes is concerned that the boundary conditions at the solid surfaces are unknown. This is hardly trivial -- decades were to elapse before the no-slip condition for a fluid at a solid wall was universally accepted. Indeed, it was Taylor's analysis of rotating cylinder flow that settled the matter. Second, the realization that the inner cylinder in rotation would be the least stable, and would produce the eddies we know as Taylor vortices, is surely the intuition of genius. Taylor vortices were not noted for another 75 years (1923). Third, Stokes is concerned with the boundary conditions at the free surface of partially filled cylinders. Fourth, he is remarking on the use of a tracer to mark the flow -- something we do easily today with a Laser Doppler Velocimeter (LDV) apparatus which measures fluid velocity by observing the Doppler shift of scattered light from small particles seeding the flow.

When the equations of motion for a viscous fluid were formulated by Navier (1823) and Stokes (1845) a considerable debate arose on how to best measure viscosity. It did not take long before experimentalists interested in this question realized that there are two forms of fluid motion, which today we would term (roughly) laminar and turbulent. Since turbulent flow is not described by simple integrals of motion, deducing the viscosity from turbulent observations usually gives anomalously high values of viscosity.

Max Margules was born in Brody (Galizien) Austria April 23, 1856 and died in 1930. He was trained in Vienna in theoretical physics but became perhaps the first theoretical meteorologist. He began meteorological studies at the end of the 1880's and worked in the subject until 1906. Margules appears to have been the first person to seriously propose constructing a rotating cylinder viscometer. In 1881 Margules wrote[3]:

"Suppose a cylinder hangs vertically on a vertical axis which rotates uniformly. Suppose the cylinder is immersed in a coaxial cylindric container, which contains the fluid to be investigated. Then, due to the friction of the fluid, the relative position of the cylinder with respect to the axis during the rotation will be different from the one in the state of rest. Now one can measure the torque by means of a simple apparatus which results in a torsion angle of equal magnitude; this way one measures the resistance of the fluid against the rotation of the cylinder. The latter motion we assume to be stationary... Even in 1930, papers in the *Physical Review* refer to the "Margules rotating cylinder type viscometer."[4]

Seven years after this publication, two young men, Arnulf Mallock and Maurice Couette began to build rotating cylinder viscometers and made preliminary announcements in London and Paris. Of these apparently only Couette knew of Margules' paper.

On November 30, 1888, Lord Rayleigh, the Secretary of the Royal Society of London communicated a paper "Determination of the Viscosity of Water" by Henry Reginald Arnulph Mallock (1851-1933).[5] Arnulph Mallock was a nephew of William Froude, the famous naval architect. He studied at Oxford and after graduating helped Froude build the original ship model tank: a trough of water used to test ship models by towing. In 1876 Mallock went as an assistant to Lord Rayleigh, being especially valuable to Rayleigh because he was a skilled instrument builder. Mallock described experiments conducted during April and May of 1888 on the viscosity of water using a pair of concentric cylinders with the outer one driven and the inner one suspended on a torsion fiber as shown in Figure 2 on the left. He ventured that the experiments might be "of some interest on account of the newness of the method employed..." In July of 1895 Lord Kelvin, President of the Royal Society, communicated a full paper by Mallock.[6] In this pioneering paper, Mallock describes (among other topics) various precautions he took so that "...the water in the annulus between E and A is very nearly in the same condition it would be if E and A were infinitely long..." The last statement is a precursor of much current discussion in classical Couette flow concerning the influence of end conditions on the flow.

Mallock's apparatus was designed to operate with three different cylinder arrangements. The first one, shown on the left, rotates outer cylinder, E, measuring the torque on the inner cylinder A. Another cylinder, G, surrounds E. The gap between E and G is filled with water, as is the inner cylinder A, and thermometers are placed in these two regions. The temperature in the annulus between A and E is taken to be the mean of the two thermometer readings. Air was trapped in a region at the bottom of A so that fluid torque was exerted only on the cylindrical wall of A. The short cylinder K is stationary. Mercury was placed between K and E in an attempt to produce end conditions with the same velocity distribution as occurs in the water being measured. The telescope T reads the displacement of the calibrated circular disk attached to the upper stem B, which supports cylinder A.

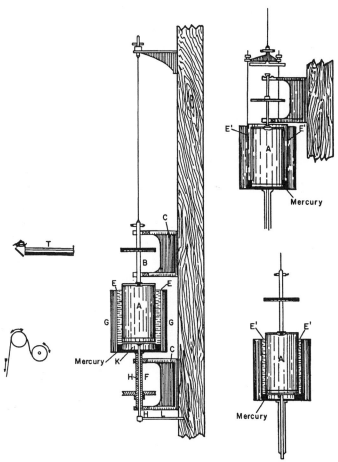

Figure 2. Arnulph Mallock's apparatus for his pioneering investigations on the flow of fluids between concentric cylinders. The parts of the apparatus are discussed in the text.

The upper right figure shows an arrangement to rotate the inner cylinder and measure torque on a suspended outer cylinder. The arrangement shown lower right was used to repeat the experiments of the outer cylinder rotation using different size cylinders. Overall, the cylinders were sizeable: $R_1 = 7.632$ cm, $R_2 = 8.687$ (for E) and 9.943 cm (for E'), h = 24.5 cm. P was a paper recorder for data.

Mallock found that when the inner cylinder was rotating, the torque and angular velocity were not linearly related and concluded (incorrectly) that such a flow was always unstable. We note in retrospect that Mallock's lowest speed of rotation (~ 2 RPM) is larger than the critical value calculated to produce Taylor vortices for his cylinder sizes. With the outer cylinder rotating he found the flow to be stable at low rotation and unstable at high rotation rates.

Mallock's experiments were watched with great interest by Kelvin, who was thinking about stability theory at the time. In a letter to Lord Rayleigh dated July 10, 1895, he wrote:[7]

"On Saturday I saw a splendid illustration by Arnulph Mallock of our ideas regarding instability of water between two parallel planes, one kept moving and the other fixed. Coaxal cylinders, nearly enough planes for our illustration. The rotation of the outer can was kept very accurately uniform at whatever speed the governor was set for, when left to itself. At one of the speeds he shewed me, the water came to a regular regime, *quite smooth*. I dipped a disturbing rod an inch or two down into the water and immediately the torque increased largely. *Smooth* regime would only be reestablished by slowing down and bringing up to speed again, gradually enough."

"Without the disturbing rod at all, I found that by resisting the outer can by hand somewhat suddenly, but not very much so, the torque increased suddenly and the motion became visibly turbulent at the lower speed and remained so..."

M. Maurice Couette (1858 - 1943) announced the first experiments with his viscometer in Paris in 1888.[8] His most important conclusion was that there are two forms of fluid motion, one given by exact integrals of the equations of motion and one, at higher speeds, which does not conform to the integrals of motion. Couette was aware of Osborne Reynold's pioneering studies on turbulence in flow through pipes which were published in 1883. The circular geometry, however, is a fundamentally different flow.

In 1890 Couette published a lengthy study of viscosity using a pair of cylinders with the outer rotating and the inner suspended on a fiber to measure torque (this was noted as a thesis of the faculty of Sciences Of Paris, No. 693).[9] The thesis, done in the laboratory of M. Lippmann, contained a study of flow through tubes to determine viscosity as well. Today such rotating cylinder viscometers are known as Couette viscometers, even though Mallock's was clearly independently developed at about the same time.

We reproduce a cross section of Couette's large and impressive apparatus in Figure 3 and a plan and elevation in Figure 4. Couette's cylinders had radii $R_1 = 14.39$ cm, $R_2 = 14.64$ cm, $h = 7.91$ cm, and had short guard cylinders at each end fixed to a tripod M. The tripod rested on three heavy piers. The outer cylinder V was rotated by means of a pulley and the inner cylinder was suspended by a steel torsion fiber. The base of the apparatus was a 50 x 50 cm square of cast iron. Small torques were measured by the deflection of the inner cylinder, larger torques were balanced by means of an Atwood's machine attached to a pulley on the suspension. Couette showed that the viscosity of water is constant until some critical rotation rate which corresponds to a Reynolds number $R_c = \Omega_c R_2 d / \nu \sim 2000$ where (in modern notation) Ω is the angular velocity of the cylinder, R_2 the radius, d the gap between cylinders and ν is the kinematic viscosity. He used his instrument to measure the viscosity of air and reported a value of 179 $\mu\rho$ at 20°C compared to 182 $\mu\rho$ obtained by Bearden in 1939.[10] He was able to record the onset of turbulence in the experiment with air as well as water.

We should observe here that both Mallock and Couette were great instrument builders. The lengthy description of the construction of Couette's viscometer is impressive even today. Their skills enabled them to build some of the most precise instruments seen to that date. Their method of approaching their research has been continued by others over the years with many instruments establishing new standards of measurement capability. Couette was also a competent theorist: he was the first to consider the eccentric cylinder problem in an effort to estimate the errors in viscosity which would result if the suspended cylinder were not concentric.

Couette's name soon came to be associated with the flow. The well known book *Hydrodynamics* which is a reprint of a 1932 National Research Council report, shows that Mallock, Couette and Margules were well-known in those days.[11] Furthermore, the literature gives frequent reference to "plane Couette flow" as the flow between two planes, with one in motion. This is also true in the gas dynamic literature.[12]

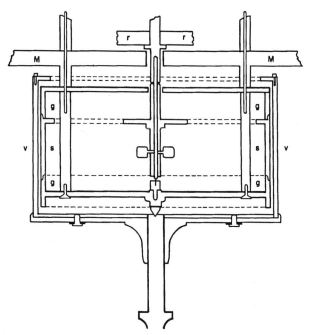

Figure 3. Cross section of Couette's apparatus, taken from his 1890 thesis. The fixed cylinders g are guards, and s is the suspended cylinder.

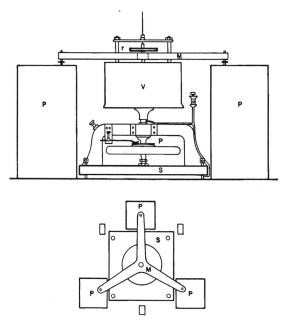

Figure 4. Plan and elevation of Couette's massive apparatus. P is a set of three heavy piers, M is an iron tripod suspension, V is the outer cylinder, a heavy brass vase, S is a 50 cm x 50 cm base, G is a greaser for oil, A is an arcade, P is a grooved pulley. The apparatus was driven by a battery-powered DC motor. The elevation is reduced by a scale of 10 and the plan by a scale of 20.

After these enterprising beginnings, the field became quiescent for almost 30 years, until Geoffrey Ingram Taylor (1886 - 1975) took up the problem. Taylor's 1923 paper contains an examination of the linear stability theory for the general case of viscous flow with both cylinders in rotation in the same and opposite directions.[13] Taylor's theoretical stability diagram for the flow was a *tour-de-force* considering the lack of computers which are so much a part of today's research. The paper also contains an account of Taylor's experimental apparatus which used ink visualization, and the results of measurements and photographs of patterns of the unstable flow for the first time. This paper, published in the *Philosophical Transactions of the Royal Society of London*, can fairly be called one of the most influential investigations of twentieth century physics. The correspondence between theory and experiment obtained by Taylor for the stability rested in an important way on the no-slip boundary condition for the flow at the boundaries. This success was taken by many as perhaps the most convincing proof of the correctness of the Navier-Stokes equations and of the no-slip boundary condition for the fluid at the cylinder walls.

In the 1950's Subrahmanyan Chandrasekhar ("Chandra"), one of the greatest theorists of all time, undertook a comprehensive study of hydrodynamic stability, and in his typical fashion made many new contributions to the field. He then synthesized the whole to write a massive treatise called *"Hydrodynamic and Hydromagnetic Stability"*[14]. His book included basic discussions of hydrodynamic stability, a major discussion of Rayleigh-Bénard convection and Taylor-Couette flows. In each case he discussed the effect of a magnetic field if a conducting fluid were used and in a number of cases examined the effect of rotation on the stability as well. In this work he addressed a number of generalizations of Taylor-Couette flow. Chandra's book brought our understanding of the experimental and theoretical situation on Taylor-Couette flow up to date, and made possible the next generation of experiments and theories which followed some years later.

These new experiments and theories stress the understanding of flows well beyond the onset of instability where finite amplitude flows, further bifurcations, chaos and turbulence occur with ever-changing flow patterns.

Chandra, a good friend of Taylor and now a Nobel laureate, is still working at the University of Chicago. He is the subject of a recent biography.[15]

3. TORQUE MEASUREMENTS

The torque G transmitted to a length h of one cylinder as the result of rotation of either is given by

$$G = \frac{4\pi\eta R_1^2 R_2^2 h (\Omega_1 - \Omega_2)}{R_2^2 - R_1^2}$$

Here R_1 and R_2 are the radii of the inner and outer cylinders, Ω_1 and Ω_2 are the corresponding angular velocities, h is the length of the suspended cylinder and η the viscosity. Thus a determination of the torque G gives an absolute measure of viscosity η, without need of calibration. In a torsion apparatus, $G = \kappa\phi$, where κ is the torsion constant and ϕ the deflection in laminar flow. G/Ω is a constant, or nearly so.

An interesting and large torque apparatus was built by G.I. Taylor about 1936 and later modified to take Pitot tube measurements of the velocity distribution in the gap (see Figures 5 and 6).[16] Taylor investigated the turbulent velocity distribution in the gap by pitot tube measurements and compared the results to his torque measurements. No one appears to have pursued such measurements further.

When efforts were first being made to understand the viscosity and superfluidity of ^{4}He, a dominant method of determining viscosity was to measure the damping of an oscillating disk. The observed damping depends on the product $(\eta\rho_n)^{\frac{1}{2}}$. Hollis Hallett realized that a rotating cylinder viscometer can give the absolute viscosity (if the flow is stable) and designed the viscometer of Figure 8 in the classically stable configuration of a rotating outer cylinder.[17] His basic design was followed by Donnelly[18] who built an apparatus with the inner cylinder rotating in the hopes of studying the Taylor-Couette problem in helium II (Figure 9). Both the Hollis Hallett and Donnelly viscometers were based, in part, on Bearden's design.

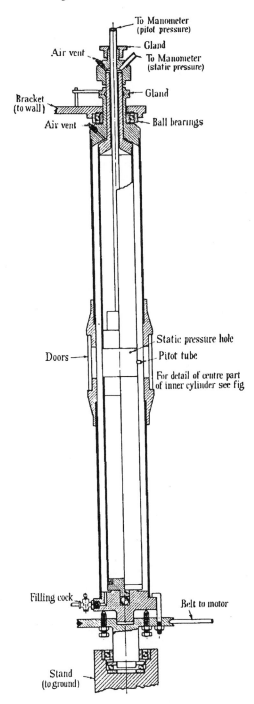

Figure 5. Apparatus built by G.I. Taylor about 1936 used to measure torque on the inner cylinder and modified to measure the velocity distributions between cylinders by means of a traversing pitot tube. The outer cylinder was 8.11 cm inside diameter and 84.4 cm long and made of brass. The diameter of the inner cylinder was 5.02 cm interchangeable with one 6.27 cm diameter.[16]

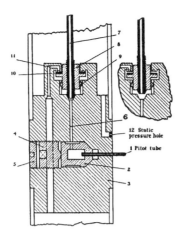

Figure 6. Detail of the Pitot tube design used by Taylor.[16] The tube had to be positioned for each radial distance and the apparatus refilled.

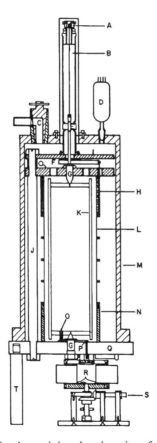

Figure 7. Bearden's viscometer for determining the viscosity of air.[10] A, adjustable torsion wire support; B, torsion wire; K, inner cylinder rotating on centers; G; L, suspended cylinder; N, guard cylinders; M, bell jar; R, magnetic drive. The diameter of the inner cylinder was 12.44224 cm and suspended cylinder was 14.95298 cm in diameter. The length of the suspended cylinder L was 26.04583 cm. The experimental details of this precision apparatus are truly remarkable. The material was Dow metal.

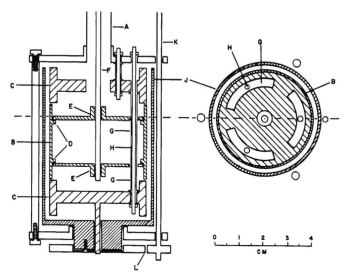

Figure 8. Viscometer designed by Hollis Hallett at Cambridge in 1953 with a rotating outer cylinder and suspended inner cylinder.[17] $R_1 = 1.9913$ cm and $R_2 = 2.0970$ cm. The inner cylinder was 3 cm long and had guard cylinders. This apparatus was used to show that helium II is unstable to rotation of the outer cylinder as well as the inner cylinder in contrast to classical fluids.

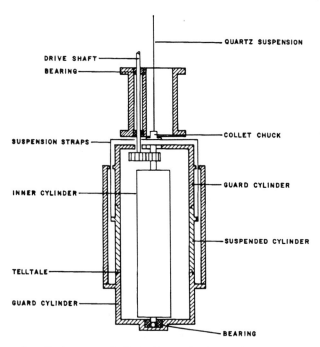

Figure 9. Diagram of small viscometer built by Donnelly for studies in classical fluids and liquid helium.[18] The radius of the outer cylinder is $R_2 = 2.00023$ cm. The length is 10 cm and the suspended cylinder 5 cm. The material was Dow metal.

Donnelly's viscometer was designed to work in both cryogenic liquids and classical liquids, as shown in Figure 10. The drive consisted of Graham variable speed transmissions and sets of reducing gears to bring motor speeds to the desired range. The torque was determined by a "null" method, where the torsion head was rotated to bring the outer cylinder to a standard position of zero torque as determined by an interferometer. This determined the angle of deflection ϕ. The lamp, mirror and photocell on the inner cylinder drive determined the period of rotation P of the inner cylinder through the use of an electronic counter. The product of ϕ P determined the viscosity of the fluid in laminar flow and the onset of instability as shown in Figure 11.

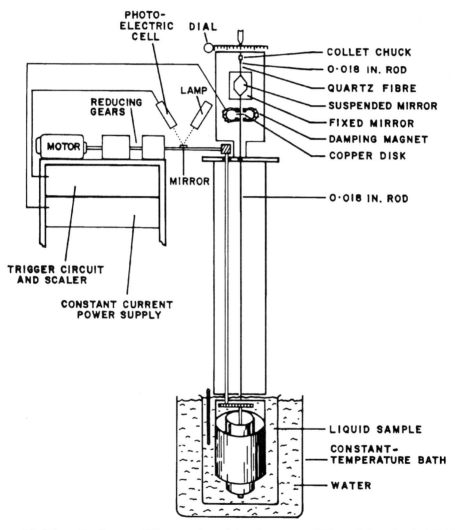

Figure 10. Schematic diagram of the operation of the viscometer of Figure 9 in a classical fluid.[18] The reference point for torque measurements was established by white light fringes from an interferometer consisting of a thick glass plate half-silvered on the front and silvered on the back face, attached to the torsion head and a suspended mirror, silvered on both sides, below the quartz suspension fiber.

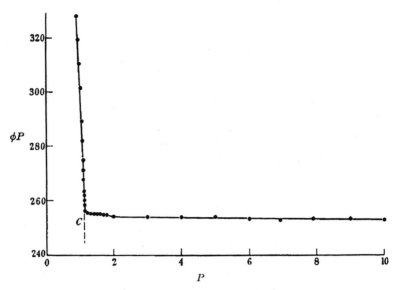

Figure 11(a). Plot of the effective viscosity ϕP as a function of the period of rotation P. The torque $G = K\phi$ where K is the torsion constant of the fiber used in ϕ the angle of deflection. The ratio G/Ω_i is proportional to η in laminar flow, but rises when instability sets in. In our early work it was easier to measure period P, where $P = 2\pi/\Omega$, and hence ϕP is proportional to effective viscosity. The onset of instability is at C which is the calculated value.[18]

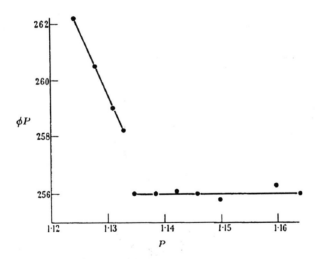

Figure 11(b). Detail of measurements about the critical speed showing that the onset of instability is characterized by the discontinuity in the effective viscosity.[18]

The success in determining the onset of Taylor vortices for ordinary fluids was exploited by Donnelly and Ozima[20] who constructed a viscometer to operate with mercury in a magnetic field. The density of mercury is so large that the outer cylinder would ordinarily float. Donnelly and Ozima overcame this by a second torsion fiber on the bottom of the apparatus (Figure 12).

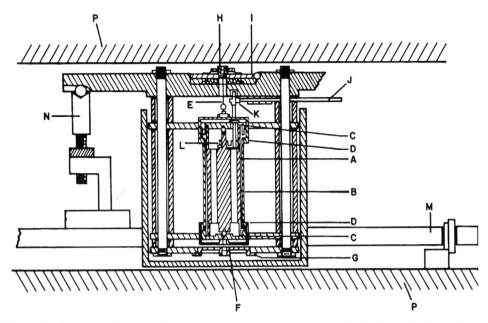

Figure 12. Apparatus for measuring torque from mercury in a magnetic field built by Donnelly and Ozima in 1961.[20] The suspension consisted of two torsion wires, the lower of which prevents the suspended cylinder from floating. A, inner cylinder; B, outer cylinder; C, guard cylinders; D, Suspension Straps; E, torsion wire with mirror at bottom; F, lower suspension wire; I, worm and gear; J, drive shaft; K, helical gears; N, levelling screws; P, cyclotron magnet pole faces.

Although the primary reason for building rotating cylinder viscometers in the first place was the determination of viscosity, the work of both Mallock and Couette showed that there is a remarkable difference in stability on rotating the outer and inner cylinders. Careful torque measurements can yield important information about the flow:

(1) determination of the onset of Taylor vortices by examination of the "effective viscosity," that is the ratio of torque to angular velocity.
(2) determination of the nonlinear growth of Taylor vortices beyond the onset of instability.[20] (Donnelly and Simon)
(3) observation of the onset of the wavy mode [21] (Allen Cole)
(4) relative stability of rotating inner and outer cylinders
(5) use with unusual fluids: mercury (Figures 8 and 12), liquid helium (Figure 8), non-Newtonian fluids.

There is an unresolved experimental problem in viscometry worth noting here.[22] The most precise viscometric determination of the viscosity of air was determined by Bearden using the instrument shown in Figure 7. He showed that the effective viscosity is independent of Reynolds number to about 1 part in 18,000. In our viscometers of Figs.8 and 12, the effective viscosity systematically increases with Reynolds number (Figure 10 (a)) -- an effect we attribute to end effects. At present it is completely unknown why Bearden's data, done with an apparatus of modest aspect ratio, shows such a constant effective viscosity. The issue is important because it lies at the heart of the use of a Couette viscometer to measure absolute viscosity.

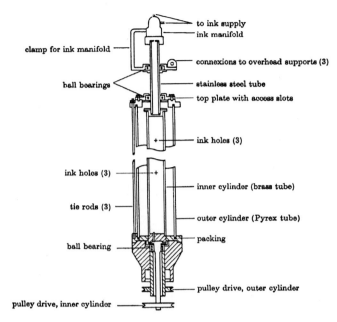

Figure 13. Apparatus built in the late 1950's by Donnelly and Fultz at the University of Chicago.[23] The radius of the glass outer cylinder is $R_2 = 6.2846 \pm .006$ cm. The tube is 94 cm long. Ink was supplied to two sets of holes through a manifold detailed in Figure 14.

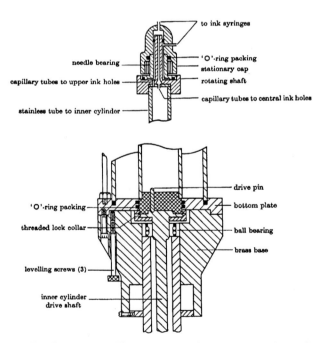

Figure 14. Details of the ink manifold and base of the apparatus shown in Figure 13.

4. FLOW VISUALIZATION BY INK

G.I. Taylor's original apparatus used ink as a tracer and Dave Fultz and I used the same technique in building our apparatus in the late 1950's. The glass outer cylinder in the apparatus of Figure 13 is remarkable for its accuracy and stability. It started as a piece of precision bore glass and was made even more precise by constructing an iron hone used to true the inside surface, followed by polishing. An average of 144 measurements of the inside diameter yielded a radius $R_2 = 6.2846 \pm 0.006$ cm. This apparatus is at the University of Texas, Austin.

The ink method can produce some beautiful photographs, such are in Chandrasekhar's book. But as a long series of measurements are made, the fluid becomes increasingly turbid and the ink can eventually cause a change in viscosity.

A modern variant on ink tracers is the Baker technique [24] where an electrical current causes a PH change in an indicator fluid mixture. A recent successful use of this technique is in a paper by Park, Barenghi and Donnelly on the subharmonic destabilization of flow near an oscillating cylinder. [25] This technique is likely underutilized: it is simple, does not change the properties of the fluid no matter how long it is used and the release of color can be electrically programmed in various useful ways.

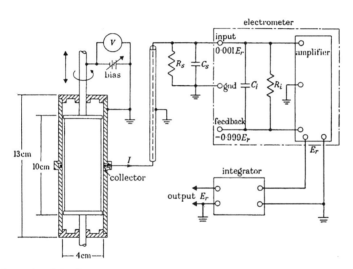

Figure 15. Schematic of the "ion" apparatus developed by Donnelly at the University of Chicago and which gives a response linear in the radial component of fluid velocity in a thin electrical boundary layer near the walls.[27]

5. FLOW VISUALIZATION BY ELECTROCHEMICAL TECHNIQUES

Figure 15 shows a simple technique for studying Taylor vortices. The essence of the method is to use a fluid of very low conductivity such as freon. Minute impurities (probably water) cause the conductivity to become finite and an electrical double layer forms on both cylinders to prevent electrical current from flowing indefinitely: the Fermi surface of the metal boundaries will not coincide with the energy of solvated impurities. When Taylor vortices begin, they make one electrical boundary layer thinner and the opposite boundary, thicker, resulting in a current flow in the circuit. The theory has been worked out by Donnelly and Tanner, and shows that the current is linearly proportional to the radial component of the Taylor flow in the boundary layers.[27] This was used to establish the "Landau law" that the equilibrium amplitude of disturbances grows as the square root of the difference between the Reynolds number and the critical Reynolds number.[28]

The use of the ion technique to visualize vortices quantitatively is shown in Figure 16.[26] An electrical elevator device drew the outer cylinder slowly upward. The current collected on an insulated ring faithfully reproduced the vortex flow. The circular collector will not, of course, reproduce wavy vortex flow. This requires a point collector. An apparatus using point collectors designed by Bob Walden for his thesis is shown in Figure 17.[29]

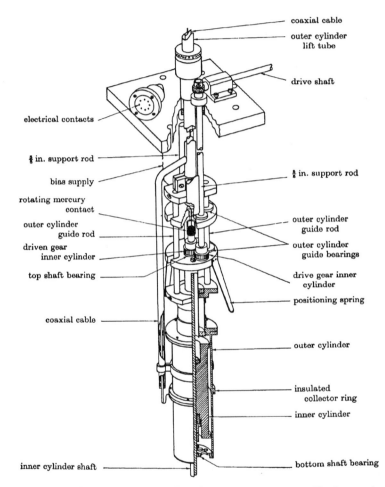

Figure 16. Implementation of the ion technique in an apparatus to scan Taylor vortices by drawing up the outer cylinder slowly.[26]

The ion technique is attractive in being low in cost and proportional to the radial component of vortex flow. It is not, however, easily calibrated for absolute response, and it is subject to drifts in sensitivity if the fluid is exposed to air.

It is amusing to note that we generally put gold plate on our apparatus to minimize stray potentials. After some months of running with the ion technique, the Taylor vortices etch the gold, leaving a permanent vortex pattern!

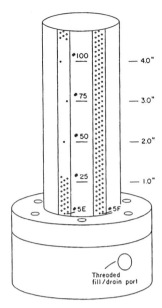

Figure 17. An ion apparatus built at the University of Oregon having about 900 local probes allowing local measurements to be made anywhere in the flow.[29]

6. FLOW VISUALIZATION BY PARTICLES

Reflecting flakes

In 1956 Schultz-Grunow and Hein published some beautiful pictures of Taylor vortices using aluminum paint pigment powder: something one can get from a paint store.[30] The aluminum flakes align in a shear flow. While the pictures were very nice, I don't think many of us appreciated right away what a revolution this really was. Here was a marker which didn't change the fluid properties with time and which allows both spatial and temporal patterns to be observed. More recently fish scales (Kalliroscope) have been used the same way. Matisse and Gorman have written a nice article on the use of Kalliroscope in fluid mechanics.[31] The flow patterns obtained from a tracer can be observed photoelectrically to produce time series of the flow and is a common technique today.[32] Time series and spatial patterns are correctly and sensitively observed, but there is no known way to deduce velocities from Kalliroscope measurements.

Two questions should be asked about Kalliroscope. First: does Kalliroscsope modify the flow? We believe it is important to measure the viscosity and temperature dependence of viscosity of the fluid mixed with Kalliroscope in the concentration actually used in the experiment. Except for flows so slow that gravitational sedimentation occurs, Kalliroscope appears to be relatively trouble free. It does have to be renewed periodically, and occasionally sticks to the boundaries of the apparatus. Second: what is the relationship of the reflectance to actual velocities in the flow? We have the impression that Kalliroscope has a sensitivity comparable to LDV but that the response saturates rapidly.

The limitation of measuring reflectance at a point can be overcome by scanning the light source and detector, or by using an array detector [33] or most recently by using a video camera and "frame grabber" board in a computer to record the entire flow and display it in many useful ways.[34] There will be examples of this technology in other papers in this volume.

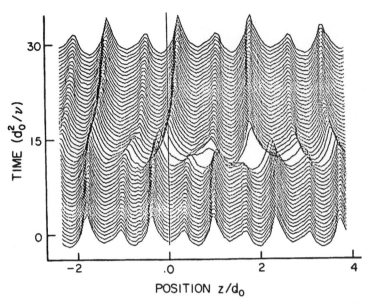

Figure 18. An example from Ref. 34 of the use of a frame grabber to produce a picture of time evolution of Taylor vortex flow with a spatial ramp illustrated in Figure 20. The travelling wave of vortex pairs in the ramped section is apparent. Vortex pairs are lost periodically in time and space.

Laser Doppler Velocimetry

Smaller particles (not flakes) are used as seeds in the flow for Laser Doppler Velocimetry. LDV observations, on the other hand, yield quantitative flow velocities at a point, but are less useful for observation of a pattern because of the large number of point measurements required to reveal the pattern. This technique has been discussed in detail elsewhere, but its chief value is quantitative flow measurements at a point and in one dimension.[35] Three dimensional instruments are coming on the market but are formidably expensive at the moment.

Particle tracking

Looking to the future, there are experimental programs going on in wind tunnel work to do "particle tracking".[36] Here the flow is seeded by reflecting particles and illuminated by laser flashes a short time apart. With fast digital processing it is possible to obtain two dimensional maps of perhaps 10,000 flow vectors at once. These results are quantitative.

7. TEST FLUIDS

Water

For a crude experiment, "water is water". For demanding applications, however, there is no substitute for using distilled water, which among other virtues has no dissolved air. When distilled water is used, simply measuring the temperature is enough to fix the kinematic viscosity.

Glycerine & water

A simple and inexpensive way to achieve higher viscosities than water is to use glycerine-water solutions. Tables of the viscosity of these mixtures are available in the AIP Physicist's Desk Reference.[37] When Kalliroscope is added to these mixtures, there is no substitute for measuring the viscosity at several temperatures and providing a way to interpolate the results. We use calibrated Cannon-Fenske viscometers, which are available in a choice of viscosity ranges. Glycerine-water solutions are hydroscopic and it is necessary to measure viscosity and temperature dependence of viscosity fairly often.

Solvents

If a measurement needs a wide range of Reynolds numbers, the lighter organic compounds are useful. We have used carbon tetrachloride, benzene, the freons, acetone, and Taylor has reported using gasoline. Practically any Newtonian fluid can be used in Taylor-Couette experiments if care is taken in determining the viscosity.

Mercury

Mercury is an attractive fluid to use for hydromagnetic experiments. While there are fears about using free mercury, in fact there is little hazard because the droplets become surface contaminated and lose their vapor pressure within a few minutes of falling on the floor. On the other hand, even a drop of mercury in contact with a hot surface can generate high vapor pressure and an accompanying health hazard.

One problem we encountered was in hydromagnetic Couette flow. We found essentially no difference between plastic cylinders and stainless steel cylinders.[38] We gained the impression that metal cylinders in contact with mercury rapidly acquire a non-conducting coating which prevents them from serving as conducting cylinders with the handbook conductivity.

Liquid sodium can be used for such experiment, but its dangers are so well known that I have not seen a reference to its use in our field.

Liquid helium and cryogenic liquids

Liquid and gaseous nitrogen and helium are attractive test fluids because of their low kinematic viscosity. Both Hallett's and Donnelly's viscometers were designed for low temperature use.

Liquid helium below the lambda transition has very special properties such as nearly infinite thermal conductivity and the ability to support novel wave forms such as second sound.

The first apparatus to use second sound in Taylor-Couette flow of helium II is shown in Figure 19.[39] Here second sound transducers were used to probe a flow of variable aspect ratio. Unfortunately, the authors were not aware of the existence of quantized vortices in their flow, and so missed an important discovery reported simultaneously by Hall and Vinen in Cambridge. For a discussion of vortices in helium II and their study by second sound and ions, see a recent book by the author.[40] A review article on Taylor-Couette flow in helium II has been prepared by Donnelly and LaMar.[41]

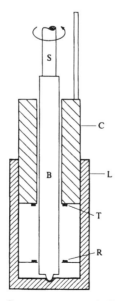

Figure 19 The first helium II Taylor-Couette apparatus built at Yale by Wheeler, Blakewood and Lane[39]. T and R are second-sound pulse transmitters and receivers. S is a drive shaft. B the inner cylinder. L the outer cylinder. C an adjustable plug.

8. DESIGN OF APPARATUS

The role of precision machining and highly precise measurements can scarcely be overemphasized in the study of Taylor-Couette flow. The care and expertise exercised by Mallock and Couette has been continued through the intervening century, and today these experiments, together with their close allies, the Bénard experiments, are examples of the very latest and most sophisticated methods in modern condensed matter physics.

Motor drives and turntables

Taylor-Couette experiments require drives capable of a wide range of speeds and extraordinary stability. Dave Fultz introduced us to the use of Graham variable speed transmissions driven by synchronous motors. When warmed up, which took about one-half hour, these drives would maintain speed accurate to better than 1/10 % indefinitely. Even with the variable speed transmission, however, trains of variable reducing gears sets were often still required.

Some laboratories have built turntables to do experiments with rotating fluids. These can be very convenient and amount to a tradeoff between the time and expense to build them, and their ease of use in practice. For example the larger tables can easily mount the apparatus and preliminary signal processing electronics whose power is supplied through large slip rings. Slip ring systems used to be messy mercury troughs, but modern coin-silver slip rings have great reliability and low noise.

Today, there is widespread use of stepping motors which can be driven from computers and can essentially be made to produce almost any rotation with high accuracy. We have even used stepping motors to produce stochastic stirring of a Bénard cell.[42] Note that stepping motors and controls are not constructed for low electromagnetic interference and can be an electrical nuisance near sensitive equipment.

A new branch of investigation in Taylor-Couette flow occurred in an investigation of the effect of modulating the rate of rotation of the inner cylinder.[43] Since Graham transmission were the best available in the 1960's we used one Graham transmission to vary the speed of another. The story of the method of doing these experiments and the ultimate resolution of experiment and theory is described in a recent survey.[42]

Variants on the basic flow configuration

More general flows can be produced than with one cylinder rotating and one at rest. For example one can rotate both cylinders in the same and opposite directions. One can add axial flow, azimuthal flow by means of suitably arranged pumps, and one can impose a radial temperature gradient between cylinders. With suitable porous construction, even radial flow can be provided as discussed by Bühler elsewhere in this volume. Cylinders can be mounted eccentrically.

External fields can be imposed. We have already shown how a conducting fluid such as mercury can be used in the presence of a magnetic field.[19] Recently, we have been studying the effects of Coriolis forces by placing the apparatus horizontally on a rotating table.[44] Generally speaking, such external fields tend to stabilize the flow.

Many concentric cylinder arrangements offer controlled stirring of solutions under predictable rates of shear. These are now finding applications for preparing solutions containing biological materials.

Spatial patterns in the flow evolve as the speed of rotation is increased. The dynamics become very complicated, as are the conditions necessary to achieve a reproducible flow, which depend on the past history in a special way.

End conditions

As Stokes imagined, the end conditions at the top and bottom are important for the flow. For the inner cylinder rotating, fixed ends, rotating ends, tapered annulus ends and a mercury bottom have all been investigated. A particularly ingenious end design is shown in Figure 20: it consists of a tapered cylinder which provides a slow spatial ramp by changing the gap between cylinders slowly. This idea has proved stimulating and useful in a variety of experimental investigations.[45,46]

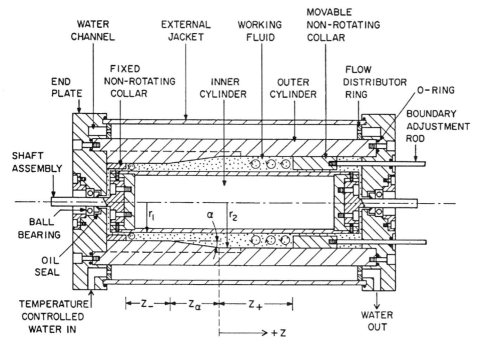

Figure 20. Tapered end developed at the University of California, Santa Barbara allowing a slow spatial ramp to be substituted for the usual abrupt end.[43]

9. EXPERIMENTAL CONTROL AND PROCEDURES

Selection of operating conditions

An important characteristic of the flow is the ratio of the radii of the cylinders. The aspect ratio is the ratio of the length of the cylinders to the gap between them. If the apparatus has a short aspect ratio, some aspects of chaotic flow may be studied; in apparatus of higher aspect ratio, turbulence is quite easily generated.

Computer control

Modern computers have been an invaluable aid to research in our subject. The ability to program the course of an experiment with instructions to all instruments on a bus, has revolutionized research. In particular, it is now easy to compute quantities on line which formerly could be done only an a main frame, if at all. This means in particular, that real time data reduction can be a reality and can warn against defective data long before one could formerly have detected a fault.

Experimental protocol

The experimental protocol, that is, the sequence of commands for data aquisition, is of utmost importance in rotating cylinder experiments. The rate of ramping to and from a given rate of rotation can influence the pattern which appears, and if done too quickly can introduce hysteresis which is absent for slower ramps. A paper reporting dramatic instances of the non-uniqueness of patterns in Taylor-Couette flow was written by Donald Coles in 1965.[47] Howard Snyder was the first to warn us that the time to equilibrate the flow after a change in angular velocity would be a fraction of the viscous diffusion time L^2/ν over the entire length of the cylinders.[48]

The effect of the rate of ramping the angular velocity to the transition to Taylor vortex flow was investigated by us based on earlier experience by Caldwell and Donnelly.[49] Examining the hysteresis which results in increasing angular velocity ramping, Park, Crawford and Donnelly [50] defined a dimensionless ramping rate a* which characterized the amount of hysteresis to be expected in a transition which should exhibit none if done sufficiently slowly. Ramping rates have been a major problem in data acquired in past years: ramps which are too fast exhibit hysteresis which is unlikely to be exactly reproduced by other investigators.

It should be emphasized that the a* criterion is applicable only to the transition to Taylor vortex flow and has not been systematically investigated for other transitions. As a result, when we conduct surveys of transitions in other situations, we routinely ramp up and back over the suspected transition, reducing the ramping rate until hysteresis is acceptably small, or the transition is recognized as intrinsically hysteretic.

Temperature control is a larger problem than might have been anticipated. The fundamental reason is that many fluids used in Taylor Couette flow change their viscosity about 2% per degree C. Both the absolute viscosity and its temperature dependence should be well understood in any experimental investigation. It is not, however, absolutely necessary to regulate the temperature of the apparatus to a high degree of accuracy providing the ambient temperature is reasonably slowly varying. It is possible to measure the temperature and correct the viscosity in real time for the temperature at the moment of data acquisition.

Thermometers are not easily made absolute. One reliable system are the Hewlett-Packard quartz thermometers, which have a calibration traceable to the NBS and which retain their calibration over long periods of time, and have a simple digital interface to computers.

If the experiment has considerable Joule heating, or strong external lighting, temperature gradients may become important and need to be recognized as a source of potential error.

10. ANALYSIS AND PRESENTATION OF DATA

Analysis of reflectance data

The fundamental observable in a Taylor-Couette experiment using Kalliroscope and a fixed reflectance detector is a time series of light intensities. Typically a 2048-point time series is taken at each increment in the control parameter space using a data acquisition frequency $50 \times f_{cyl}$, where f_{cyl} is the inner-cylinder frequency. From the time series one can readily determine four quantities that can be used to assess the flow dynamics: *average relative reflectance, variance of the reflectance, spectral number distribution*, and *spectral mode number*.

The time series is averaged at each point in parameter space to determine the average relative reflectance. A sharp drop in relative reflectance indicates the transition from the base flow to secondary flow. To determine the onset of time-dependence in a secondary flow one inspects the variance of the reflectance at each point. In the base flow state, and in Taylor vortex flow, for example, the variance is low; when the flow makes a transition to time-dependence such as wavy vortex flow one observes a marked rise in the variance.[51] While the wavy vortex flow is easily seen in Fourier spectra, the variance is a good general purpose detector for any transition to time dependence. One still needs to classify the time-dependent flow state as ordered or disordered.

The principal tool here is the Fast Fourier Transform (FFT) of the time series. Time-dependent flows are either periodic or ordered, or they display some degree of aperiodicity or disorder. An FFT of the reflectance time series from a periodic or quasiperiodic state is known a produce or power spectrum with frequencies that correspond to actual frequencies in the flow [52]. Savas [53] and Schwarz [54] have shown that there is also a correspondence between the dynamics of Kalliroscope platelets and the dynamics of the flow when the flow is disordered.

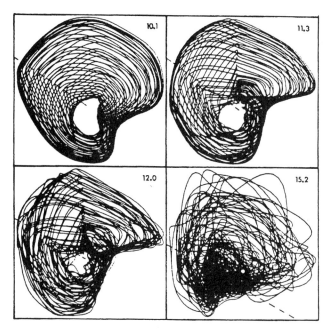

Figure 21. Two-dimensional portraits $V(t+\tau)$ versus $V(t)$ constructed for two Reynolds numbers ($R/R_c=10.1$ and 11.3) corresponding to modulated wavy vortex flow and two Reynolds numbers (R/R_c 12.0 and 15.2) corresponding to chaotic flow. (Brandstater and Swinney [57])

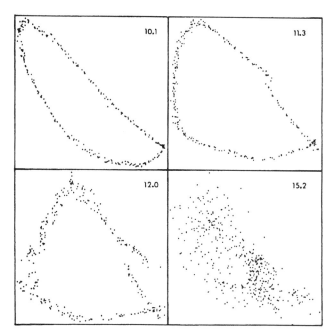

Figure 22. Poincaré sections given by the intersection of orbits of three-dimensional phase portraits [where the third axis is $V(t+\tau)$] with planes normal to the paper through the dashed lines in Figure 21. [The Delay times τ at the four Reynolds numbers 162, 144, 144 and 108 ms, respectively; about 40 of the 300 orbits observed at each Reynolds number shown in Figure 21. (Brandstater and Swinney [57])

As the noise level in the power spectrum increases we can employ two methods which can be used to measure relative changes in the power spectral noise. The first statistic is the spectral number distribution (SND) which is simply a reordering of the power spectrum whereby the number of spectral components above a certain power is plotted against that power.[55] We then choose a cutoff which we call the spectral noise. The spectral noise is chosen to be the power at which the number of components drops below 25% of the total number of components. The choice of cutoff, though arbitrary, has been shown to be a robust method of measuring bifurcations to and from disordered states. Note however, that the actual value of the spectral noise is a meaningless number; what matters is relative change in the spectral noise. Another useful statistic is the spectral mode number (SMN).[56] The SMN is

$$D = \left(\sum_{i=1}^{n} P_i\right)^2 / \sum_{i=1}^{n} P_i^2$$

where P_i is the power of an n-component power spectrum. For a pure sine wave $D = 1$ and for white noise $D = n$. As in the spectral noise from the SND, we can use the SMN to detect bifurcations among temporally ordered and disordered states.

Lyapunov exponents, maps, strange attractors.

Brandstater and Swinney[57] have undertaken a careful study to characterize both the bifurcation to chaos in Taylor-Couette flow and the dynamics of the chaotic flow. Their data analysis requires the low noise data provided by LDV measurements. What one gains with the sorts of techniques that Brandstater and Swinney apply is information related to the actual dynamics of a particular flow regime. As a result they can come closer to determining whether the noise seen in a power spectrum is indicative of deterministic and low dimensional dynamics (ie. chaos), or whether the flow is high dimensional (ie. turbulent). This sort of information is important because it has implications to studies of the transition to turbulence and to nonlinear dynamics. Power spectra, phase space portraits and circle maps obtained from velocity time-series data indicate that the nonperiodic behavior is deterministic, that is, it is described by strange attractors. The technical demands of this type of analysis in Taylor-Couette flow are severe and the analysis methods sophisticated. This paper is a clear guide to this type of analysis. Figs 21 and 22 give examples of the kind of data which can be obtained with sufficient care.

11. TURBULENCE STUDIES

The outer cylinder in rotation exhibits a direct transition to turbulence, which merits far more attention than it has had. There are many topics in turbulent flow that can be examined: Harry Swinney's work at this conference is a fine modern example with special attention being paid to intermittentcy in the counter-rotating cylinder case.

New possibilities are the examination of the effect on the transition to turbulence of "riblets" and other boundary modification methods. Finally, the use of liquid helium allows the very highest Reynolds numbers to be achieved.

REFERENCES

1. Isaac Newton, *Mathematical Principles* ed. F. Cajori, p. 385, University of California Press, Berkeley (1946).
2. G. G. Stokes, *Mathematical and Physical Papers*, Vol 1, p. 102 Cambridge University Press (1880); Vol 5, (1905).
3. Max Margules, "Über die bestimmung des reibungs- und gleitungs-coëfficienten Aus eben bewegungen einer flüssigkeit" ", Wiener Berichte, (Second series) **83**, 588-602 (1881).
4. H.R. Lillie, "The Margules method of measuring viscosities modified to give absolute values", Phys. Rev. **36** 347-362, (1930).
5. A. Mallock, "Determination of the viscosity of water", Proc. Roy. Soc. **45**, 126-132 (1888).

6. A. Mallock, "Experiments on fluid viscosity" Phil. Trans. Roy. Soc. **187**, 41-56 (1896).

7. Lord Rayleigh "Further remarks on the stability of viscous fluid motion", Philos. Mag. vol. **28,** 609 (1914). [*Scientific Papers*, Lord Rayleigh, #388, Dover Publications 1964].

8. M. Couette, "Sur un nouvel appareil pour l'étude Du frottement des liquides", Comptes Rendues **107**, 388-390 (1888); "La viscosité des liquides", Bull. des Sciences Physiques 4 - 62, 262-278 (1888), Soc. Francoise de Physiques, Paris, Bull des Scéances 60-61, 108-109 (1889).

9. M. Couette, "Études sur le frottment des liquides", Ann. Chim. Phys. Ser VI Vol 21 433 - 510 (1890).

10. J.A. Bearden, "A precision determination of the viscosity of air", Phys. Rev. **56**, 1023-1040, (1939)

11. H.L. Dryden, F.P. Murnaghan, H. Bateman, *Hydrodynamics* Dover Publications, New York (1956), [Republication of Bulletin 84 of the National Research Council, *Report of the Committee on Hydrodynamics*, 1932.]

12. L. Talbot, Editor, *Rarefied Gas Dynamics* Academic Press, New York, (1961).

13. G.I. Taylor, "Stability of a viscous liquid contained between two rotating cylinders", Phil. Trans. Roy. Soc. **A, 223**, 289 (1923).

14. S. Chandrasekhar, *Hydrodynamic and Hydromagnetic Stability*, Oxford, Clarendon Press (1961).

15. K.C. Wali, *Chandra, A Biography of S. Chandrasekhar*, University of Chicago Press (1991).

16. G.I. Taylor, "Fluid friction between rotating cylinders I - Torque measurements", Proc Roy Soc A, **157**, 546 - 564, (1936)

    G.I. Taylor, "Fluid friction between rotating cylinders II - Distribution of velocity between concentric cylinders when outer one is rotating and inner one is at rest", Proc Roy Soc A, **157**, 565 - 578 (1936)

17. A.C. Hollis Hallet, "Experiments with a rotating cylinder viscometer in liquid helium II," Proc. Camb. Phil. Soc. **49**, 717-727 (1953).

18. R.J. Donnelly, "Experiments on the stability of viscous flow between rotating cylinders, I. Torque measurements" Proc Roy Soc A **246**, 312-325 (1958).

19. R.J. Donnelly, M. Ozima, "Experiments on the stability of flow between rotating cylinders in the presence of a magnetic field." Proc Roy Soc A **266** 272-286 (1962).

20. R.J. Donnelly and N.J. Simon " An empirical torque relation for supercritical flow between rotating cylinders" J. Fluid Mechanics **7**, 401 - 418 (1960).

21. J.A. Cole, "Taylor-vortex instability and annulus-length effects," J. Fluid Mech. **75**, 1-15 (1976).

22. R.J. Donnelly and M.M. LaMar, "Absolute measurement of the viscosity of classical and quantum fluids by rotating-cylinder viscometers," Phys. Rev. A **36** 4507 (1987).

23. R.J. Donnelly, D. Fultz, "Experiments on the stability of flow between rotating cylinders II. Visual observations." Proc Roy Soc A **258** 101-123 (1960).

24. D.J. Baker, "A technique for the precise measurement of small fluid velocities" J. Fluid Mech. **26**, 573-575 (1966).

25. K. Park, C.F. Barenghi, and R.J. Donnelly "Subharmonic destabilization of Taylor vortices near and oscillating cylinder" Phys Lett 78A 152-154 (1980).

26. R.J. Donnelly, "Experiments on the stability of flow between rotating cylinders IV. The ion technique." Proc Roy Soc A **283** 509-519 (1965).

27. R.J. Donnelly and D.J. Tanner, "Experiments on the stability of flow between rotating cylinders V. The theory of the ion technique." Proc Roy Soc A **283** 520 - 530 (1965).

28. R.J. Donnelly, K.W. Schwarz, "Experiments on the stability of viscous flow between rotating cylinders, VI. Finite-amplitude experiments." Proc Roy Soc A **283**, 531-546 (1965).
29. R.W. Walden, "Transition to turbulence in Couette flow between concentric cylinders " PhD thesis, University of Oregon, (1978)
30. F. Schultz-Grunow and H. Hein, "Beitrag zur Couettestromung" Z. Fur Flugwissenschaften **4** 28-30 (1956).
31. P. Matisse and M. Gorman, "Neutrally buoyant anisotropic particles for flow visualization," Phys. Fluids **27**, 759-760 (1984).
32. R.J. Donnelly, K. Park, R. Shaw and R.W. Walden, "Early non-periodic transitions in Couette flow" Phys Rev Lett **44**, 987-989 (1980).
33. G. Crawford, "Transitions in Taylor wavy-vortex flow" PhD thesis, University of Oregon, (1983).
34. Li Ning, G. Ahlers and D.S. Cannell "Wavenumber selection and travelling vortex waves in spatially ramped Taylor-Couette flow", Phys Rev Lett (1991).
35. F. Durst, A. Melling and J.H. Whitelaw, *Principles and practice of laser-doppler anemometry*, Academic press, New York (1981)
36. D.C. Bjorkquist, "Particle image velocimetry for determining structures of turbulent flows." Flow Lines *(a publication of TSI Inc.)* **6** 3-8 (1991).
37. R.J. Donnelly, in *A Physicist's Desk Reference* Ed. Herbert L. Anderson, AIP New York, 1989, p.201.
38. R. J. Donnelly and D.R. Caldwell, "Experiments on the stability of hydromagnetic Couette flow" J. Fluid Mech. **19**, 257 - 263 (1964).
39. R.G. Wheeler, C.H. Blakewood and C.T. Lane, "Second sound attenuation in rotating helium II" Phys Rev **99**, 1667-1672 (1955).
40. R.J. Donnelly, *Quantized vortices in helium II*, Cambridge University Press, Cambridge (1991).
41. R.J. Donnelly and Michelle M. LaMar, "Flow and Stability of Helium II Between Concentric Cylinders", J. Fluid Mech. **186**, 163-198 (1988).
42. R.J. Donnelly in *Nonlinear Evolution of Spatio-Temporal Structures in Dissipative Systems*, ed. F.H. Busse and L. Kramer, Plenum Press, NY (1990).
43. R.J. Donnelly, "Experiments on the stability of viscous flow between rotating cylinders, III. Enhancement of stability by modulation." Proc Roy Soc A **281**, 130 - 139 (1964).
44. R.J. Wiener, P.W. Hammer, C.E. Swanson and R.J. Donnelly, " Stability of Taylor-Couette flow subject to an external Coriolis force." Phys. Rev. Lett. **64**, 1115 (1990).
45. M.A. Dominguez-Lerma, D.S. Cannell and G. Ahlers, "Eckhaus boundary and wave-number selection in rotating Couette-Taylor flow" Phys Rev A **34**, 4956 - 4970, (1986).
46. Li Ning, G. Ahlers and D.S. Cannell " Wavenumber selection and ravelling vortex waves in spatially ramped Taylor-Couette flow" Phys Rev Letters
47. D. Coles, "Transition in circular Couette flow" J. Fluid Mech. **21**, 385-425, (1965).
48. H.A. Snyder "Change in wave-form and mean flow associated with wavelength variations in rotating Couette flow" J. Fluid Mech **35** 337-352 (1969).
49. D.R. Caldwell and R.J. Donnelly, "On the reversibility of the transition past instability in Couette flow", Proc. Roy. Soc A**267** 197-205 (1962).
50. K. Park, G.L. Crawford and R.J. Donnelly "Determination of transition in Couette flow in finite geometries", Phys Rev Lett **47**, 1448-1450 (1981).
51. P.W. Hammer, R.J. Wiener, C.E. Swanson and R.J. Donnelly, "Bifurcation phenomena in nonaxisymmetric Taylor-Couette flow" (Phys Rev A15, in preparation, 1992).
52. M. Gorman and H.L. Swinney, "Visual observation of the second characteristic mode in a quasiperiodic flow", Phys. Rev. Lett. **43**, 1871 (1979)

53. Ö. Savas, "On flow visualization using reflective flakes", J. Fluid Mech. 152, 235 (1985)
54. K. Schwarz, "Evidence for organized small-scale structure in fully developed turbulence", Phys. Rev. Lett. 64, 415 (1990).
55. W.L. Ditto, M.L. Spano, H.T. Savage, S.N. Rauseo, J. Heagy and E. Ott "Experimental observation of a strange nonchaotic attractor", Phys Rev Lett **65**, 533-536 (1990).
56. J. Crutchfield, D. Farmer, N. Packard, R. Shaw, G. Jones and R.J. Donnelly "Power spectral analysis of a dynamical system", Phys Lett **76A**, 1-4 (1980).
57. A. Brandstater and H.L. Swinney, "Strange attractors in weakly turbulent Couette-Taylor flow", Phys Rev A**35** 2207-2220 (1987).

MODE COMPETITION AND COEXISTENCE IN TAYLOR-COUETTE FLOW

John Brindley[1] and Frank R. Mobbs[2]

[1]Department of Applied Mathematical Studies
[2]Department of Mechanical Engineering
University of Leeds, Leeds LS2 9JT

1. INTRODUCTION AND BACKGROUND

The geometrical simplicity of the Taylor-Couette configuration, embodies rotational, reflectional and translational symmetry. It has enabled theoreticians, under some conditions, to construct, by rigorous deduction from the Navier-Stokes equations, generic mathematical models for the bifurcational behaviour of the system. Thus certain simple symmetry-breaking "modes" have been identified, and their bifurcation from steady Couette flow, or from steady Taylor vortices, has been located in parameter space. [1-5]

Our particular concern here is with modal coexistence at large amplitude, and in a rigorous sense the analysis of [1-5] is not valid for our situation. The predictions made there have, however, received experimental support in circumstances where their analysis *is* valid, and they provide an ideal conceptual framework within which to view and interpret our results.

Briefly, the object of the theoretician has been to construct, as nearly rigorously as possible, a model system describing the time evolution of the flow. The model should be capable of a full bifurcational analysis, so that the distribution in parameter space of the various qualitative flow patterns may be established. Interest centres on the search for points, P_o, in parameter space, of maximum degeneracy, in the neighbourhood of which rigorous centre manifold/normal form reductions yield equations whose solutions exhibit, in various sectors of parameter space centred on P_o, all the types of qualitative behaviour possible in the system. The concept of "hidden organising centres" (points of degeneracy) which are inaccessible in the actual model equations but which are made accessible by the addition of (non-physical) terms of small amplitude [5], is perhaps of particular value to our interpretation of results. The simplest degeneracy occurs when two distinct new "modes", steady or time-dependent, bifurcate from a given flow at the same point in parameter space. The existence of such degeneracies has not been proved when the outer of the two cylinders remains at rest. Indeed for small gaps, coincident bifurcation for Couette flow of two or more modes does not occur [6]; bifurcation to steady Taylor vortices always precedes any bifurcation to a time-dependent mode. The question remains open, however, for wide gaps. In each of the cases in which the cylinders co-rotate or counter-rotate such degeneracy has been conclusively established; an excellent agreement between theory and experiment has led to good understanding of some of the wide range of qualitatively different flows seen in those cases [7,8].

A final remark on earlier work must concern the question of "state selection". The bifurcation from Couette flow to steady Taylor vortices, which always precedes any time-dependent behaviour when the outer cylinder is at rest, is "degenerate" in the sense that the spatial structure of the vortex flow is non-unique; Taylor vortices whose aspect ratio lies within a finite range may exist stably. It is well understood theoretically [9], and elegant theory and experiment has illuminated the phenomenon in short cylinders [10], whose dimensions exert a quantizing effect on the admissible aspect ratios of cells. In long cylinders, with which we have been almost exclusively concerned, end effects are still important in this way, and the excessive multiplicity of possible steady cell patterns leads to major problems of interpretation of experimental data.

Virtually all the results described here have come from image processing techniques, described fully in [11], which have enabled us to obtain time series information simultaneously from up to 40 points in the flow, distributed either azimuthally or in meridional planes. This data makes it possible to analyse spatial structure in a very thorough way, and the "phase" information so obtained has formed a vital addition to the more commonly obtained "time series at a point" data in interpreting this spatial structure. The main result is that many, if not most, of the time-dependent flows, however complex their visual appearance, may be understood as superpositions of two or more simple modes, each having a different spatial structure itself compatible with boundary constraints and symmetries. This coexistence of course requires a balanced nonlinear interaction between the modes themselves and between each mode and some "mean" flow. A number of time-dependent modes of different character may be distinguished, some of them global (in the sense that the end conditions of the apparatus have a significant effect), others local, and dependent on the aspect ratio of a particular Taylor cell in which they occur.

This aspect ratio, λ, whose value is a consequence of the state selection process, has an influence of crucial importance in the evolution of time-dependence, and we choose to regard it as a parameter (even though its value is internally selected rather than externally imposed) in our presentation and discussion of results. It, together with the ratio, η, of cylinder radii, the non-dimensional length, Γ, of the cylinder and the Taylor number, T, form a four-dimensional parameter space within which the dynamics of time-dependent behaviour may be investigated and classified. Reproducibility of phenomena and comparability of results between experiments <u>requires</u> equality of values for <u>all four parameters</u>, though in some regions of parameter space the dependence on one of them may be weak.

The physical character of the modes themselves is dominated by a wave-like structure in the azimuthal direction, each mode having an integer wave number. The modes appear to be localised (in the sense that they have a maximum amplitude there) at one or other of the outward and inward flow regions between cells or at the cell cores. Some are advected with the speed of the cell core, others with speed not directly identifiable with a single point of the flow but compatible with speeds in the inflow or outflow regions.

2. A PARAMETER-SPACE MAP FOR ONSET OF TIME-DEPENDENCE

The availability of several different assemblies of apparatus has permitted the accumulation of a great deal of data describing the onset and consequent evolution of time-dependent behaviour. Our first objective, in this section, is to draw on this data to establish a comprehensive and comprehensible "map" of the behaviour in the appropriate parameter-space.

Many authors have drawn attention to the dependence on the radius ratio, η, or on the apparatus aspect ratio, Γ, of the critical Taylor number for onset of "waves"; fewer have remarked on the dependence on cell aspect ratio, λ. We have found that <u>all</u> these parameters are crucially important, not only in influencing the value of T at the initial onset of time-dependent wavy behaviour, but also in determining the temporal behaviour and spatial form of the flow structures which subsequently develop. Many of the observations are suggestive of two or more independent but interacting physical instability processes, and we shall develop this idea later. It is convenient first to devote three subsections to results which expose clearly the dependence on η, Γ and λ respectively, before attempting to draw the whole picture together in a coherent way.

(i) <u>Dependence on η</u>

It is clear that in laboratory experiments it is not possible to vary η continuously, but results from a wide range of values of η obtained in different experimental series enable us to view the onset of time-dependence as a function of η and λ with Γ kept constant. At each value of η a cusped curve, as in Figure 2, is observed. As η increases, the position of the cusp migrates to the left, to smaller values of λ. This means that for narrow gap configurations the left hand branch of the cusp is very difficult to access. Although the form of this idealized picture remains broadly the same, its details are very sensitive to the values of Γ.

It is important to note here, as will be made clear later, that this boundary of onset of time-dependence, defined in terms of T, η, Γ and λ, is in fact made up of parts relating to quite different physical modes, whose spatial structures and temporal frequencies are quite distinct.

(ii) <u>Dependence on Γ</u>

Our results suggest two distinct influences of Γ on the dynamics. First is the direct inhibiting effect of, especially, rigid end boundaries on the flow in adjacent Taylor cells. This effect forces the value of T_w to rise from a few percent higher than T for large Γ to many times T_c for small Γ even in narrow gap configurations.

The second, more subtle but perhaps more important, influence of Γ is on the <u>state selection problem</u>. By this we mean the selection of the number of Taylor cells in the apparatus. Many beautiful experiments and associated theory by Brooke Benjamin and Mullin [10] have made clear the character of the bifurcation problem for Γ and have shown the existence of multi-equilibrium states. For larger Γ, the range of states (defined by different cell totals) available, according to any simple linear theory of instability, increases, and the migration of the flow from one state to another by creation or annihilation of cells becomes more common. Indeed the attainment and subsequent retention of a particular state, even at fixed T, is a far from trivial task in the laboratory if Γ is large. For this reason we have found it convenient, and it is apparently also physically sensible, to present most of our results in terms of the value of λ, which is a measure of the state actually occupied by the system but which is not a directly controllable laboratory parameter; it is, however, useful to display some results showing the effect of varying Γ for the same value of λ.

Thus in Figure 1 we show the onset of time-dependent behaviour of the cell boundaries, displayed against Γ, for particular values of η and λ. For Γ "small" (i.e. $\Gamma < 12$), the onset of time-dependence for $\lambda > 1$ occurs at a very much higher value of T than the expected value for long cylinders, where, for this value of η, the value of $\frac{T_w}{T_c}$ is not much greater than unity. At "large" Γ ($\Gamma > 16$), on the other hand, the value of T_w is much more in line with "classical" ideas. The values of frequencies at onset of time-dependence show dramatic differences. The left-hand branch of the cusped time-dependence onset curve (Fig.2), for which λ is < 1, is quite insensitive to the value of Γ, and it seems plausible at this stage that the mode or modes responsible for the low frequency right-hand branch are global in some sense, and feel the end effects, whilst the left-hand branch represents a phenomenon more local in character (though even here the coherence of phase information suggests strong inter-cell coupling).

(iii) <u>Dependence on λ</u>

Wherever possible, experiments have been carried out at fixed λ once that value has been attained in some way, e.g. by varying the "ramping" rate of Ω. The striking characteristic of the results for onset of time-dependence is the strong, "peaking" at some value of λ. The onset curves are strongly suggestive of at least two distinct modes or physical processes of instability. Typical results are shown in Figure 2. These are wide gap results ($\eta = 0.497$) and a wide range of Γ values has been used. It is clear that the left-hand branch of the time-dependent onset curve is virtually independent of Γ, as we have remarked earlier. The right-hand branch is dependent on Γ, and Figure 3, which shows the values of the dominant frequency peak at onset, gives a clear indication that two separate modes are involved even on this apparently single branch. For $\Gamma > 17$, the onset frequency ω_w is clearly defined and is identifiable with the "classical" wavy mode. For smaller values of Γ a quite different frequency, ω_2, appears at onset. The left-hand branch of the time-dependent onset curve has associated with it a different frequency again, ω_1. In §3 we present information on phase and spatial structure which throws light on this complex behaviour. For narrow gap cases (e.g. $\eta = 0.85, 0.908$) we find almost no change of T_w over the range of λ investigated when Γ is > 30. There is, however, rapid increase in T_w for $\lambda < 1$ when Γ is < 12.

3. SPATIO-TEMPORAL BEHAVIOUR; FREQUENCY AND PHASE INFORMATION

The foregoing qualitative overview of the onset of time-dependent behaviour provides a framework within which to review three particular sets of results, corresponding to radius ratios $\eta = 0.497$, $\eta = 0.713$ and $\eta = 0.84$ respectively.

(i) $\eta = 0.497$

Figures 2,3, summarise information on values of T and on the dominant frequencies at onset of time-dependent behaviour in a number of experiments with different values of Γ. The universality in the left-hand branch is clear, as is the onset frequency. The right-hand branch of the onset curve is seen to be a composite of two parts, one dominant for "small" Γ, the other for "large" Γ. The independence of the two parts is strikingly illustrated by the frequency results, which are totally different. The frequency ω_w is clearly identifiable with the "classical" wavy mode in which the cell boundaries are perturbed by in-phase travelling waves, whilst evidence from spectra obtained concurrently from different spatial locations suggest that the frequency ω_1 is associated with modes advected with the vortex core velocity, and the frequency ω_2 with the outward jet region. Results obtained for a range of Taylor number at each of two values of λ, suggest that each of these modes can occur almost independently of the existence of the others. Thus, for λ "small", we find that ω_1 appears, followed by ω_w, whilst, for larger, λ ω_2 occurs first. For

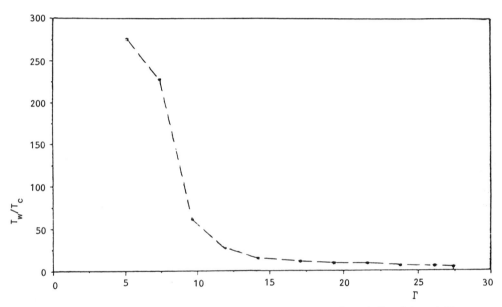

Fig.1. Variation of Tw with Γ for fixed $\eta = 0.79$ and fixed $\lambda = 1.16$

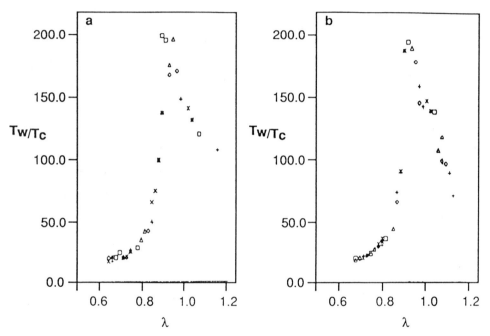

Fig.2. Value of T_w/T_c at onset of time-dependence against λ for a range of values of Γ; (a) $11<\Gamma<13$, (b) $15<\Gamma<17$. In all cases $\eta = 0.497$

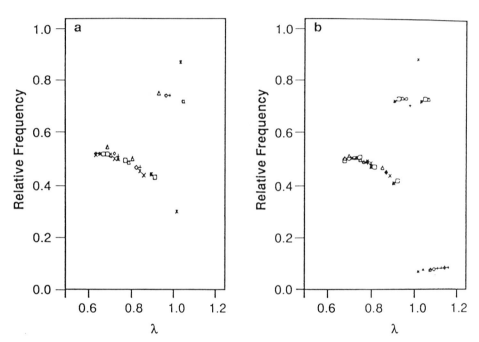

Fig.3. Values of frequency at onset of time-dependence for a range of values of Γ as in Fig.2; (a) $11<\Gamma<13$, (b) $15<\Gamma<17$. In all cases $\eta = 0.497$

still larger λ the first onset of time-dependence is through the frequency ω_w. Identification of the sources of the frequencies suggests a renaming of ω_w, ω_1 and ω_2 as ω_{IJ}, ω_c and ω_{OJ} respectively and it is through this self-evident notation that we shall refer to the equivalent frequencies at the other values of η.

(ii) $\eta = 0.713$

We turn in this case to an examination of phase or wave-number behaviour. The most striking characteristics always resemble Figure 4; regions in parameter space where the spectra show sharp peaks and coherent dependence on T are seen to exhibit erratic wave-number behaviour. The wave-number associated with a given frequency peak is virtually always non-integer, and varies significantly both with T and in space. Moreover the wave-number associated with a harmonic of the dominant frequency is not the harmonic of the wave-number associated with that frequency.

These wave-number characteristics are typical of experiments at several η values, and, though the details are Γ-dependent, the same qualitative behaviour exists for all relatively large Γ.

(iii) $\eta = 0.804$

At this value of η, the right-hand branch of the onset curve is always accessed first, and, for large Γ, the classical wavy mode appears first. Even in this case, however, the frequency and wave-number evolution with increasing T is far from simple, and Figure 4 contains much information crucial to the understanding of the main argument of this paper. In it we display the frequency and wave-number results obtained from a line of pixels orientated along the vortex core, and to simplify the picture we have applied a cut off to the spectra as high as 0.99. In other words, only the highest peak in the spectrum has been retained in each case.

From the onset of waves, at T/T_c 5, until T/T_c 100, the dominant frequency, first falling with T and then levelling out (labelled $3\omega_c$) is seen. The accompanying wave-number information for $5 < T/T_c < 100$ shows a consistent wave-number very close to 3. It is clear that the dominant frequency is $3\omega_c$, and the consistent calculation from phase information of a wave-number 3 indicates conclusively a wave advected with the vortex core speed, Ω_c.

A quite different state of affairs is seen where T/T_c is > 100; at these values a new frequency (labelled $3\omega_{IJ}$) is dominant. It is identifiable with information on wave speed and/or dominant frequencies given by other experimenters near this value of η [12,13], but it is not a multiple of the vortex core speed Ω_c. The phase information associated with this frequency is quite unexpected. The inferred wave-number varies wildly with T and does not take integer values, even in an approximate sense. No simple concept of a wave advecting with a speed identifiable with the frequency $3\omega_c$ is possible on the basis of information obtained from the vortex core.

A clue to the understanding of this puzzling information is provided by examining information obtained from a line of pixels in the inward jet region of the Taylor vortex. Again, for the range $5 < T/T_c < 100$, we see that $3\omega_c$ is the dominant frequency, and, for $T/T_c > 100$, the dominant frequency switches to $3\omega_{IJ}$ (the precise point of switch over dominant frequency is not identical at core and inward jet; a lower cut off shows that both frequencies, $3\omega_c$ and

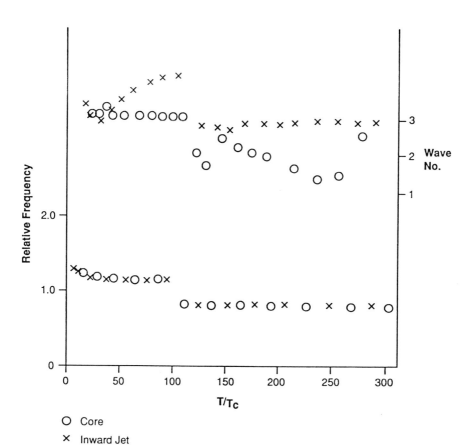

Fig.4. Frequency and apparent wave number measured simultaneously near a cell core - o, and near an inward jet - x. Scale for frequency on left hand axis; for wave number on right hand axis. $\eta = 0.804$.

$3\omega_{IJ}$, exist together in the spectrum over a wide range of T, but the ratio of the peak strength associated with each of them changes with T). The behaviour of wave-number is, however, reversed, in that, for $5 < T/T_c < 100$, the wave-number is non-integer varying, whilst for $T/T_c > 100$ it is very steady and close to 3. The information is consistent with a wave having a wave-number 3, advected with a speed "Ω_{IJ}". Unlike Ω_c, the speed Ω_{IJ} is not associated with any obvious location in the flow.

The results described are typical of an extensive range of data relating to other values of η, Γ and λ. In all, four time-dependent modes (each essentially a travelling wave) may be identified:

(i) <u>Outward Jet Mode</u>

In this mode only the outward jet oscillates; the inward jet boundary remains plane. Adjacent outward jets may be in-phase or anti-phase. For wide gaps this mode replaces the outward jet dominated wavy mode (described at (iii) below) as λ increases, but its amplitude also decreases as λ increases. As η increases the range of dominance of this mode is gradually eroded by the extension of the outward jet dominated wavy mode to higher λ and the extension of an inward jet dominated wavy mode to lower λ; it vanishes altogether for narrow gaps.

(ii) <u>Inward Jet Mode</u>

In this mode only the inward jet oscillates. It occurs over a limited range of λ close to unity for wide gaps and low Γ.

(iii) <u>Outward Jet Dominated Wavy Mode</u>

In this mode both outward and inward jets oscillate with maximum amplitude usually in the outward jet at the outer cylinder. It is not present for $\eta = 0.497$; it appears at very low λ at $\eta = 0.54$, and extends to higher values of λ as η increases. For narrow gaps the outward jet amplitude decreases with increasing λ until the waves on both jets are approximately of equal amplitude.

(iv) <u>Inward Jet Dominated Wavy Mode (the classical wavy mode)</u>

In this mode the motion is dominated by high amplitude oscillations of the inward jet near the inner cylinder; for wide gaps it occurs at high values of λ, gradually extending to lower values of λ as Γ is increased. The relative amplitude of the outward jet oscillation increases as λ increases. The mode extends to lower values of λ as η increases and forms the classical wavy mode. It is convenient to summarise the onset frequencies as follows

(a) <u>Wide gaps</u> (e.g. $\eta = 0.497$). The outward jet dominated wavy mode, the outward jet mode, and the inward jet mode all have frequencies which are integer multiples of the core speed. The inward jet dominated wavy mode frequency is not so related.

(b) <u>Narrow gaps</u> (e.g. $\eta = 0.804$). For λ small, both core and boundaries have the same frequency, which is an integer multiple of the core speed. This frequency is retained by the core up to high values of λ but the boundaries then have a different frequency. This second frequency, which is not an integer multiple of core speed, is common to core and boundaries at very high λ.

To conclude this rather complex set of results on spatio-temporal behaviour we repeat several vital points.

(i) Simple frequency behaviour, with sharp spectral peaks, is not always, or even usually, accompanied by simple, easily comprehended, azimuthal phase behaviour.

(ii) Though the same frequency peaks are identifiable in spectra obtained simultaneously from different locations in a Taylor cell, the wave-numbers associated with those peaks vary in space and with Taylor number.

(iii) Several physically distinct time-dependent modes are identifiable, and it is sometimes possible to identify frequencies of wave-numbers with a simple modal advection process.

(iv) More usually it appears that two (or more) modes coexist, either retaining their own frequencies, or in some frequency locked state.

4. A SIMPLE CONCEPTUAL MODEL

We have argued earlier [11] that the phase, or wave-number behaviour may be understood in terms of simple linear superpositions of two (or more) modes, each having a different integer as its azimuthal wave-number but having a common frequency. Thus any field variable, χ, is written as

$$\chi = a_1 \sin m_1(\theta + m_2 \omega t) + a_2 \sin m_2(\theta + m_1 \omega t)$$
$$= p \cos m_1 m_2 \omega t + q \sin m_1 m_2 \omega t$$
$$= r \sin(m_1 m_2 \omega t + \phi)$$

where $p = a_1 \sin m_1 \theta + a_2 \sin m_2 \theta$, $q = a_1 \cos m_1 \theta + a_2 \cos m_2 \theta$

$r = \sqrt{p^2 + q^2}$, $\phi = \sin^{-1} \frac{p}{r} = \cos^{-1} \frac{q}{r}$

and the amplitudes a_1, a_2 vary in space and with the parameters T, η and λ. As an almost trivial example we can take $m_1 = 1$, $m_2 = 2$, $a_1 = a_2 (=$ constant$)$. Then the apparent "wave-number" of the whole field, m, is given by $\frac{\partial \phi}{\partial \theta}$, and ϕ has a phase jump of π at $\theta = \frac{2\pi}{3}$.

Again, if we choose $a_1 = 2a_2$ ($=$ constant), and $m_1 = 1$, $m_2 = 2$, we find an apparent wave number of 1.35. Clearly if $a_1 \gg a_2$, the apparent wave-number is very close to m_1, and vice versa.

Now suppose that a_1 and a_2 have some variation with r and z, in other words that the "eigenfunctions" associated with the m_1 and m_2 "modes" have some spatial structure. At once we see that the apparent wave-number becomes a function of position, approaching the value m_1, say, at any point in the flow where the m_1 mode strongly dominates the m_2 mode in amplitude.

On this basis we can understand the curious behaviour discussed in detail in 3 above, in which a wave-number near 3 exists coherently near the vortex core for $5 < T/T_c < 100$, associated with the frequency $3\omega_c$, and the wave-number 3 exists coherently near the inward jet for $T/T_c > 100$, associated with a frequency $3\omega_{IJ}$. Thus, for $5 < T/T_c < 100$, the dominant physical disturbance is a wave of wave-number 3, advected with the core speed and having a maximum strength at the core. In the inward jet this mode is less strong, and it is contaminated by one or more Fourier components to produce a non-integer wave number. Variations in the relative sizes of a_1 and a_2 as T changes lead to variations in this apparent non-integer wave-number in experiments at different values of T.

For $T/T_c > 100$, on the other hand, we have a dominant physical mode localised near the inward jet and advected with a coherent speed ω_{IJ}. In this case the disturbance of the core is not totally dominated by the inward jet mode, and its non-integer and varying wave-number arises from contamination by other modes.

It appears that, for substantial ranges of T, two (or more) modes of this kind, each with a localised spatial structure, coexist. Their different physical character permits locking of their frequencies, presumably when integers m_1, m_2 can be found for which $m_1\omega_2$ and $m_2\omega_{IJ}$ are sufficiently close together, though at other times they can coexist independently at different frequencies. We must stress again that, though our conceptual model is linear, the coexistence observed in the actual experiment is nonlinear; we are observing finite amplitude equilibrium coexistence of modes. Any adequate mathematical model must correctly predict this nonlinear coexistence, together with the apparent exclusion of other modes. The wide separation in values of T at onset of the two modes probably precludes any analytic approach at the parameter values of this example, but, as we discuss further elsewhere [14], a full exploration of the (T, η, λ, Γ)space may reveal the general distribution of points of simultaneous instability of two or more modes near which analysis is possible [4].

5. SUMMARY AND DISCUSSION.

The use of image processing facilities, permitting simultaneous observations at a number of spatial locations, has yielded data from which, in particular, phase information may be extracted. This information has strongly suggested a common occurrence in parameter space of multi-modal behaviour, i.e. the coexistence of two (or more) wave-like modes at integer wave-number but with different spatial structure. These modes may have the same or different frequencies and may be located (in the sense that their amplitudes are a maximum there) at the core of the parent Taylor cell, or in the inward jet or outward jet regions of the cell. Four different modes, clearly distinguishable by spatial structure and phase and frequency behaviour have been identified and described.

Reference has already been made in the introduction to the valuable foundations to a theoretical understanding already laid by a number of authors [1-5]. On the whole these approaches have exploited the observed results [7] that, with counter-rotating cylinders (introducing yet another control parameter!), two distinct time-dependent modes can bifurcate simultaneously from the basic Couette flow. Our results suggest that, though such a simultaneous bifurcation cannot be obtained experimentally when the outer cylinder is at rest, nevertheless, at finite amplitude coexistence of modes is common. It seems possible that there may exist hidden "organising centres", in the four-dimensional parameter space of T, η, λ, Γ, which may be crucially important to the successful theoretical modelling of the observed behaviour.

Though in an experimental sense it would not be possible, in a mathematical model we may seek a point of intersection in parameter space of all three instability surfaces, and by perturbing about this point of maximum degeneracy, examine the modal interaction phenomena.

Finally, we have concentrated in this paper solely on azimuthal phase information. Recent analysis of axial phase information yields results suggestive of similar multiplicity of axial modes. This is hardly surprising, in view of the multiplicity of possible unimodal steady states known to exist concurrently at a given value of T [10]. Detailed results will be presented fully in a subsequent paper.

References

1. Chossat,P. and G.Iooss. Japan J.App.Math. 2, 37-68, 1985
2. Iooss,G. J.Fluid Mech. 173, 273-288,1986
3. Chossat,P.,Y.Demay and G.Iooss Arch.Rat.Mech.Anal. 99,213-248,1987
4. Golubitsky,M. and W.F.Langford. Physica 32D,362-392,1988
5. Hill,A. and I.Stewart. Three-mode interactions with O(2) symmetry and a model for Taylor-Couette flow, preprint, Mathematics Institute, Univ. of Warwick,1990
6. Davey,A., R.di Prima and J T Stuart. J.Fluid Mech.31,17-52,1968
7. Andereck,C.D., S.S.Liu and H.L.Swinney. J.Fluid Mech.164,155-183,1986
8. Langford,W.F., R.Tagg, E.Kostalich, H.L.Swinney and M.Golubitsky. Phys.Fluids.31,776-785,1986
9. Ahlers,G., D.S.Connell, M.A.Dominguez-Lerma and R.Heinrichs. Physica D.23,202-219,1986
10. Brooke-Benjamin,T. and T.Mullin. J.Fluid Mech.121,219-230,1982
11. Brindley,J. and F.R.Mobbs. in Advances in Turbulence, ed.G.Compte-Bellot and J.Mathieu. Springer-Verlag,7-16,1987
12. Park,K. and K.Jeong. Phys.Fluids 27,2201-2203,1984
13. King,G.P., Y.Li, W.Lee, H.L.Swinney and P.Marcus. J.Fluid Mech. 141,365-390,1984
14. Brindley,J. and F.R.Mobbs unpublished m/s.

LOW-DIMENSIONAL SPECTRAL TRUNCATIONS

FOR TAYLOR-COUETTE FLOW

K.T. Coughlin and P.S. Marcus

University of California at Berkeley
Department of Mechanical Engineering
Berkeley, California

Abstract

Experiments have shown that low–dimensional attractors exist in chaotic fluid flows, implying that only a small number of degrees of freedom are active. Current spectral computational techniques demand that the solution be represented as a sum of separable functions of space and time; hence, many more modes than the dimensionality of the attractor are needed to resolve the solution. The question we address in this paper is whether one can exploit the fact that the dynamics is low dimensional to substantially reduce the number of computational modes, without sacrificing quantitative accuracy. To do so would be of great practical importance. Using the formalism of inertial manifold theory, we discuss a novel procedure for constructing a low–dimensional truncation of the Navier–Stokes equations with no adjustable parameters, in which an initially small set of basis functions is augmented as needed during time integration. The method will be applied to temporally quasiperiodic and chaotic Taylor–Couette flows, with the choice of basis functions based on previous numerical work. Implementation is in progress.

1 INTRODUCTION

Recently, there has been a rapid advance in the study of strongly nonlinear systems, primarily through the use of numerical simulation. However, it has become clear that many interesting problems in which a wide range of length and time scales are active will not be computationally accessible in the foreseeable future, despite improvements in computer hardware. To make further progress, methods of solving partial differential equations (PDE's) will be needed that are substantially more efficient than those currently in use. For finite difference computations, the use of adaptive grid solvers[1] (which vary the grid resolution over the domain in order to focus computational work on the most complex regions), has been an important innovation. In this paper, we discuss in very preliminary terms the development of an adaptive spectral method, *i.e.* a spectral, initial–value algorithm in which an initial, optimal set of basis functions is chosen, with more basis functions added as required during the calculation.

Traditional spectral methods represent the solution to a PDE as a sum of known basis functions (such as sines and cosines) multiplied by time–dependent amplitudes; the latter are the computational variables[2]. They are most useful in problems where the geometry of the domain possesses a high degree of symmetry; most often in closed flows. Frequently, these problems also have interesting regimes which are described by relatively low–dimensional attractors, which suggests that there may exist a preferred description in which the number of computational variables N needed is comparably low. Assuming that prior analysis of the flow has led to a small set of physically well-motivated basis functions (which are adequate in a limited regime of parameter space — similar to an amplitude expansion), we can augment the set by generating additional basis functions from the nonlinear terms in the Navier–Stokes equation. This is necessary because nonlinear interactions generically introduce directions in phase space not spanned by the original set, and in strongly nonlinear problems these will be non-negligible. Because the fundamental low dimensionality of the attractor is not affected by nonlinearity, this procedure should converge, but it is important to note that it is *not* true that a PDE with an attractor of low dimension D will generally be well represented using $N \sim D$ variables.

As an example, consider traveling waves, which are space-time periodic (in x say), and thus have a one dimensional attractor. They can be generically described as a spectral sum (assuming an infinite or periodic domain)

$$V(\mathbf{r},t) = \sum_{m=-\infty}^{\infty} b_m e^{im(x-ct)} \tag{1}$$

where c is the phase speed of the wave. The coefficients b_m (which may be functions of the other spatial coordinates) are non-vanishing for an infinite number of m's, and no coordinate transformation can change this fact. The system point lives on a one dimensional curve in phase space where there are no constraints on the representation of the solution. However, spectral numerical methods *always* demand that the solution be represented as a sum of the form

$$V(\mathbf{r},t) = \sum_{m=-\infty}^{\infty} f_m(t) g_m(\mathbf{r}) \tag{2}$$

where the g_m are members of a complete set, and may themselves be written as sums of simple functions. When separability is imposed on the computational variables we find we need $N \gg D$ for adequate resolution. This is so regardless of whether the attractor is smooth or 'strange'.

On the other hand, it is intuitively clear that for most purposes an infinite number of coefficients is unnecessary; for a smooth solution of the form in equation 1 the magnitudes of the coefficients decrease exponentially for large m, and a finite number of them will be enough. Similarly, given that all temporally organized flows and many weakly turbulent flows possess a high degree of spatial order, it is reasonable to hypothesize that, with an astute choice of basis functions, the solutions should be well represented with N of order $10D$ to $100D$. The problem then reduces to computing the 'right' basis functions, a highly nontrivial task. In this we rely on a good understanding of the physics of the problem, through full numerical simulations at a few parameter values.

Steady or time periodic flows, no matter how spatially complex, have linear eigenmodes which are temporally simple. The spatial structure of these eigenmodes is a good representation of the the physical instabilities in the flow, they can also be computed in a straightforward manner; hence, they are a sensible choice for the starting components of a spectral basis set. We adapt the set of basis functions as follows: In the usual spectral representation, a complete set of known basis function are truncated to a finite, incomplete set. The nonlinear terms in the PDE generate components which lie outside the space spanned by the truncated set. These are ignored, but a quantitative estimate can be made of the error incurred in doing so. If the error becomes too large, the calculation can easily be refined by including more basis functions. Suppose, however, that one has a finite set of functions which are computed in some arbitrary manner so that they are not an obvious subset of a complete and infinite set of basis functions. If, by some quantitative diagnostic these are determined to be insufficient for accurate computations, then a method must be found of generating additional functions to augment the basis. We outline such a method below, drawing on the conceptual structure of inertial manifold theory as developed by Jolly *et. al.* , Titi, and Foias *et. al.* [3,9,10]

The rest of the paper is organized as follows. In section 2 we review the Taylor–Couette flows that are relevant to our discussion, and describe the physical problem that will be used as a test case for our method. In section 3 we present a brief overview of inertial manifold theory, and describe our algorithm, which we refer to as an adaptive spectral method. Section 4 contains our discussion and conclusions.

2   TAYLOR–COUETTE FLOW

We consider the set of states found in narrow gap systems where the outer cylinder is held fixed and the inner cylinder speed increased. The bifurcation parameter is the Reynolds number $R \equiv \Omega a(b-a)/\nu$, where Ω is the inner cylinder frequency, ν is the kinematic viscosity of the working fluid, and a and b are the inner and outer cylinder radii respectively. We nondimensionalize setting the gap width, the inner cylinder speed and the fluid density equal to one. In our computations $a/b = .875$ and we impose strict periodicity in the axial and azimuthal directions with axial wavelength λ and azimuthal wavenumber s.

At $R \equiv R_c$ there is a bifurcation to axially periodic, axisymmetric Taylor vortex flow (TVF), consisting of toroidal vortices wrapped around the inner cylinder. At R a few percent above R_c, a supercritical Hopf bifurcation leads to wavy vortex flow (WVF), in which azimuthally traveling waves appear on the vortices. These are characterized by azimuthal wavenumber m_1 and phase speed c_1. At still higher R the periodic flow becomes modulated (modulated wavy vortex flow or MWV), with the power spectrum characterized by the presence of two incommensurate frequencies. The flow is quasiperiodic relative to the laboratory frame, but periodic in the frame rotating with the original traveling wave. The azimuthal periodicity of the flow may change in this transition. In experimental work Gollub & Swinney[4] found, as R is further increased, a non–hysteretic transition from MWV to chaos. Further experimental work[5] showed that a low–dimensional attractor exists for the chaotic flow. Because of the rotational symmetry MWV is a limit cycle in the correct frame; hence, the transition directly to chaos is not understood by current theories.

In previous work[6], we solved the Navier-Stokes equations using a pseudo-spectral initial value code, and showed that several distinct branches of modulated waves exist, and that several routes to chaos are exhibited by this system. Relevant to this paper is a numerically discovered mode competition between two simultaneously growing Floquet modes (linear eigenmodes of the unstable WVF), which we have termed 'ZS' and 'GS' modes. Our observations have been experimentally confirmed[7].

2.1 Instability of WVF

We have discussed elsewhere the role of the outflow jet between adjacent Taylor vortices in this sequence of instabilities. The radial outflow produces a correspondingly strong azimuthal jet, which becomes sharper and narrower as R increases, for both TVF and WVF. The WVF becomes unstable to a Floquet mode, which has space-time dependence of the form $e^{i(m_2\theta - \omega t)}\mathbf{f}(r, \theta, z, t)$. Here m_2 is the azimuthal wavenumber of the modulation, ω is real at onset, and $\mathbf{f}$ has the space-time periodicity of the unstable WVF. This functional form is consistent with the patterns observed by Gorman & Swinney, and the theoretical work of Rand[8].

In the frame rotating with the initial WVF, the MWV is time periodic and it is trivial to separate the steady-state flow (defined as $\bar{\mathbf{v}}$) from the fluctuating (or quasiperiodic) piece ($\mathbf{v}^{qp}$). The latter is equal to the Floquet mode near onset. Spatially, the Floquet mode is concentrated in the region of the outflow jet and consists of a set of small vortices, which are best visualized by plotting the azimuthal vorticity ω_θ^{qp}, as in figure 1. Contours of ω_θ^{qp} are plotted on an unrolled cylindrical surface parallel to the inner cylinder at mid-gap for the ZS (1a) and GS (1b) modes. In both cases, the fluctuating field consists of compact, three-dimensional vortex pairs sitting exactly on top of the outflow. For the ZS mode in figure 1a, $m_2 = 3m_1$, and for the GS flow in figure 1b $m_2 = m_1$. As a function of time, the vortex pairs drift along the outflow at a mean speed equal to the modulation frequency (relative to the frame rotating with the underlying WVF) divided by the wavelength $2\pi/m_2$ of the Floquet mode. The visual signature of the ZS mode is small, rapidly moving ripples at the outflow, and of the GS mode a periodic flattening of the outflow contour. The space time symmetry of these two modes is identical, thus mode competition can occur.

As a function of R, we see the following sequence of flows[7] (with $s = 4$ and $\lambda = 2.5$): At $R/R_c = 8.50$ there is a ZS equilibrium with the GS mode evident as a decaying transient. At $R/R_c = 9.80$ the GS mode is stable, and transients are dominated by the decaying ZS mode. There is an intermediate range of R near $9R_c$ where both modes are present with positive growth rate, and the flow appears to be chaotic. The long term temporal behavior of the flow over this range is as yet undetermined. We infer that the growth rate of the ZS mode is positive below about $R/R_c = 9.0$ and negative above, while the growth rate of the GS mode is negative below roughly $R/R_c = 9.5$ and positive above. Presumably, in an extended parameter space there is a codimension two point at which both their eigenvalues cross the imaginary axis simultaneously. To find such a point through the computations is prohibitively expensive; hence this information is not useful here.

3 LOW–DIMENSIONAL MODEL

We are interested in the physics and mathematics of this problem in itself: if, when and how the mode competition leads to chaos, whether there is mode locking, what the effects of the rotational symmetry are on the dynamics, *etc*. Fortunately, it is also an ideal test problem for the development of adaptive spectral methods: it is a natural candidate for the spectral decomposition described above, and we can make precise quantitative comparison between the new method and both experiments and traditional spectral computations. In addition, the behavior of the system is not known *a priori* so we can be sure that we are not building the results into our model. Before describing the algorithm, we briefly outline some ideas from inertial manifold theory which are used in our discussion[9].

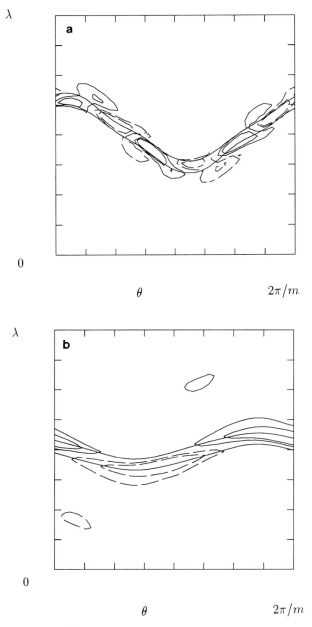

Figure 1. Contours of ω_θ^{qp} in the (θ, z) plane at midgap for a) the ZS mode at $R/R_c = 8.5$, $m_2 = 3m_1$, contour interval $= \pm 0.1$, and b) the GS mode at $R/R_c = 9.8$, $m_2 = m_1$, contour interval $= \pm 0.3$. Dashed contours represent negative values, solid contours are positive. The outflow jet has been indicated by the solid contour on which $v_\theta = 0.2$.

3.1 Inertial manifold theory

An inertial manifold $\mathcal{M}$ for a PDE, if it exists, is a a smooth, finite dimensional manifold in the (infinite dimensional) phase space of the system which attracts trajectories exponentially. When $\mathcal{M}$ exists, asymptotically the entire dynamics of the PDE is described by a finite-dimensional system of ordinary differential equations (ODE's), called the inertial form, which is obtained by restricting the equations of motion to $\mathcal{M}$. If a global attractor $\mathcal{A}$ exists then $\mathcal{A} \subset \mathcal{M}$. In general $\mathcal{A}$ may not be smooth, while $\mathcal{M}$ is smooth by definition, and $\mathcal{A}$ may attract trajectories arbitrarily slowly. Thus $\mathcal{M}$ is the more practical object to work with. The goal of the theory is to construct an approximate inertial manifold (AIM) explicitly, project the equations onto this manifold, and solve the simpler system. Existing work suggests that such a procedure can lead to significant improvements in the truncation error[10]. Unlike center manifold equations, the inertial form may be valid for many bifurcations, as long as the global attractor remains in the space defined by the AIM.

Formally, one separates the PDE into a linear and a nonlinear term (which we assume without loss of generality to be quadratic):

$$\dot{u} = \mathcal{L}u + \mathcal{F}(u,u). \tag{3}$$

We define $\mathcal{P}$ as the projection onto a finite subspace of the full phase space $\mathcal{H}$, and $\mathcal{Q} \equiv \mathcal{I} - \mathcal{P}$ (where $\mathcal{I}$ is the identity operator) as the complementary projection. Setting $p = \mathcal{P}u$ and $q = \mathcal{Q}u$ we obtain:

$$\dot{p} = \mathcal{P}\mathcal{L}p + \mathcal{P}\mathcal{F}(p+q, p+q) \tag{4}$$

$$\dot{q} = \mathcal{Q}\mathcal{L}q + \mathcal{Q}\mathcal{F}(p+q, p+q). \tag{5}$$

Note that if the basis functions p, q are chosen to be eigenfunctions of $\mathcal{L}$, then the projection operators can be dropped from the linear terms. If $\mathcal{M}$ is of dimension less than or equal to the dimension of $\mathcal{P}\mathcal{H}$, then formally it can be defined as the graph[9] of a function $q = \Psi(p)$. Thus the p's are the coordinates on $\mathcal{M}$ and the q's are the coordinates over the rest of the space. Clearly, how the space is to be divided into p's and q's is an important consideration, which is in general problem dependent.

To construct the inertial form one must calculate the function $\Psi(p)$ which models the action of the neglected modes, through the nonlinearity, on the modes that are kept. To calculate it exactly requires solving the full problem. Several methods of constructing an approximating function $\bar{\Psi}(p)$ have been proposed[9], which rely on iterating some form of equation 5 with $\dot{q}$ set equal to zero, and are thus defined satisfactorily only for steady states. An approximate function $q = \bar{\Psi}(p)$ defines the AIM, and the inertial form is constructed by replacing equation 4 with

$$\dot{p} = \mathcal{P}\mathcal{L}p + \mathcal{P}\mathcal{F}(p+\bar{\Psi}(p), p+\bar{\Psi}(p)) \tag{6}$$

and dropping equation 5. For example, in a traditional spectral method the solution u is represented as a truncated sum of N basis functions. These are the p's, the neglected basis function with index greater than N are the q's, and $\mathcal{M}$ is approximated by $q = \bar{\Psi}(p) = 0$.

3.2 Adaptive spectral methods

In this section we discuss our proposed method in the context of the Taylor–Couette problem described above. We use the formal language of inertial manifold theory, but forgo any attempts to be rigorous.

Let $\mathbf{u}_0$ be the equilibrium WVF solution to the Navier–Stokes equations in the appropriate frame (so it is steady state). For an arbitrary solution $\mathbf{u}$ we set $\mathbf{u} = \mathbf{u}_0 + \mathbf{u}'$. We define the operator $\mathcal{L}(\mathbf{u}_0)$ to be the Navier–Stokes linearized around $\mathbf{u}_0$, and the bilinear operator $\mathcal{B}(\mathbf{u}, \mathbf{v}) \equiv (\mathbf{u} \cdot \nabla)\mathbf{v}$. With this definition, $\mathcal{L}(\mathbf{u}_0)$ has two growing eigenmodes, which are the GS and ZS Floquet modes. Let $\mathcal{P}$ be the projection onto the space spanned by these modes (which has dimension 4), $\mathcal{Q}$ be the complementary projection, and set $p = \mathcal{P}\mathbf{u}'$ and $q = \mathcal{Q}\mathbf{u}'$.

We first define our AIM by setting $q = \bar{\Psi}(p) = 0$, and solve

$$\dot{p} = \mathcal{L}(\mathbf{u}_0)p + \mathcal{P}\mathcal{B}(p,p) \tag{7}$$

using an initial value algorithm. We also monitor the size of the neglected term $\mathcal{Q}\mathcal{B}(p,p)$, which is a quantitative measure of how well this AIM approximates the true local (in time) solution. When this term becomes significant, we define a new basis function, orthogonal to the original set but capturing the structure along the escaping direction. We then continue the computation using the augmented basis set, redefining $\mathcal{P}$ and $\mathcal{Q}$ appropriately.

Enlarging the basis set $p \to p' = p + \phi(p)$ in equation 4 (where $\phi(p)$ is the piece of $\mathcal{B}(p,p)$ that does not lie in span$\{p\}$) is similar to defining a new AIM by $q = \bar{\Psi}(p) = \phi(p)$. Thus we are allowing the initial value computation to locally define the best approximation to $\mathcal{M}$. Also, because we keep the dynamical equation for $\phi(p)$, our method is effectively following the deformations of this AIM as a function of time. When necessary, another basis function will be added, which corresponds to further refinement of the AIM. Note there is no need to assume here a clean separation of the large, slow scales from the short, fast scales.

Beyond the first iteration the situation is complicated somewhat by the fact that the linear operator $\mathcal{L}(\mathbf{u}_0)$ no longer commutes with the projection operators. The construction of these operators, and the matrix of coupling coefficients entailed by $\mathcal{B}$, is computationally intensive, however it needs to be done only once for each round of time integration. The time integration itself is trivial. The fact that the spatial structures present in the physical flow are well represented by the Floquet modes implies that the process should converge rapidly to a 'good enough' set of coordinates (or basis functions) p. This method can also be used to compute the unstable steady state WVF in this parameter regime, and in that case should be exact since the only unstable directions are those defined by the Floquet modes.

4 DISCUSSION

We have presented an algorithm for computing accurate solutions to the Navier–Stokes equations for Taylor–Couette flow, using a very small set of ordinary differential equations, the implementation of which is still in progress. We have described a physical flow, accessible in the laboratory, which will be used as a test case for the numerical procedure. We refer to the procedure as an adaptive spectral method: spectral because we convert the PDE to a set of ODE's by representing it as a sum of spatial functions multipled by time dependent amplitudes, and adaptive because we follow the calculation as it evolves along new directions in phase space and use these to construct new basis functions.

We would like to emphasize that, unlike the approaches developed by Aubrey, Lumley and others[11], we are not trying to derive from the full problem a simpler, approximate model with qualitatively similar dynamics. Our aim in this work is to maintain the high accuracy of full simulation while vastly increasing the efficiency of our computations. We do not claim that our method is rigorously justified; in fact, the mathematical underpinnings of our algorithm could be considered somewhat tenuous. It is thus essential to be able to test results from the truncated system against both full simulations and laboratory experiments. We will then be in a position to state definitively whether this procedure is viable. We hope that this work will lead to generally useful algorithms for studying systems which combine nontrivial spatial structure with complex, but low-dimensional, temporal dynamics.

REFERENCES

1. M. J. Berger and J. Oliger, **J. Comp. Phys.** 53:4, (1984).
2. D. Gottleib and S. A. Orszag, "Numerical Analysis of Spectral Methods", SIAM, Philadelphia, (1977).
3. P. Constantin, C. Foias, B. Nicolaenko and R. Témam, "Integral Manifolds and Inertial Manifolds for Dissipative Partial Differential Equations", Applied Mathematics Sciences, No. 70, Springer, Berlin, (1988).
4. J. P. Gollub and H. L. Swinney, Onset of Turbulence in a rotating fluid, **Phys. Rev. Lett.** 35:927 (1975).
5. A. Brandstater and H. L. Swinney, Strange attractors in a weakly turbulent Couette-Taylor flow, **Phys. Rev. Lett.** 35:2207 (1987).
6. K. T. Coughlin and P. S. Marcus, Modulated waves in Taylor-Couette flow; Parts 1 and 2, to appear in **J. Fluid Mech.**, (1991).
7. K. T. Coughlin, P. S. Marcus, R. P. Tagg and H. L. Swinney, Distinct quasiperiodic modes with like symmetry in a rotating fluid, **Phys. Rev. Lett.** 66:1161 (1991).
8. M. Gorman, H. L. Swinney and D. Rand, Doubly periodic circular Couette flow: experiments compared with predictions from dynamics and symmetry, **Phys. Rev. Lett.** 16:992 (1981).
9. The presentation here is drawn primarily from M. S. Jolly, I. G. Kevrekidis, and E. S. Titi, Approximate inertial manifolds for the Kuramoto-Sivashinsky equation: Analysis and computations, **Physica D** 44:38 (1990).
10. E. S. Titi, On approximate inertial manifolds to the Navier-Stokes equations, **J. Math. Anal. App.** 149:540 (1990).
11. N. Aubrey, P. Holmes, J. L. Lumley, and E. Stone, The dynamics of coherent structures in the wall region of a turbulent shear layer, **J. Fluid Mech.** 192:115 (1988).

THE COUETTE TAYLOR PROBLEM IN THE SMALL GAP APPROXIMATION

Yves Demay, Gérard Iooss, and Patrice Laure
Institut Non Linéaire de Nice, UMR CNRS 129, Université de Nice
Parc Valrose, 06034 NICE Cedex, France

1 INTRODUCTION

We consider the classical Couette-Taylor problem in the limiting case when the radii ratio ($\eta = \frac{R_1}{R_2}$) is very close to 1 and the critical Reynolds number is very large. It is well know that the critical modes which destabilise the Couette flow are either stationary axisymmetric modes or oscillatory non axisymmetric modes with an integer azimutal wavenumber. Langford *and al.* [1] observe when η tends to 1 that the most critical oscillatory modes come altogether (Figure 4 in [2]). In this paper, our main motivation is to analyse the transition between axisymmetric and non axisymmetric modes at this limit and to determine the selected azimuthal wavelength. By this way, we give some informations on the occurrence of oscillatory motion when the two cylinders move in the opposite direction. The governing equations considered here are deduced from the Navier-Stokes equations by a suitable choice of scale taken at the limit $\eta = 1$. The new dimensionless parameters are the inner and outer Taylor numbers T_i, and, the basic flow is the planar Couette flow. The linear analysis points out the possibility of oscillatory instabilities for negative value of T_2. The bifurcated structures are studied by means of Ginzburg-Landau type of equations. Thus, we can take into account the interaction between critical modes with spatial wave numbers close to the critical one. In addition, we give precisely the coefficients occuring these equations. At the critical point, these coefficients can be obtained in the same way as those of classical amplitude equations ([3]).

2 GOUVERNING EQUATION

Let us denote respectively by R_1, R_2 and Ω_1, Ω_2 the radii and rotation rates of the inner and outer cylinders. We consider the case when the inner and eventually the outer Reynolds numbers $\mathcal{R}_j = \frac{R_j \Omega_j d}{\nu}$, $j = 1$ and 2, (ν being the kinematic viscosity) are very large when η is close to 1. We use the same scales as Tabeling [4] and the parameters are T_j, $j = 1$ and 2, defined by

$$T_j = \frac{R_j \Omega_j d}{\nu} \sqrt{2(1-\eta)}, \quad j = 1, 2 \tag{1}$$

where $d = R_2 - R_1$ and T_1 is positive while T_2 may be negative.

Ordered and Turbulent Patterns in Taylor-Couette Flow
Edited by C.D. Andereck and F. Hayot, Plenum Press, New York, 1992

More precisely, we choose a new set of dimensionless variables defined by the relation

$$x = \frac{r}{d} - \frac{R_1 + R_2}{2d}, \quad y = \theta\sqrt{\frac{2R_1}{d}}, \quad z = \frac{Z}{d} \tag{2}$$

where r is the distance from the axis of cylinders, θ the azimuthal coordinate and Z the axial coordinate. In addition, the scaling factors for velocity are taken in a such way that the incompressibility condition is recovered in the new cartesian coordinates :

$$v_x^* = v_z^* = \frac{\nu}{d}, \quad v_y^* = \frac{\nu}{d}\sqrt{\frac{R_1}{2d}}. \tag{3}$$

After suppression of terms of order $O(1-\eta)$ in the Navier-Stokes equations written in cylindrical coordinates, we have a solution which corresponds to the Couette solution given by

$$V_0 = (0, v_0(x), 0), \text{ with } v_0(x) = (T_2 - T_1)x + \frac{T_2 + T_1}{2} \tag{4}$$

and the perturbation U now satisfies ([5])

$$\begin{cases} \frac{\partial U}{\partial t} = \Delta_{xz}U - v_0(x)\frac{\partial U}{\partial y} + \begin{pmatrix} v_0(x)u_y + \frac{u_y^2}{2} \\ -v_0'(x)u_x \\ 0 \end{pmatrix} - (U.\nabla)U - \begin{pmatrix} H_x(x,z) \\ H_y(y) \\ H_z(x,z) \end{pmatrix}, \\ \frac{\partial H_x}{\partial x} = \frac{\partial H_x}{\partial z}, \frac{\partial H_y}{\partial x} = \frac{\partial H_y}{\partial z}, \\ \nabla.\, U = 0, \end{cases} \tag{5}$$

with the boundary conditions

$$U|_{x=\pm 1/2} = 0, \tag{6}$$

where Δ_{xz} is the usual Laplace operator in the two coordinates x and z. System (5) differs from the system used by Tabeling [4] by the "pressure" term. In fact, it is shown at the limit $\eta \to 1$ that there is an additional variable H_y on its second component which is only a function of y (in [4] H_y is 0, which is wrong for y dependent motion)

Let us finally observe that the system (5) is translation invariant in y and z directions, and is invariant under $z \mapsto -z$ symmetry.

3 LINEAR STABILITY ANALYSIS

We look for eigenmodes of the form

$$U = \hat{U}(x)e^{i(\alpha z + \beta y)} \tag{7}$$

belonging to some eigenvalue σ. Let us remark that the eigenvalues have the following properties due to the form of the linearized operator from equation (5) and the symmetry $z \mapsto -z$:

$$\sigma(-\alpha, \beta, T_1, T_2) = \sigma(\alpha, \beta, T_1, T_2) \text{ and } \sigma(\alpha, -\beta, T_1, T_2) = \overline{\sigma(\alpha, \beta, T_1, T_2)} \tag{8}$$

The most unstable mode is given by the eigenvalue σ_0 with largest real part, and the neutral stability curve in the (T_1, T_2) plane, corresponds to the function

$$T_{1c} = \min_{\alpha,\beta} T_1(\alpha, \beta, T_2) \tag{9}$$

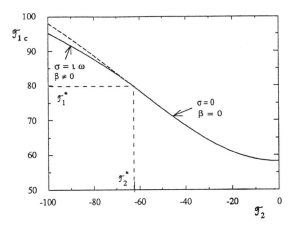

Figure 1. Neutral stability curve.

with T_1 solution of $\Re(\sigma_0(\alpha,\beta,T_1,T_2)) = 0$ for fixed values of T_2, α and β. This gives in particular the critical wavelengths α_c (Table 1), β_c (Figure 2) and the neutral curve $T_{1c}(T_2)$ is plotted in Figure 1. The critical curve is obtained for $\beta = 0$ and $\sigma_0 = 0$ up to a value T_2^* (~ -62.5) of T_2. For $T_2 < T_2^*$, the minimum corresponds to positive value of β and a complex eigenvalue. In order to explain this transition occuring at point (T_2^*, T_1^*), we analyse more precisely the dependency of critical variables with respect to β. We solve numerically the equation $\Re(\sigma_0(\alpha,\beta,T_1,T_2)) = 0$ for various values of T_2, and, the following functions

$$T_1'(\beta, T_2) = \min_\alpha T_1(\alpha, \beta, T_2), \qquad (10)$$

are plotted in Figure 2. Using the properties (8), the Taylor expansion of σ_0 in the neighbourhood of $(\alpha_c, \beta = 0, T_{1c})$ yields for $T_2 > T_2^*$

$$\sigma_0 = ia_1\beta + a_2(T_1 - T_{1c}) - a_3(\alpha^2 - \alpha_c^2)^2 - a_4\beta^2 - ia_5\beta(\alpha^2 - \alpha_c^2) + \ldots \qquad (11)$$

where the coefficients a_i are real. In this way, we show that at leading order, T_1' verifies the relation

$$T_1' = T_{1c} + \frac{a_4}{a_2}\beta^2 + \ldots \qquad (12)$$

Consequently, we can deduced from figure 2 that the coefficient a_4 changes its sign at $T_2 = T_2^*$, and for $T_2 < T_2^*$ the criticality is now reached for non zero azimuthal wavelength β. Moreover, due to coefficients a_5 and a_1, σ_0 becomes complex as soon as β is non zero. Let us finally remark that for the monotonic transition the curves plotted in Figure 2 are very flat, so $T_1'(\beta)$ is very close to T_{1c} for small values of β.

4 WEAKLY NON LINEAR ANALYSIS

We present the envelope (or Ginzburg-Landau) equations which describe the spatio-temporal evolution of the amplitude of the perturbation at criticality. The form of Taylor expansion of these equations is obtained by means of invariance properties, and in addition the numerical computation of main coefficients are been made. As only the sign and the weight of each

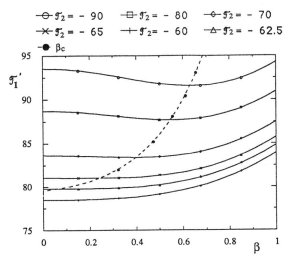

Figure 2. Graphs of $T_1'(\alpha_c, \beta, T_2)$ for fixed values of T_2.

coefficient with respect to others are important, we only present the relevant equations obtained after changes of variable and scales. In all the cases, they are given in a rigid frame moving in y direction at velocity proportional to a_1 ($= i\partial\sigma_0/\partial\beta$). This allows us to eliminate the advection term ($\partial A/\partial y$) which means that any disturbance of the bifurcated solution propagates along the mean flow at constant speed. Let us note that this term comes from the anisotropy of the problem. The computations show that the bifurcation towards rolls are becoming subcritical at $T_2 \sim -48.5$. Then, in the monotonic case we have two situations of high codimension which correspond to the annulation of third order nonlinear coefficient and the annulation of coefficient a_4 as noted before. We finally exhibit four different GL equations as T_2 decreases.

4.1 Steady bifurcation.

Let first consider the case where the critical eigenvalue and the critical wavelength β are null. The eigenmode is $U_0(x)\exp(\alpha_c z)$ and its complex conjugate $\bar{U}_0\exp(-\alpha_c z)$. If we denote by A the amplitude of the perturbation, we obtain when the first nonlinear term is negative ($T_2 > -48.5$) the same type of equation as already presented by Tabeling [4]

$$\partial_t A = \mu A + \partial_{z^2}^2 A + \partial_{y^2}^2 A + ia\partial_{yz}^2 A - A^2\bar{A} \tag{13}$$

where the bifurcation parameter μ is proportional to $T_1 - T_{1c}$; values of coefficient a are given in Table 1.

Equation (13) has non trivial solutions of the form :

$$A_0 = Q\exp i(qy + pz + \Omega t) \text{ with } Q^2 = \mu - p^2 - q^2 \text{ and } \Omega = -apq \tag{14}$$

where p measures the dilatation (or the compression) of rolls and q the modulation in azimuthal direction. Following [4] and [6], one obtains an equation describing the dynamics of the phase variable ϕ of a solution $A = A_0(qy + pz + \Omega t + \phi(z, y, t))$. A generalization of the

Table 1. Values of coefficient a occuring in equation (13).

T_2	α_c	T_{1c}	a
0.	3.13	58.21	.15
-10.	3.14	58.86	.21
-20.	3.15	60.98	.29
-30.	3.19	64.31	.39
-40.	3.24	68.55	.51
-45.	3.27	70.91	.58

Tabeling's approach yields,

$$\frac{\partial \phi}{\partial t} + v_y \frac{\partial \phi}{\partial y} + v_z \frac{\partial \phi}{\partial z} = \alpha_{20} \frac{\partial^2 \phi}{\partial y^2} + \alpha_{11} \frac{\partial^2 \phi}{\partial y \partial z} + \alpha_{02} \frac{\partial^2 \phi}{\partial y^2} \tag{15}$$

with

$$v_y = ap, \quad v_z = aq,$$

$$\alpha_{20} = 1 - 2\frac{q^2}{Q^2}, \quad \alpha_{11} = -\frac{4pq}{Q^2}, \quad \alpha_{02} = 1 - \frac{2p^2}{Q^2}.$$

On the physical grounds, v_y and v_z mean that any disturbances along the y- (resp. z-) direction propagate in the z- (resp. y-) direction. In the moving frame at velocity $v = (0, v_y, v_z)$, the two conditions

$$\alpha_{20} > 0 \text{ and } 4\alpha_{20}\alpha_{02} - \alpha_{11}^2 > 0$$

giving the stability of the solution (14) yields

$$\mu \geq 3(q^2 + p^2) \tag{16}$$

which is the classical Eckhaus criterion.

The numerical computations show that for $T_2 \sim -48.5$ the first nonlinear coefficient of the amplitude equation vanishes. Thus, it is necessary to expand the previous equation up to fifth order. As this coefficient is negative, we obtain

$$\partial_t A = \mu A + \partial_{z^2}^2 A + \partial_{y^2}^2 A + i a \partial_{yz}^2 A + \epsilon A^2 \bar{A} - A^3 \bar{A}^2$$
$$+ i f_1 A^2 \partial_z \bar{A} + i f_2 A \bar{A} \partial_z A + g_1 A^2 \partial_y \bar{A} + g_2 A \bar{A} \partial_y A + ... \tag{17}$$

where μ and ϵ are the two small parameters describing this codimension two bifurcation; other coefficients are given in Table 2.

Table 2: values of coefficients occuring in equation (17).

T_2	α_c	T_1	a	f_1	f_2	g_1	g_2
-48.5	3.305	72.62	.64	-.54	-.70	17.02	1.38

If we neglect the spatial dependency with respect to y, we recover the analysis presented in Refs [7] and [8] for the degenerate case. Here, computations are more complicated, and the non trivial solution (14) verifies the relations

$$Q_\pm^2 = \frac{\epsilon + pf \pm \sqrt{\Delta}}{2}, \quad \Omega = -(apq + Q^2 gq)$$

with $f = f_1 - f_2$, $g = g_1 - g_2$ and $\Delta = (\epsilon + pf)^2 + 4(\mu - p^2 - q^2)$. We show that the solution Q_-^2 may exist in a specified domain under the surface ($\mu_c = p^2 + q^2$), but it is always unstable. However, the coefficients of the phase equation (15) around the solution Q_+^2 are now

$$\alpha_{20} = 1 - \frac{2q^2}{\delta}(\frac{1}{X} - \frac{gg_1}{\delta} + \frac{g^2 q^2}{X\delta^2})$$

$$\alpha_{11} = -\frac{q}{\delta}[-ag + 2f_2 + 4\frac{p}{X} + g_1 g \frac{-2p+fX}{\delta} - 2g^2 pq^2 \frac{-2p+fX}{X\delta^2}]$$

$$\alpha_{02} = 1 - \frac{-2p+fX}{2\delta}[-\frac{2p+X(f_1+f_2)}{X} + g^2 q^2 \frac{(-2p+fX)}{X\delta^2}]$$

where $X = Q_+^2$ and $\delta = \sqrt{\Delta}$. It is difficult to exhibit a so nice relation as the Eckhaus criterion, but numerical trials show that the surface delimiting the fields where the solution Q_+^2 is stable is close to the paraboloid (16). It is under this paraboloid in three sectors ($p > 0$ and $q \sim 0$; $p \sim 0$ and either $q > 0$ or $q < 0$) starting from the point $(\mu, p, q) = (0, 0, 0)$, and above it in the other cases.

The next codimension two situation occuring at $T_2 \sim -62.5$ is related to the following envelope equation

$$\partial_t A = \mu A + \partial_{z^2}^2 A - \nu \partial_{y^2}^2 A - \partial_{y^4}^4 A + ia\partial_{yz}^2 A + \epsilon A^2 \bar{A} - A^3 \bar{A}^2$$

$$+ if_1 A^2 \partial_z \bar{A} + if_2 A \bar{A} \partial_z A + g_1 A^2 \partial_y \bar{A} + g_2 A \bar{A} \partial_y A + ...$$

(18)

where now the two small parameters are ν and μ. As ϵ is rather important, the rolls are very subcritical and we have to take into account terms μA^2 and μ^2 in the expansion (18). Thus, the complete study of this case has not been made.

Table 3. values of coefficients occuring in equation (18).

T_2	α_c	T_1	ϵ	a	f_1	f_2	g_1	g_2
-62.5	3.462	79.77	.50	.64	-.31	-.56	19.36	1.02

4.2 Hopf bifurcation

For $T_2 < T_2^*$, we have four critical modes $U_0(x) \exp i(\alpha_c z + \beta_c y)$, $U_1(x) \exp i(-\alpha_c z + \beta_c y)$ and complex conjugates. The Ginzburg-Landau equations have the following form :

$$\partial_t A = a_2 \mu A + i\omega A + 2b_1 \alpha_c \partial_z A + 4a_3 \alpha_c^2 \partial_{z^2}^2 A + a_4 \partial_{y^2}^2 A + 2ia_5 \partial_{yz}^2 A + bA^2 \bar{A} + cAB\bar{B}$$

(19)

$$\partial_t B = a_2 \mu B + i\omega B - 2b_1 \alpha_c \partial_z B + 4a_3 \alpha_c^2 \partial_{z^2}^2 B + a_4 \partial_{y^2}^2 B + 2ia_5 \partial_{yz}^2 B + bB^2 \bar{B} + cA\bar{A}B$$

where A and B are the amplitudes of the perturbation associated to critical modes U_0 and U_1 respectively. Neglecting spatial dependencies, equation (19) possesses two types of non trivial solutions, the travelling waves ($A \neq 0$ and $B = 0$) or standing waves($|A| = |B|$). The former is stable with respect to spatially homogeneous perturbation when $\Re(b) < 0$ and $\Re(c) - \Re(b) < 0$, the latter when $\Re(b) < 0$ and $\Re(c) - \Re(b) > 0$. Unfortunately, the numerical computations show that neither of these solutions is stable.

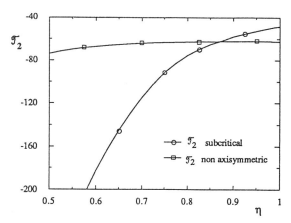

Figure 3. Evolution versus η of $\mathcal{T}_2$ giving the subcriticality (O) of Taylor vortices and the interaction between axisymmetric and first non axisymmetric critical modes($\square$)

5 CONCLUSION

The determination of the two points of high codimension for $\mathcal{T}_2$ equal to -48.5 or -62.5 can be deduced from previous works on the Couette-Taylor problem by a limit process as η tends to 1. As shown on Figure 3, the former value corresponds to the outer Taylor number for which the Taylor vortices are becoming subcritical, and the latter to the interaction between the axisymmetric mode and the first non axisymmetric mode. But with the small gap formulation, we have now a smooth transition between steady axisymmetric critical modes and non axisymmetric oscillatory modes, and the azimuthal wavelength increases continuously from 0 as $\mathcal{T}_2$ decreases from $\mathcal{T}_2^*$. This fact was a priori not obvious with a formulation using outer and inner Reynolds numbers as parameters ([1]). Moreover, this work gives an example of bifurcations with anisotropic properties which are characterized by the presence of advection and cross derivative terms in the amplitude equations and the fact that any disturbance azimuthal direction involves a temporal modulation.

References

[1] W. Langford, R. Tagg, E. Kostelich, H. Swinney and M. Golubitsky, Phys. Fluids, 31, 776 (1987).

[2] M. Golubitsky, W.F. Langford, Physica D 32, 362-392 (1988).

[3] Y. Demay and G. Iooss, J. Méca. Théo. Appli., n^0 spécial, 193 (1984).

[4] P. Tabeling, J. Physique-lettres, 44, 16, 665 (1983).

[5] Y. Demay, G. Iooss and P. Laure, Eur. J. Mech. B/Fluids, in print (1992).

[6] J. Lega, Thesis, Nice University (1989).

[7] W. Eckhaus and G. Iooss, Phys. D 39, 124 (1989).

[8] A. Doelman, Thesis, Utrecht University (1990).

STRUCTURE OF TAYLOR VORTEX FLOW AND THE INFLUENCE OF SPATIAL AMPLITUDE VARIATIONS ON PHASE DYNAMICS

D. Roth[1], M. Lücke[1], M. Kamps[2], and R. Schmitz[3]

[1]*Institut für Theoretische Physik, Universität des Saarlandes*
D–6600 Saarbrücken, FRG
[2]*Höchstleistungsrechenzentrum, Stabsstelle Supercomputing*
D–5170 Jülich, FRG
[3]*Forschungszentrum Jülich GmbH, Institut für Festkörperforschung*
D–5170 Jülich, FRG

In the first part the spatial structure of Taylor vortices at moderate supercritical Reynolds numbers is analysed in detail using 2–dimensional numerical simulation data. For radius ratios $\eta = 0.75, 0.88$ the dependence of axial Fourier modes on the Reynolds number is presented in the case where only the inner cylinder rotates as well as for co — and counterrotating outer cylinder. In the second part the dynamical response of the vortex structure to small perturbations is investigated, in particular for the experimentally relevant situations where the amplitude of the flow is not constant over the whole systems. A simple generalized diffusion equation for the phase is derived. The space dependence of the amplitude $R(z)$ influences the time evolution of the phase of the structure. In the experimentally most common case of a finite system with rigid end plates the amplitude enhancement near the ends (Ekman vortex) leads to a substantial increase of the long–time relaxation rate of the phase. Quantitative agreement between phase description and numerical simulation of the full Navier Stokes equations is found and qualitative agreement with experimental results.

The Taylor–Couette system shows a rich variety of interesting features and allows quantitative experimental as well as theoretical studies. Extensions of G.I. Taylor's fundamental work[1] as changes in the geometry, different fluids, additional forces (heat, rotation, pressure, modulation) have discovered new patterns and bifurcation scenarios. But there are also tasks left in the original setup. Two of them we want to attack in this paper: First we will study the nonlinear structure of the Taylor vortices in detail which has not yet been done for independently driven outer and inner cylinders. Secondly we will investigate the influence of end plates involving Ekman vortices on the phase diffusion response of the Taylor rolls.

NOTATION AND INVESTIGATION METHOD

Two co–axial cylinders of length L, radii R_1 and R_2, gapwidth $d = R_2 - R_1$, rotating with angular velocities Ω_1 and Ω_2 are considered. The system properties are given by a radius ratio $\eta = R_1/R_2$, a rotation ratio $\mu = \Omega_2/\Omega_1$, and a control parameter $\epsilon = \Omega_1/\Omega_{1c} - 1$ relative to the critical velocity Ω_{1c} for the onset of Taylor vortices at $\mu = 0$. The fluid velocity is decomposed in cylindrical coordinates $\mathbf{u} = u\mathbf{e}_r + v\mathbf{e}_\varphi + w\mathbf{e}_z$.

The full axisymmetric Navier–Stokes equations are solved numerically using a SOLA version of the MAC finite difference scheme with an iterative solver for the pressure[2]. Boundary conditions at top and bottom ends are periodic for the structure investigation and rigid in the phase diffusion case.

STRUCTURE OF STATIONARY TAYLOR VORTICES

Since there is no closed expression describing the shape of Taylor rolls in the η, μ, ϵ–parameter space, one needs a lot of single data to cover the whole range of interest. Most authors[3,4] in the literature have restricted their analyses to $\mu=0$, the case when only the inner cylinder is rotating. Recently $\mu \neq 0$, in particular $\mu<0$, the counterrotating case, has been attracting much interest. Hence we present here the μ–dependence of some characteristic values, as the axial Fourier coefficients of the fields and the radial position of the roll centers. Those data can be useful to determine

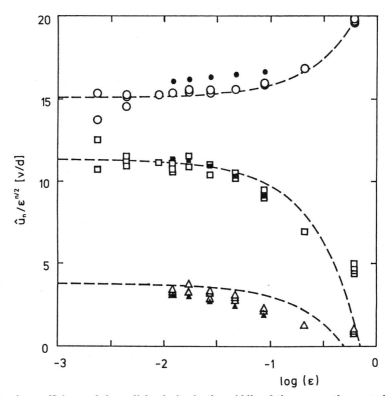

Fig. 1. Fourier coefficients of the radial velocity in the middle of the gap vs the control parameter ϵ. Open (full) symbols refer to radius ratio $\eta = 0.75$ (0.893), circles belong to n = 1, squares to n = 2, and triangles to n = 3. Dashed lines are fits to the $\eta = 0.75$–data according to eq. (1).

the validity range of model representations of the Taylor vortices like the Ginzburg–Landau equation (GLE) or few–mode Galerkin models, and to compare the change of structure caused by external forces as pointed out in the introduction.

According to Davey[5] we decompose the radial velocity

$$u(r,z) = \sum_{n=1}^{\infty} \hat{u}_n(r) \cos(nkz) \qquad (1a)$$

where the Fourier coefficients $\hat{u}_n(r)$ have the form

$$\hat{u}_n(r) = \epsilon^{n/2} A_n(r) [1 + \epsilon B_n(r)]. \qquad (1b)$$

Similar expressions hold for v, w and the pressure p. Fig. 1 recalls[4] the way how we get the amplitudes A_n and B_n. For $\eta = 0.75$ and $\eta = 0.88$ the Fourier coefficients are plotted versus ϵ. The different values belong to various k. Davey's formula (1) holds quite well and the results depend weakly on η. The A_n and B_n found by fitting the numerically obtained data to eq. (1b) are compiled in table 1. The quotient A_2/A_1 is rather large which turns out to be the typical situation for counterrotating cylinders (see below). Consequently an accurate description of Taylor vortex flow requires more than only one Fourier mode. That is also obvious from the inflow/outflow asymmetry. A second remark on table 1 is that the correction terms B_n are large in comparison to 1 except for the case of w_1. Due to this fact the w–component is the best choice for a single–mode representation of Taylor rolls.

Fig. 2 shows the μ–dependence of the Fourier coefficients. Since the radial shape changes with μ we have taken the position where $|\hat{u}_n(r)|$ is maximal. These maximum

Table 1. Coefficients obtained by fitting the first three Fourier coefficients $\hat{u}_n$, $\hat{w}_n$, $\hat{p}_n$, for small ϵ by $\epsilon^{n/2} A_n(1 + \epsilon B_n)$. Units are d for r, ν/d for the velocities, and $\rho \nu^2/d^2$ for p.

	$u(r_1 + \tfrac{1}{2})$	$w(r_1 + \tfrac{1}{4})$	$p(r_1 + \tfrac{1}{4})$
A_1	15.2	15.8	−214
B_1	0.5	−0.07	−0.34
A_2	11.1	−6.0	26
B_2	−1.4	2.3	4.3
A_3	3.5	1.4	−44
B_3	−2.0	−0.25	−1.3

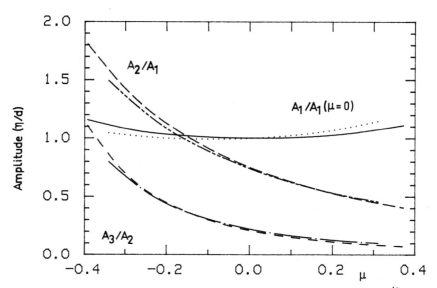

Fig. 2. μ–variation of the Fourier coefficients of u in the small ϵ limit, i. e. $\hat{u}_n \simeq \epsilon^{n/2} A_n$. For the first coefficient $A_1(\mu)/A_1(\mu = 0)$ is taken, the higher ones (n = 2,3) are divided by $A_1(\mu)$. The radial position is where u(r) is extremal. Lines types are (i) $\eta = 0.75$ n=1: full line, n=2: long dashed, n=3: short dashed and (ii) $\eta = 0.88$ n=1: dotted, n=2:dash–dot–dotted, n=3: dash–dotted.

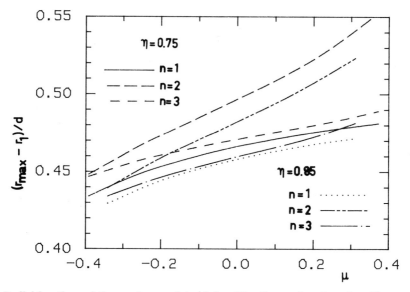

Fig. 3. Radial locations of the maximum of $A_n(r)$ (see Fig. 2) as a function of μ. Line types as in Fig. 2.

62

locations are drawn in Fig. 3. Lowering η shift them inwards, while the coefficients themselves hardly change. However, there are small differences in the scaling: $A_1(\eta = 0.88, \mu = 0) = 16.5$, while $A_1(\eta = 0.75, \mu = 0) = 15.7$ (see Fig. 1). The first coefficient is almost μ–independent, while the higher ones increase with lowering μ. Thus the contribution by higher axial Fourier modes to Taylor vortex flow increases strongly when lowering μ from the $\mu = 1$ limiting case. This in turn restricts the application range of the lowest–order GLE as a quantitative method to describe Taylor vortices already at $\mu = 0$ and even more so at negative μ. In contrast to that Rayleigh–Bénard convection — corresponding for a Prandtl number of 1 to the $\eta \to 1$, $\mu \to 1$ limit of the Taylor system — is much better represented by just one Fourier mode.

PHASE DYNAMICS

To explain what we are dealing with when speaking about phase dynamics we shortly describe an experiment done by Gerdts[6]. He used a Taylor system ($\eta = 0.5$, $\mu = 0$) with a movable top plate. First, stationary Taylor rolls were prepared. Then the top plate was moved downwards a little. This causes the vortices to adjust to their new positions basically without changing their strength. By looking at a particular vortex one can see the center position relax into its final location with an exponential long–time behavior. This suggests[7] a treatment with a diffusion equation for a phase variable connected to the vortex location

$$\partial_t \psi = D_0 \partial_x^2 \psi. \qquad (2)$$

Here $D_0 = \xi_0^2/\tau_0 \simeq 1.92$ is the diffusion constant for an infinitely long system. In Fig. 4 the experimental long–time relaxation rates[6] divided by those expected from eq. (2) are plotted for different lengths and control parameter values (open symbols). They all lie well above unity tending to increase with decreasing ϵ. Here we want to show that this peculiar increase is related to the Ekman enhancement of the Taylor vortex flow amplitude near the end which is not properly accounted for by eq. (2).

To do so we start with the usual GLE for the amplitude A of a velocity component[8],

$$u(r,z,t) = A(z,t)e^{ik_c z}\hat{u}(r) + \text{c.c.} \qquad (3a)$$

which reads

$$\tau_0 \partial_t A(z,t) = \left[\xi_0^2 \partial_z^2 + \epsilon - g|A(z,t)|^2\right] A(z,t). \qquad (3b)$$

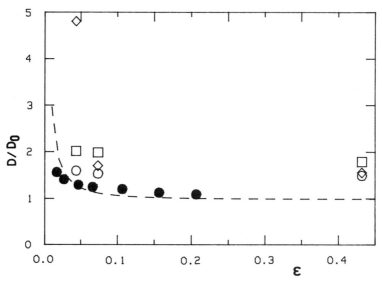

Fig. 4. ϵ–dependence of the effective diffusion constant D divided by the infinite–system value D_0. The dashed line is the result of formula (6). Filled circles come from the numerical solution of the full NSE with L = 25, $\eta = 0.75$. Open circles (diamonds, squares) are the $\eta = 0.5$ experimental findings of Gerdts[6] for L = 20.3 (12.4, 8.3).

A decomposition of $A = Re^{i\phi}$ into phase ϕ and modulus R yields the generalized diffusion equation

$$R^2(z)\, \partial_t \varphi(z,t) = D_0 R^2(z)\, \partial_z^2 \varphi(z,t) + D_0 [\partial_z R^2(z)] \partial_z \varphi(z,t) \qquad (4)$$

for the deviation $\varphi(z,t) = \phi(z,t) - \phi(z, t=\infty)$ from the final stationary phase. In eq. (4) we have assumed that the timescale of the amplitude relaxation is short compared to the relaxation time $L^2/(\pi^2 D_0)$ of the phase. Thus we drop the time dependence of R and consider it to be already stationary during the long–time relaxation of φ to zero. R(z) can be taken as input, for instance from the envelope of the measured velocity profile. Eq. (4) reduces to eq. (2) if R(z) is constant. Thus we want to emphasize that a nonuniform envelope of the velocity profile R(z) changes the time evolution of the phase of the vortex chain.

After inserting $\varphi(z,t) = \sum_n \varphi_n(t)\sqrt{2}\sin(n\pi z/L)$ eq. (4) reduces to an algebraic eigenvalue problem[4]. The smallest eigenvalue gives the long–time behaviour of φ. The very first approximation of this eigenvalue can be obtained by truncating the infinite dimensional system of equations to a 1x1–system. Then we get the effective diffusion constant D with

$$\frac{D}{D_0} \simeq \frac{R_0 + 3R_2}{R_0 - R_2} \quad ; \quad R_j = \frac{1}{L} \int_0^L dz\, R(z) \cos \frac{j\pi z}{L}. \qquad (5)$$

Within this drastic truncation D depends on the size of the second axial Fourier–cosine projection of R(z), R_2. Thus $D > D_0$ if R_2 is positive.

We now turn to investigate the influence of Ekman vortices in a finite system. As sketched out in Fig. 5, the rigid boundary condition on both ends causes Ekman vortices leading to an enhancement of the modulus R(z) near the ends. Here $R_2 > 0$, so we expect $D > D_0$. In order to see analytically the influence of the length L and the control parameter ϵ on the effective diffusion constant we model R(z) for convenience with two exponentials

$$R(z) \simeq A_0 + B\left[\exp(\tfrac{-z}{\ell}) + \exp(\tfrac{L-z}{\ell})\right] \qquad (6a)$$

rather than the correct Jacobian double periodic function. We take the ϵ–dependence of the bulk amplitude A_0 and of the penetration length ℓ of the Ekman vortex from the GLE (3), while the strength B of the Ekman vortex is fitted empirically. Then we get

$$A_0 = \sqrt{\epsilon/g} \quad ; \quad \ell \simeq \xi_0/\sqrt{\epsilon} \qquad (6b)$$

$$A_0 + B \simeq B_0 + \epsilon B_1 \quad ; \quad B_0,\, B_1 \text{ constant.} \qquad (6c)$$

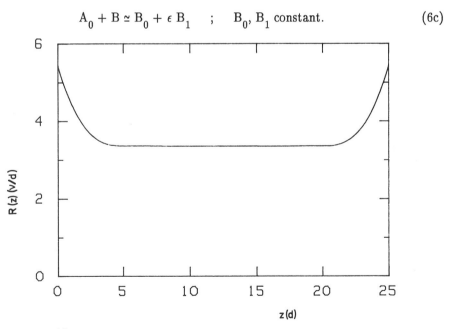

Fig. 5. Modulus R(z) of the amplitude of the radial velocity field in a system with two rigid ends, L = 25, ϵ = 0.047, η = 0.75.

Inserting the representation (6) into the approximation (5) gives an expression for D/D_0 in the presence of rigid end plates

$$\frac{D}{D_0} \simeq 1 + \frac{8 B \ell}{A_0 L} = 1 + \frac{8\xi_0 g^{1/2}}{L} (B_0 \epsilon^{-1} {-} g^{-1/2} \epsilon^{-1/2} + B_1). \qquad (7)$$

The correction to 1 shows an offset for larger ϵ due to B_1. Within the 1x1 matrix truncation D diverges for $\epsilon \to 0$. With increasing truncation order of the eigenvalue problem, however, there is a rounding off. To check our simple model results (5–7) we performed a numerical simulation ($L = 25$, $\eta = 0.75$, $\mu = 0$) of the *full* NSE with the correct boundary conditions[4]. These results are shown in Fig. 4. The numerical data (filled circles) agree surprisingly well with our simple model formula (7). Note in particular the substantial increase of D/D_0 to about 1.6 near $\epsilon = 0$. The experimental points of Gerdts do not fit quantitatively, but in their general tendency. One must keep in mind that it is more difficult to control the boundary conditions in experiments than in numerical simulations. Further experimental work should be done to test our predictions.

REFERENCES

1. G. I. Taylor, Phil. Trans R. Soc. Lond. A **223**, 289 (1923).
2. C. W. Hirt, B. D. Nichols, and N. C. Romero, Los Alamos Scientific Laboratory Report LA–5852, 1975. See also M. Lücke, M. Mihelcic, and K. Wingerath, Phys. Rev. A **31**, 396 (1985).
3. H. A. Snyder and R. B. Lambert, J. Fluid Mech. **26**, 545 (1966); J. P. Gollub and M. H. Freilich, Phys. Fluids **19**, 295 (1977); H. Fasel and O. Booz, J. Fluid Mech. **138**, 21 (1984); T. Berland, T. Jøssang, and J. Feder, Phys. Scr. **34**, 427 (1986); R. M. Heinrichs, D. S. Cannell, G. Ahlers, and M. Jefferson, Phys. Fluids **31**, 250 (1988); H. Kuhlmann, D. Roth, and M. Lücke, Phys. Rev. A **39**, 745 (1989).
4. M. Lücke and D. Roth, Z. Phys. B **78**, 147 (1990).
5. A. Davey, J. Fluid Mech. **14**, 336 (1962).
6. U. Gerdts, Ph. D. Thesis, Kiel 1985 (unpublished).
7. P. Tabeling, J. Phys. (Paris) Lett. **44**, L665 (1983); for a review concerning phase dynamics see: H. Brand, in: *Pattern, Defects, and Material Instabilities*, D. Walgraef and N. M. Ghoniem (ed.), NATO ASI Series E, vol. **183**, Kluwer, Dordrecht, (1990), p. 25.
8. R. Graham and J. A. Domeradzki, Phys. Rev. A **26**, 1572 (1982).

CHAOTIC PHASE DIFFUSION

THROUGH THE INTERACTION OF PHASE SLIP PROCESSES

Hermann Riecke[1]   Hans-Georg Paap[2]

[1]Department of Engineering Sciences   [2]Physikalisches Institut
and Applied Mathematics   der Universität Bayreuth
Northwestern University   D-8580 Bayreuth, Fed. Rep. Germany
Evanston, IL 60208, USA

1. INTRODUCTION

Chaotic dynamics in extended pattern forming systems like Rayleigh-Bénard convection or Taylor vortex flow has drawn considerable interest recently. Most intensely investigated has been the complex Ginzburg-Landau equation describing travelling waves near onset[1-3]. A remarkable feature of these waves is, that they can be unstable to spatial perturbations for all wave numbers for which they exist, i.e. they are Benjamin-Feir unstable[4]. This is to be contrasted to the situation obtained for steady patterns close to threshold: in spatially homogeneous quasi-one-dimensional systems they are always stable within a band of wave numbers. Its limits are given by the Eckhaus instability which triggers a 'phase-slip process' which destroys (or creates) a roll (or vortex) pair. The situation changes, however, if the system is allowed to be spatially inhomogeneous. More precisely, if a control parameter $\mathcal{R}$ varies in space such that it becomes subcritical, $\mathcal{R} < \mathcal{R}_c$, in part of the system ('subcritical ramp'), then the stable band is reduced and - in the limit of infinitely slow variations - shrinks to a single wave number[5,6]. Of particular interest in the present context is the fact, that the solution need not be stable for the wave number singled out in this way. It has been predicted that this will lead to dynamics resulting from the persistent creation (or destruction) of vortices[7]. This prediction, which was made quantitative by applying the phase diffusion approach to Taylor vortex flow[8,9] (TVF), was recently confirmed in a Taylor system in which the straight cylinders have been replaced by tapered ones[10]. The resulting dynamics turns out to be periodic and the corresponding frequency agrees well with the results obtained from the phase equation[9]. In the present contribution we address the following questions. Why is the experimentally observed dynamics periodic? Could one set up a system in which the dynamics becomes more complicated? What kind of dynamics can be found?

An important detail of the experimental set-up is the sudden transition from the tapered to the homogeneous section. This singles out a special location which could trigger the phase-slip processes to occur always at the same place leading to periodic dynamics. This motivates the present investigation of set-ups without any such preferred location. Since the main ingredients required for the dynamics are phase diffusion and phase-slip processes (Eckhaus instability), it is - as a first step - not necessary to investigate the full Taylor system. This is important to notice, since the phase diffusion approach does not suffice to describe the evolution of the system through the phase-slip processes and one has to resort to a full simulation of the basic equations. This would be prohibitively expensive for TVF considering the fact that the Navier-Stokes equations would have to be simulated for a large aspect ratio system over many phase slip processes. We therefore consider a simple reaction-diffusion model which exhibits these basic features and, indeed, find low-dimensional chaotic dynamics and spatial-temporal chaos. These results together with quantitative calculations based on the phase equation for TVF suggest that chaotic dynamics can also be obtained in Taylor vortex flow with only the inner cylinder tapered. We have reported some of these results in a previous publication[11].

2. A REACTION-DIFFUSION MODEL

For a model to be suitable for the present study it has to exhibit a steady supercritical bifurcation to spatially periodic states and may not be derivable from a potential in order to allow persistent dynamics. In addition, it should have two relevant control parameters to admit a tuning of the wave number selected by subcritical ramps. This is satisfied by a reaction-diffusion model for the 'concentrations' u and v with simple cubic nonlinearities[12],

$$\partial_t u = \partial_x^2 u + u - u^3 - R_1 v, \quad \partial_t v = D \partial_x^2 v - av - av^3 + R_2 u. \qquad (1)$$

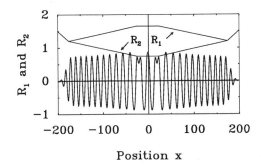

Fig.1.'Concentration' $u(x)$ and control parameters $R_1(x)$ and $R_1(x)$ for a ramp with linear 'tapering'. Note that the two phase slips occur close to the corners of the ramp at $x = \pm 25$.

The Turing instability to a steady pattern occurs when the coupling coefficients satisfy $R_1 R_2 < R_{1c} R_{2c} \equiv (D+a)^2/4D$. In the following $D = 4$ and $a = 1.2$ throughout. In this model the only relevant secondary instability of the pattern is the Eckhaus instability. In the presence of *slow* spatial variations of the control parameters $R_i = R_i(X), X = \epsilon x$, the slow dynamics of this system can be described by the phase equation[12]

$$\partial_T \phi = D_\parallel(q, R_i) \partial_X^2 \phi + r_1(q, R_i) \partial_X R_1 + r_2(q, R_i) \partial_X R_2, \qquad (2)$$

where ϕ is the phase of the pattern, $T = \epsilon^2 t$, and the local wave number is given by $q = \partial_X \phi$.

In the following we investigate (1) for two kinds of ramps in the control parameters $R_1(x)$ and $R_2(x)$. Both of them are reflection symmetric and become subcritical for large $|x|$. The first set-up has a homogeneous section in the center which connects to a linear 'tapering' at the sides. Similar to the experiments the transition betwen the two sections is sudden as shown in fig.1. In the second set-up parabolic ramps are used, which are perfectly smooth in the relevant part of the system.

3. RAMPS WITH CORNERS

To show the effect of a sudden transition between the homogeneous and the tapered section we first consider ramps of the form

$$R_i = R_< + s_1|x|, 3L/4 < |x| < L, R_i = R_m + s_{2,i}|x|, h < |x| < 3L/4, R_i = R_{>,i}, |x| < h. \quad (3)$$

In all subsequent simulations $R_i(L) = 1.8$ and $R_i(3L/4) = 1.2$ are fixed and $s_1, s_{2,i}$ and $R_{>,i}$ are varied such that $R_1(x=0) + R_2(x=0) = 2.4$. The relevant control parameter is then $\mathcal{R} \equiv R_1(0)$. Fig.1 shows such a set-up for $2L = 399$. Of interest is now the location where the phase slips occur relative to the corners of the ramp.

Within the phase equation the expansion (or compression) of the pattern is largest at the corner where the inhomogeneous terms, which tend to induce a drift of the pattern, make a jump[11]. One would therefore expect that the phase slips will always occur there. However, close to the onset of the Eckhaus instability the relevant perturbation mode has a very large wave length and the instability sets in only when the unstable wave number is reached over a length of that size. Consequently, the phase slip occurs at a distance related to this wave length. Beyond onset the wave number is driven further into the unstable regime and the perturbation wavelength decreases: the phase slip is expected to move closer to the corner. This is in fact observed in the experiments[10].

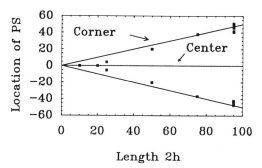

Fig.2. Location where the phase slips occur (symbols) as a function of the length $2h$ of the homogeneous section.
Solid lines: transition to the 'tapered' section, dotted line: middle of the system.

Fig.2 shows the result of our simulation for various lengths $2h$ of the homogeneous section at fixed $\mathcal{R} = 1.65$. The symbols denote the location where the phase slips (PS) appear. The middle of the system is indicated by the dotted line and the transition to the 'tapered' section by solid lines. The length is chosen as $2L = 1599 + 2h$. For small h the phase slips occur in the middle of the system whereas for large values they arise close to the corners. The dynamics was found to be periodic except for $h = 95$, where a competition between two locations ($x \approx 40$ and $x \approx 50$) leads to more complicated behavior. The latter is related to the fact that the phase slips occur only at the extrema of the pattern (cf. fig.1). It is seen that the corner, in fact, singles out special positions and the dynamics is determined by a small number of phase slips centers. In particular, the pattern is destabilized only at these isolated places and not over a large part of the system.

4. CHAOS BY SMOOTH PARABOLIC RAMPS

To obtain richer dynamics the phase slips should occur at many locations. This is best achieved by using a smooth ramp which homogeneously stretches the pattern. Motivated by a simplified phase equation, we investigate the following family of parabolic ramps[11],

$$R_i(x) = R_< + s\mid x\mid, 3L/4 <\mid x\mid < L, \quad R_i(x) = R_{>,i} + c_i x^2, \mid x\mid < 3L/4, \qquad (4)$$

with $2L = 399$. As reported previously, the steady pattern becomes Eckhaus unstable for $\mathcal{R} = 1.45$ and phase slips occur periodically at symmetrically related, alternating locations. With increasing $\mathcal{R}$ they shift off-center and above $\mathcal{R} = 1.527$ undergo a period-doubling cascade involving both the location and the time between successive phase slips. The cascade can also be seen when monitoring the local value of one of the 'concentrations' at a fixed position. Fig.3a shows the power spectrum of such a time series for $\mathcal{R} = 1.52767$, corresponding to the period 16 in the period-doubling cascade. For larger values of $\mathcal{R}$ the dynamics becomes chaotic, exhibits a period 3 (with two subsequent period-doublings) and yields the spectrum shown in fig.3b at $\mathcal{R} = 1.65$.

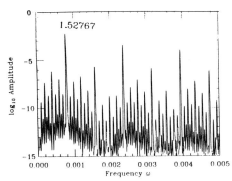 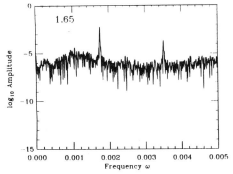

Fig.3. Power spectrum of $u(t)$ at $x = 20$.
a) Period 16 at $\mathcal{R} = 1.52767$, b) strongly chaotic at $\mathcal{R} = 1.65$.

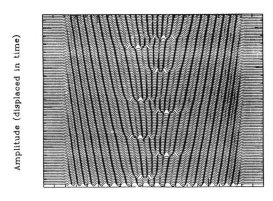

Fig.4. Space-time diagram for the 'concentration' $u(x,t)$ at $\mathcal{R} = 1.7$.
Note the variation in the time period between successive phase slips.

The dimension of the attractor obtained from the time series is found to be $d = 1.7$ for $\mathcal{R} = 1.6$ and $d = 2.5$ for $\mathcal{R} = 1.7$. Despite these strongly chaotic signatures the pattern looks rather orderly most of the time. Fig.4 gives a space-time diagram for $\mathcal{R} = 1.7$. The pattern drifts outward to the subcritical regime and new 'elementary cells' are perpetually created in the center by phase slips. It is the strongly varying time between such phase slips which constitutes the chaotic behavior. For this system length the pattern becomes Eckhaus unstable only over a length of about 3 wavelengths. We have also investigated the case $2L = 1599$ for which this length increases to about 20 and the dynamics becomes more complicated[11]. Remarkably, we found there that the pattern always undergoes *symmetrical* phase slips ('tip splitting', cf. fig.1). This preference is not contained in the Ginzburg-Landau equations commonly used in investigations of spatio-temporal chaos, which are based on multiple space scales[1–3].

5. CHAOTIC DYNAMICS IN TAYLOR VORTEX FLOW?

To obtain smooth ramps in the Taylor system which are similar to those discussed in section 4 one could in principal choose parabolically shaped 'cylinders'. Alternatively, one could turn to spherical Couette flow, which for small gap widths can be considered as a specifically ramped Taylor system. In fact, numerical simulations have hinted at the possibility of complicated dynamics in this system, if the solutions are restricted to be axially symmetric[13]. They are, however, often unstable to spirals.

Without reflection symmetry, one can easily obtain smooth ramps which become subcritical at both ends by tapering both cylinders linearly[8]. In fact, it is even sufficient to taper only the inner cylinder. Using the phase diffusion equation, which we have derived directly from the Navier-Stokes equations[8,9], we have calculated the drift frequency ω of TVF for this ramp as well as the corresponding wave number profile $q(x)$ as a function of the Taylor number $\mathcal{T}(x = 0)$. The radius ratio $\eta(x = 0)$ is 0.85 and the slope of the inner cylinder is 0.01. The value of the local phase diffusion coefficent $\mathcal{D}(q(x))$ determines the (local) stability of the pattern with respect to the Eckhaus instability. The results are shown in fig.5. One can see that even in this simple geometry the minimal value $\mathcal{D}_{min}$ decreases with growing $\mathcal{T}$, reflecting the fact that the wave number selected by the ramp moves toward the limit of the stable band. In fact, for $\mathcal{T} = 650$ the Eckhaus boundary is reached and the pattern becomes locally unstable.

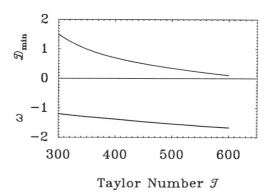

Fig.5 Taylor vortex flow with tapered *inner* cylinder.
Drift frequency ω and minimal value $\mathcal{D}_{min}$ of the phase diffusion coefficient as a function of the Taylor number $\mathcal{T}(x = 0)$. Note that $\mathcal{D}_{min}$ reaches 0 for $\mathcal{T} \approx 650$ indicating convective instability. Chaotic dynamics is expected sufficiently above this value of $\mathcal{T}$.

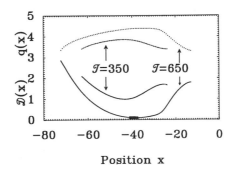

Fig.6 Wave-number $q(x)$ selected by the ramp and corresponding values of the diffusion coefficient $\mathcal{D}(x)$ for $\mathcal{T} = 350$ and $\mathcal{T} = 650$. The heavy line marks where the pattern becomes unstable for $\mathcal{T} > 650$. When this region is large enough chaotic dynamics is expected.

Fig.6 shows the profiles $q(x)$ and $\mathcal{D}(x)$ for two values of $\mathcal{T}$. In the symmetric systems discussed above $\mathcal{D} = 0$ would imply the occurence of phase slips. In the present system this is not necessarily the case, since the pattern has an overall drift. Thus the condition $\mathcal{D}(q) = 0$ gives only the condition for *local convective* instability. To obtain non-trivial dynamics the growth rate has to be large enough for the full phase slip process to occur before the growing perturbation is swept into the stable regime downstream. In addition, the phase slips have to occur sufficiently close in time in order for them to be able to interact while they are being swept away. Thus one has to address the questions of local and global absolute instability[14]. The situation seems, however, not straight forward, since the Eckhaus instability does not lead to a nonlinear state with a well-defined group velocity and the local growth rate is not constant but depends (through the wave number) on the state of the system. Here no detailed discussion of this interesting issue is attempted.

6. CONCLUSIONS

In the present contribution we have investigated the dynamics resulting from subcritical ramps which select an Eckhaus-unstable wave number. We found that the interaction of phase slip processes occuring at different locations can lead to interesting dynamics. Corners in ramps induce phase slip centers and the phase slips occur at a small number of different places. Their interaction can lead to non-trivial dynamics. This may be the origin of the dynamics observed in recent experiments on TVF under the influence of a Coriolis force[15]. Smooth ramps can lead to a homogeneous stretching of the pattern which destabilizes the pattern over a large portion of the system. In the model system, this was seen to lead to strongly chaotic dynamics. We expect such dynamics also to occur in various simple set-ups of the Taylor system (even if *only* the inner cylinder is tapered). There the instability leading to the transition to chaotic dynamics is convective, which raises additional interesting questions.

Acknowledgments: It is a pleasure to acknowledge discussions with V. Croquette and in particular J. Theiler, whose code was used to calculate the dimensions. This work was partially supported by DARPA under grant AFOSR F49620-87-C-0117, by the Deutsche Forschungsgemeinschaft (DFG) and by NATO through grant CRG 900276. Computer time was provided through the NAS program of the NASA Ames research center.

REFERENCES

1. C.S. Bretherton and E.A. Spiegel, Phys. Lett. **96A** (1983) 152;
   H.T. Moon, P. Huerre and L.G. Redekopp, Phys. Rev. Lett. **49** (1982) 458.
2. Y. Kuramoto, Prog. Theor. Phys. **71** (1984) 1182.
3. P. Coullet and J. Lega, Europhys. Lett. **7** (1988) 511;
   P. Coullet, L. Gil and J. Lega, in *Advances in Fluid Turbulence*,
   Eds. G. Doolen, R. Ecke, D. Holm, V. Steinberg, North-Holland 1989.
4. J.T. Stuart, R.C. DiPrima, Proc. R. Soc. Lond. **A 362** (1978) 27.
5. L. Kramer, E. Ben-Jacob, H. Brand and M.C. Cross, Phys. Rev. Lett. **49** (1982) 1891.
6. D.S. Cannell, M.A. Dominguez-Lerma and G. Ahlers, Phys. Rev. Lett. **50** (1983) 1365.
7. L. Kramer and H. Riecke, Z. Phys. **B59** (1985) 245.
8. H. Riecke and H.-G. Paap, Phys. Rev. Lett. **59** (1987) 2578.
9. H.-G. Paap and H. Riecke, Phys. Fluids **A 3** (1991) 1519.
10. L. Ning, G. Ahlers and D.S. Cannell, Phys. Rev. Lett. **64** (1990) 1235.
11. H. Riecke and H.-G. Paap, Europhys. Lett. **14** (1991) 433.
12. P.C. Hohenberg, L. Kramer and H. Riecke, Physica **15D** (1985) 402.
13. F. Bartels, J. Fluid Mech. **119** (1982) 1.
14. J.M. Chomaz, P. Huerre and L.G. Redekopp, Phys. Rev. Lett. **60** (1988) 25.
15. L. Ning, G. Ahlers and D.S. Cannell, this conference.

PHASE DYNAMICS IN THE TAYLOR-COUETTE SYSTEM

Mingming Wu and C. David Andereck

Department of Physics
The Ohio State University
Columbus, Ohio 43210

INTRODUCTION

In principle, the flows in the Taylor-Couette system can be understood as solutions of the Navier-Stokes equation [1, 2, 3]. Unfortunately the complexity of the equation often makes it difficult to compare with laboratory results, thus necessitating the use of model equations [4, 5, 6, 7]. A typical example is the successful use of amplitude equations [4] for the states close to the onset of the first supercritical bifurcations. The amplitude equation is derived from the basic equations by the expansion of a small amplitude of the structure. In the amplitude equation, the phase variable and amplitude are two independent variables. For a case where the wavelength has slow time and space variation, the amplitude is slaved to the phase variable and the amplitude equation can be simplified to a phase equation [5, 6, 7]. For a flow pattern that is far above its onset, the amplitude equations are no longer valid, but the equations for the phase variables still are.

In this paper, we give an overall summary of our experimental studies on phase dynamics in the Taylor-Couette system. Section 2 gives the experimental setup and the data acquisition technique. Section 3 contains the experimental results on the phase dynamics of Taylor vortex flow, wavy vortex flow and turbulent wavy vortex flow, along with their theoretical models.

EXPERIMENTAL SETUP

Our experiment is conducted in two concentric cylinders with the outer one fixed. The radius of the inner cylinder is $r_i = 5.262 cm$, the outer one $r_o = 5.965 cm$. The cylinder rotation rate is controlled by a Compumotor stepper motor(model M83-93) which is precise to 0.001 Hz. A PDP-11/73 is interfaced through the Compumotor indexer to control the stepper motor. The working fluid region is bounded at both ends by Teflon rings. The upper ring touches neither the outer nor the inner cylinder, and is controlled by a traversing mechanism. The ring is able to oscillate along the axial direction over a maximum distance of 1cm under the control of a Compumotor stepper motor. The distance between the Teflon rings initially (before the modulation is added) is 49.5cm, and therefore the average aspect ratio $\Gamma = \frac{L}{r_o - r_i} = 70.4$. The working fluid is a solution of double distilled water and 44% glycerol by weight. 1% by volume of Kalliroscope AQ1000 is added for visualization. The flow pattern is viewed with a 512 × 480 pixel CCD camera which is connected to an image processor. The image data file can be saved in a micro-computer and later transferred to a VAX 8650 for further analysis.

EXPERIMENTAL AND THEORETICAL STUDIES OF THE PHASE DYNAMICS IN THE TAYLOR-COUETTE SYSTEM

The controlling parameter of the Taylor Couette system is the Reynolds number ($\propto \Omega$, the inner cylinder rotation frequency). When R exceeds a threshold value R_c, the spatially uniform circular Couette flow (CCF) changes to the axially periodic Taylor vortex flow(TVF). The first time-dependent regime, wavy vortex flow(WVF), in which an azimuthal wave is superimposed on the TVF, occurs for slightly greater R in a large radius ratio system. If R is increased further, the flow will pass the weakly turbulent region and reach the turbulent wavy vortex flow (TWVF). The azimuthal wavy boundary lines persist in the TWVF as in the case of WVF, but each vortex contains well developed turbulent flow [8].

The phase variables of interest are directly related to the positions of the vortex boundaries. For TVF, one phase variable ψ, the axial phase variable, is used to describe the position variations of the vortex boundary. For WVF, the azimuthal phase variable ϕ is introduced to describe the azimuthal wave motion. In the case of TWVF, it is not evident how many phase variables are involved. Our study only concerns the axial phase variable ψ, which relates to the average vortex boundary position in the axial direction.

In order to study the phase dynamics of the flow patterns, a sinusoidal forcing is applied to the upper ring of the system. Responses of the vortex boundaries to the modulation are recorded by the computer. The following gives a detailed experimental and theoretical descriptions of the phase dynamics in the three flow regimes.

<u>Taylor Vortex Flow</u>

Assume z_n^0 and z_n are the nth vortex boundary positions before and after the modulation is added. Then $z_n^0 = \frac{n\pi}{\tilde{q}}$ and $\psi(z_n, t) = (z_n^0 - z_n)\tilde{q}$, where $\tilde{q}$ is the axial wavevector. We adjust the cylinder rotation frequency until a stationary TVF pattern is formed. The flow pattern is set for 30 minutes before the upper ring oscillation begins. The image processor starts to take data 2 hours after the modulation is added. The vertical line profile of the flow pattern is taken every 2 minutes for 5 hours. The vortex boundary position $z_n(z_n^0, t)$ is obtained by locating the minima of the light intensity profile. Fig. 1 is a typical data set. The response of each node line is a sinusoidal function of time. The amplitude (in ln scale) and the phase of the node line motion are linearly related to the axial position (Fig. 2a , Fig. 2b).

The above observations are understood by a simple diffusion model:

$$\frac{\partial \psi}{\partial t} = D_\| \frac{\partial^2 \psi}{\partial z^2} \qquad (1)$$

where $D_\|$ is the diffusion coefficient in the axial direction.

We found that the phase at $z = 0$ exactly follows the top collar's motion due to the fact that the period of the modulation $T \gg d^2/\nu$, the diffusion time through a vortex. (typically, $T = 3040$sec. and $d^2/\nu = 12.3$sec.) This provides the following boundary condition:

$$\psi|_{z=0} = \psi_0 \sin(\omega t) \qquad (2)$$

Here ψ_0 and ω are the modulation amplitude and frequency. Solving Eqn. 1 with the above boundary condition, we obtain:

$$z_n = z_n^0 - a\sin(\omega t - \delta)/\tilde{q} \qquad (3)$$

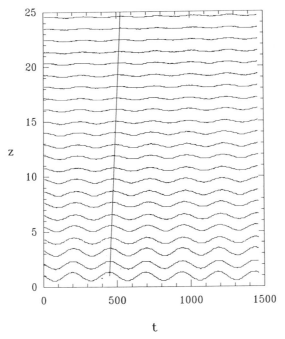

Figure 1. Node line locations of TVF subjected to the periodic boundary modulation, t represents time and z is the distance from the top collar. The solid line traces the shift in phase from vortex to vortex.

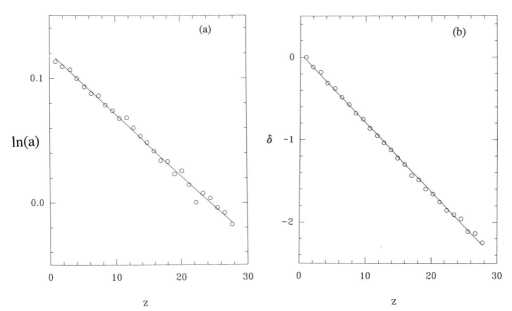

Figure 2. Circles are the experimental data and the solid lines are from the fit to the diffusion model. (a) a is the amplitude of each node line motion. (b) δ is the phase shift of the node line motion.

where $a = (\psi_0/\tilde{q})e^{-\alpha z_n^0}$, $\delta = \beta z_n^0$, a and δ are the amplitude and the phase of the node line motion, and:

$$\alpha = \beta = \sqrt{\frac{\omega}{2D_\parallel}} \qquad (4)$$

The fact that $\ln a$ and δ are linearly related to z_n^0 is consistent with our experimental results shown in Fig. 2a and Fig. 2b. The slopes of the lines give us values of α and β and hence the value of $D_\parallel$ from Eqn. 4. The variations of α and β from our experiment were within 20%. A more detailed description of the phase dynamics in TVF can be found in Ref. [9].

Wavy Vortex Flow

The phase dynamics of the wavy vortex flow is complicated by the introduction of the second phase variable ϕ, thus leading to richer possibilities. Our experimental study has been limited to the dynamics of ψ. A somewhat different approach will be needed to study the dynamics of ϕ and further work on this matter is important for fully understanding the phase dynamics in WVF.

In contrast with the Taylor vortex flow, the vortex boundary position in the lab frame $z_n(z,t)$ now changes on two time scales, the slow time T_1 (corresponding to the slow boundary perturbation, which has a typical period of several minutes) and the fast time T_2 (corresponding to the azimuthal wave motion of period $\simeq 1$ sec.). Averaging out the azimuthal wave motion, we have $\psi = \tilde{q}(\bar{z}_n(T_1) - \bar{z}_n^0)$, where $\bar{z}_n^0$ is the average vortex boundary position without modulation. In order to obtain $\bar{z}_n(T_1)$, we take about 100 consecutive line profiles (covering about 10 azimuthal waves) in 7 sec., find the intensity minima along each vertical line, then average out the azimuthal wave motion for each node line. We repeat the above process at a time interval of 20 or 30 sec. depending on the modulation period T. In general, 10 data points are taken in a period T.

$\bar{z}_n(T_1)$ is measured for flow states with different m. A m=3 state is obtained by increasing ϵ slightly above the onset of TVF, $\epsilon = 0.140$. A typical result is shown in Fig. 3a. The amplitude of $\bar{z}_n(T_1)$ decreases exponentially along the system axis, and there is a linear phase shift between oscillations of neighboring vortices. Upon further increasing the rotation frequency, the system reaches a new stable state with $m = 7$ at $\epsilon = 0.813$. In order to obtain $m = 7$ with the same vortex number N, the rotation frequency is changed rapidly from $\epsilon = 0.140$ to $\epsilon = 0.813$. For the same modulation period $T = 53.3$, $\bar{z}_n(T_1)$ reveals an axially propagating wave in the middle section of the cylinders, as shown in Fig. 3b.

The above observations can be explained within the theoretical framework of the coupled phase equations proposed by Brand and Cross [7]. Based on their symmetry analysis, ψ and ϕ are governed by:

$$\begin{aligned}\frac{\partial \psi}{\partial t} &= D_\parallel \frac{\partial^2 \psi}{\partial z^2} + C_\parallel \frac{\partial \phi}{\partial z} \\ \frac{\partial \phi}{\partial t} &= D_\perp \frac{\partial^2 \phi}{\partial z^2} + C_\perp \frac{\partial \psi}{\partial z}\end{aligned} \qquad (5)$$

where z is along the axial direction, $D_\parallel$ and $D_\perp$ are the diffusion coefficients for the axial and azimuthal directions, $C_\parallel, C_\perp \propto q_y$ and represent the coupling between the azimuthal wave motion and the axial vortex position change. These coefficients can be derived from the amplitude equation near the onset of TVF.

For the weak coupling case where $C_1 C_2 \ll |D_1 - D_2|\omega$, the coupled equations can be simplified to a diffusion model. Fig. 3a shows that for a small q_y, the small $C_\parallel, C_\perp$ case, the phase dynamics is similar to the case of TVF. The diffusion coefficient can be evaluated by Eqn. 4 and it is ~ 1.4.

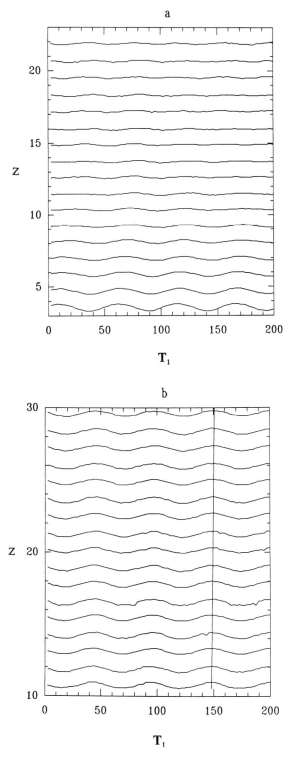

Figure 3. Responses of the axial phase variable $\bar{z}_n(T_1)$ to a modulation of period $T = 53.3$. Number of vortices N=60. (a) m=3, $\epsilon = 0.140$; (b) m=7, $\epsilon = 0.813$. The straight line traces the shift in phase from vortex to vortex.

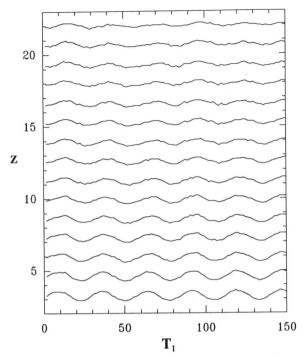

Figure 4. Responses of the axial phase variable $\bar{z}_n(T_1)$ to a modulation of period $T = 27.6$. The flow state has $\epsilon = 19.7$, N=52.

For the strong coupling case where $C_\| C_\perp \gg 2\omega(D_\| + D_\perp)$, Eqn. 5 is decoupled into a simple traveling wave equation:

$$\frac{\partial^2 \psi}{\partial t^2} - c^2 \frac{\partial^2 \psi}{\partial z^2} = 0 \qquad (6)$$

where $c = \sqrt{C_\| C_\perp}$. This is confirmed by our observation in the large q_y (m=7) flow pattern and its traveling wave behavior is shown in Fig. 3b. For more details on the phase dynamics of WVF, see Ref. [10].

Turbulent Wavy Vortex Flow

The average vortex boundary position $\bar{z}_n$ of TWVF is measured with the same data acquisition technique as in the wavy vortex flow. A typical time series of $\bar{z}_n(T_1)$ is shown in Fig. 4. The amplitude drops off exponentially and the phase shift of neighboring vortex boundaries has a linear dependence on the axial position. This indicates that the phase dynamics in TWVF can be described by a simple diffusion model. A diffusion coefficient is evaluated by Eqn. 4 from the slope of the line (phase vs. axis). It is about 18 in this case and about 10 times larger than that of TVF.

CONCLUSION

We have studied the phase dynamics both experimentally and theoretically in the TVF, WVF and TWVF. The phase dynamics in TVF and TWVF can be described by a simple diffusion model, while that of WVF is governed by a coupled diffusion model.

We thank H. Brand for persistent encouragement during this project and Z. H. Wang, F. Hayot and I. Mutabazi for helpful discussions. This work was supported by the Office of Naval Research, under contract N00014-86-K-0071 and grant N00014-89-J-1352.

References

[1] P. S. Marcus, J. Fluid Mech. **146**, 45(1984)

[2] P. S. Marcus, J. Fluid Mech. **146**, 65(1984)

[3] M. Nagata, J. Fluid Mech., **169**, 229(1986), and **188**, 585(1988)

[4] A. C. Newell, J. A. Whitehead, J. Fluid Mech., **38**, 279(1969) For application to the Taylor-Couette system see R. Graham and J. A. Domaradzki, Phys. Rev. A **26**, 1572(1982) and references therein

[5] Y. Pomeau and P. Manneville, J. Phys. Lett. **40**, L609 (1979)

[6] P. Tabeling, J. Phys. Lett. **44**, L665 (1983)

[7] H. Brand and M. C. Cross, Phys. Rev. **A**, **27**, 1237(1983)

[8] R. W. Walden, R. J. Donnelly, Phys. Rev. Lett. **42**, 301(1979)

[9] M. Wu, C. D. Andereck, Phys. Rev. **A**, Rapid Comm., **43**, 2074(1991)

[10] M. Wu, C. D. Andereck, Phys. Rev. Lett. (1991), in press

SPIRAL VORTICES IN FINITE CYLINDERS

E. Knobloch and R. Pierce

Department of Physics
University of California
Berkeley, CA 94720

In their (1986) paper on counter-rotating Taylor-Couette flow, Andereck et al (1986) report the existence of spiral vortex flow in which an upward-travelling right-handed spiral flow is replaced, at irregular intervals, by downward-travelling left-handed spiral flow (see their Fig. 4). The observations indicate that the apparatus was in fact at all times filled by left spirals near the bottom and right spirals near the top, with the boundary between the two types of spirals moving slowly up the cylinder, thereby expanding the region filled by left spirals and contracting that filled by right spirals, before moving back down the cylinder. Similar behavior was observed by R. Tagg (private communication) near the codimension-two point where the primary Hopf bifurcation is superseded by a steady state bifurcation to Taylor vortices as the outer rotation rate is decreased. Andereck et al (1986) suggest that the observed reversals may be somehow associated with the presence of ends on the cylinders. In this paper we describe the results of an analysis that lends support to this conjecture. Our approach makes use of equivariant bifurcation theory, and takes as its starting point the analysis of an idealized system in which the cylinders are considered to be infinitely long and the system translation-invariant in the axial direction as well as invariant under reflections in any orthogonal plane. The ends are introduced by bringing them, so to speak, from infinity and determining their effect on the idealized system in a perturbative fashion. This effect comes through the breaking of the translation symmetry in the axial direction. We find that, depending on the cylinder aspect ratio, a new type of solution may be present that looks qualitatively like the state reported by Andereck et al (1986). Numerous verifiable predictions about the appearance and disappearance of this alternating spiral vortex state as a function of the aspect ratio and the Reynolds number (equivalently, the rotation rate of the inner cylinder) are made.

For the idealized system we employ periodic boundary conditions in the axial direction. This procedure introduces the symmetry group O(2) into the governing equations, in addition to the symmetry SO(2) arising from the azimuthal symmetry of the apparatus. Here O(2) is the symmetry of rotations and reflections of a circle, while SO(2) denotes only the rotations.

For sufficiently counter-rotating cylinders, the basic flow, called Couette flow, loses stability at a Hopf bifurcation. Because the unstable mode has a non-zero axial wavenumber k, this bifurcation breaks the O(2) symmetry. Consequently the multiplicity of the imaginary eigenvalue at the bifurcation is not one but two (see, e.g., Crawford and Knobloch 1991). If we fix the outer rotation rate, the Hopf bifurcation takes place when the inner rotation rate exceeds a critical value given by R_c, where R is the Reynolds number based on the inner cylinder rotation speed. Then for $|R - R_c| \ll 1$, the solutions of the Navier-Stokes equations must be of the form

$$W(x, r, \phi, t) = \{v(t)e^{ikx+im\phi} + w(t)e^{ikx-im\phi} + c.c.\} f_{km}(r), \tag{1}$$

where k is the axial wavenumber that first becomes unstable, m is the corresponding azimuthal wavenumber, and $f_{km}(r)$ represents the radial eigenfunction for the quantity W. The latter represents the velocity on the axial direction, but could equally well represent the radial velocity or the departure of the azimuthal flow from Couette flow. Observe that the complex amplitudes v, w represent, respectively, the amplitude of left- and right-handed helical waves, hereafter called spiral vortices. The equations for the amplitudes v, w follow from the requirement that they inherit the symmetries of the system. On the space $\{(v,w)\} \equiv \mathbf{C}^2$ these act as follows:

$$\begin{array}{lll} \text{reflection}: & x \to -x & (v,w) \to (\overline{w}, \overline{v}) \\ \text{translations}: & x \to x + l & (v,w) \to (e^{ikl}v, e^{ikl}\overline{w}) \\ \text{rotations}: & \phi \to \phi + \theta & (v,w) \to (e^{im\theta}v, e^{-im\theta}w) \end{array} \tag{2}$$

The most general equations for (v, w) that commute with this action of O(2)×SO(2) take the form:

$$\dot{v} = g(\lambda, |v|^2, |w|^2)v, \tag{3a}$$
$$\dot{w} = \overline{g}(\lambda, |w|^2, |v|^2)w, \tag{3b}$$

where $g(0,0,0) = i\omega_0$ and ω_0 is the Hopf frequency at $\lambda \equiv R - R_c = 0$. It follows that near onset

$$\dot{v} = \left[\lambda + i\omega + a|w|^2 + b(|v|^2 + |w|^2)\right]v + O(5) + \lambda O(3) \tag{4a}$$
$$\dot{w} = \left[\lambda - i\omega + \overline{a}|w|^2 + \overline{b}(|v|^2 + |w|^2)\right]v + O(5) + \lambda O(3), \tag{4b}$$

where $O(n)$ denotes terms of order n in $|v|, |w|$.

In terms of the real variables given by $v = r_1 e^{i\phi_1}$, $w = r_2 e^{i\phi_2}$, equations (4) become:

$$\dot{r}_1 = \left(\lambda + a_r r_2^2 + b_r(r_1^2 + r_2^2)\right) r_1 + O(5) + \lambda O(3) \tag{5a}$$
$$\dot{r}_2 = \left(\lambda + a_r r_1^2 + b_r(r_1^2 + r_2^2)\right) r_2 + O(5) + \lambda O(3) \tag{5b}$$

together with two decoupled equations for $\dot{\phi}_1, \dot{\phi}_2$. This decoupling is a consequence of the translation symmetry in the axial direction and the rotation symmetry in the azimuthal direction. The solutions to equations (5) depend on the coefficients a_r, b_r, the real parts of the coefficients a, b in equations (4). There are three types of solutions, $(r_1, r_2) = (0, 0)$ corresponding to Couette flow, $(r, 0)$ or $(0, r)$ corresponding to left and right spirals, respectively, and (r, r) corresponding to ribbons. Note that left spirals travel downwards, while right spirals travel upwards. For the following we refer to the spirals as travelling waves (TW) and to the ribbons as standing waves (SW). Their respective properties are summarized in Fig. 1 (cf. Demay and Iooss 1984, Golubitsky and Stewart 1985, Knobloch 1986).

The coefficients a_r, b_r were calculated from the Navier-Stokes equations by Demay and Iooss (1984). Using their results one can show that the initial Hopf instability can give rise to either stable spirals (TW) or stable ribbons (SW) depending on the physical parameters. Indeed, until the prediction of $m = 2$ stable ribbons, no such state had been observed (see Tagg et al 1989, 1990).

It is clear that a pure spiral cannot exist in a finite cylinder. One might suspect therefore that despite the success of the infinite cylinder theory and its ability to predict new states it is possible that the presence of even distant ends can have nontrivial dynamical consequences. We assume that the primary effect of distant ends is to break the translation symmetry in the axial direction. Thus we add to equations (5) the terms $(d\overline{w}, \overline{d}\overline{v})^T$, these being the dominant terms that preserve the equivariance under reflection and rotation, but break the equivariance under translations:

$$\dot{v} = \left[\lambda + i\omega + a|w|^2 + b(|v|^2 + |w|^2)\right]v + d\overline{w} \tag{6a}$$
$$\dot{w} = \left[\lambda - i\omega + \overline{a}|w|^2 + \overline{b}(|v|^2 + |w|^2)\right]w + \overline{d}\overline{v}. \tag{6b}$$

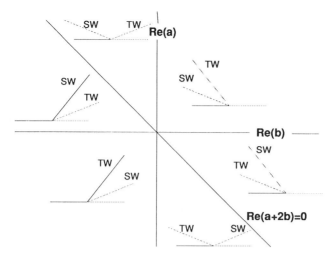

Figure 1. Bifurcation diagrams ($r_1^2 + r_2^2$ vs. λ) for the perfect problem in the (a_r, b_r) plane. Black lines denote stable branches, gray lines unstable branches.

Here $d \equiv |d|e^{i\alpha}$ is a complex coefficient that can be determined by solving the linear stability problem in a finite cylinder. For the following we assume that the quantities a_r, b_r, and $a_r + 2b_r$ are nonzero, so that the small perturbation due to the ends does not change their sign or magnitude. Thus we take a_r, b_r, etc. to be given by their values for the infinite cylinder. In the following we assume that $a_r < 0$, $b_r < 0$ so that in the infinite system the Hopf bifurcation gives rise to stable spirals.

Explicit expressions for the coefficients can be obtained, for example, from the coupled complex Ginzburg-Landau equation for left and right spiral vortices and valid in a long but finite cylinder (cf. Edwards 1990):

$$A_\tau = \left\{(1+i\gamma)\partial_X^2 + s\partial_X + \Lambda + \tilde{a}|B|^2 + \tilde{b}(|A|^2 + |B|^2)\right\} A \tag{7a}$$

$$B_\tau = \left\{(1-i\gamma)\partial_X^2 - s\partial_X + \overline{\Lambda} + \overline{\tilde{a}}|B|^2 + \overline{\tilde{b}}(|A|^2 + |B|^2)\right\} B. \tag{7b}$$

These equations are obtained by looking for solutions of the Navier-Stokes equations of the form:

$$W(x, r, \phi, t) = \epsilon Re\left\{A(X,\tau)e^{ikx+im\phi+i\omega t} + B(X,\tau)e^{ikx-im\phi-i\omega t} + c.c.\right\} f_{km}(r) + O(\epsilon^2), \tag{8}$$

where X, τ are slow space and time variables,

$$X = \epsilon x, \qquad \tau = \epsilon^2 t, \tag{9}$$

and $\epsilon \equiv (R - R_c)^{1/2}$. In equations (7) s denotes the quantity

$$s = \frac{1}{\epsilon}\left(\frac{\partial \omega}{\partial k}\right)_c; \tag{10}$$

consequently at $R = R_c$, the group velocity $\frac{\partial \omega}{\partial k}$ must be $O(\epsilon)$ in order that s be $O(1)$, an assumption implicit in the derivation of (7). In the Taylor-Couette system $\frac{\partial \omega}{\partial k}$ can in fact change sign, and parameter regimes in which $s = O(1)$ do exist (R. Tagg, private communication). When $\frac{\partial \omega}{\partial k} = O(1)$ equations (7) are replaced by modified equations derived by Knobloch and De Luca (1990). Equations (7) must be supplemented by boundary conditions at the two ends $X = \pm L$. Here $L = O(1)$ so that the aspect ratio, Γ, of the cylinder is $2L/\epsilon \gg 1$. Note that this procedure links the aspect ratio to the degree of supercriticality, $R - R_c$. The boundary conditions that are consistent with the reflection and rotation symmetries of the system must be of the form (cf. Cross 1988):

$$A - \epsilon(\mu A_X + \nu \overline{B}_X) = 0 \quad B - \epsilon(\mu B_X + \nu \overline{A}_X) = 0 \quad \text{at} \quad X = L$$
$$A + \epsilon(\overline{\mu} A_X + \overline{\nu} B_X) = 0 \quad B + \epsilon(\overline{\mu} B_X + \overline{\nu} A_X) = 0 \quad \text{at} \quad X = -L, \tag{11}$$

where μ, ν are complex coefficients, assumed to be known.

Following Dangelmayr and Knobloch (1990) we let

$$A = \epsilon^{1/2} A_0 + \epsilon^{3/2} A_1 + \cdots, \qquad B = \epsilon^{1/2} B_0 + \epsilon^{3/2} B_1 + \cdots \tag{12}$$

and let $T = \epsilon \tau$ be a superslow time. We also write:

$$\Lambda = \Lambda_c + \epsilon \Delta + \cdots, \tag{13}$$

where Λ_c represents the shift in the instability threshold arising from the presence of the ends. At $O(\epsilon^{1/2})$ we obtain:

$$[(1 + i\gamma)\partial_X^2 + s\partial_X + \Lambda_c] A_0 = 0, \qquad A_0(\pm L) = 0 \tag{14a}$$
$$[(1 - i\gamma)\partial_X^2 - s\partial_X + \overline{\Lambda}_c] B_0 = 0, \qquad B_0(\pm L) = 0 \tag{14b}$$

so that

$$\Lambda_c = (1 + i\gamma)\left[\left(\frac{\pi}{2L}\right)^2 + \left(\frac{s}{2(1 + i\gamma)}\right)^2\right]. \tag{15}$$

The corresponding eigenfunctions are given by

$$A_0 = v(T) \exp\left[\left(\frac{-s(1 - i\gamma)}{2(1 + \gamma^2)}\right) X\right] \cos\left(\frac{\pi X}{2L}\right)$$
$$B_0 = w(T) \exp\left[\left(\frac{s(1 + i\gamma)}{2(1 + \gamma^2)}\right) X\right] \cos\left(\frac{\pi X}{2L}\right), \tag{16}$$

where the complex amplitudes v, w are functions of T. The evolution equations for v and w are obtained from the solvability conditions at $O(\epsilon^{3/2})$. One obtains equations (6) with the following expressions for the coefficients:

$$\lambda = \Delta_r - \left(\frac{\pi^2}{2L^3}\right)\mu_r \qquad \omega = \Delta_i - \gamma\left(\frac{\pi^2}{2L^3}\right)\mu_r \tag{17a}$$

$$(a, b) = K(\tilde{a}, \tilde{b}), \qquad K \equiv \frac{3\pi^4(1 + \gamma^2)^5 \sinh\left[\frac{sL}{1+\gamma^2}\right]}{sL^5\left(\left(\frac{\pi}{L}\right)^2(1 + \gamma^2)^2 + s^2\right)\left(\left(\frac{2\pi}{L}\right)^2(1 + \gamma^2)^2 + s^2\right)} \tag{17b}$$

$$d = -\left(\frac{\pi^2}{4L^3}\right)(1 + i\gamma)(\nu e^{\frac{sL}{1+i\gamma}} + \overline{\nu} e^{\frac{-sL}{1+i\gamma}}). \tag{17c}$$

Thus near the onset of instability the partial differential equations (7,11) reduce to the equations deduced from symmetry principles alone.

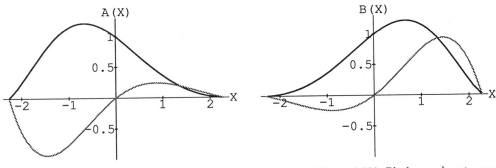

Figure 2: The linear eigenfunctions (16), $\Gamma = 40$, $L = 2.28$, $s = 2.32$, $\gamma = 0.999$. Black = real part, gray = imaginary part.

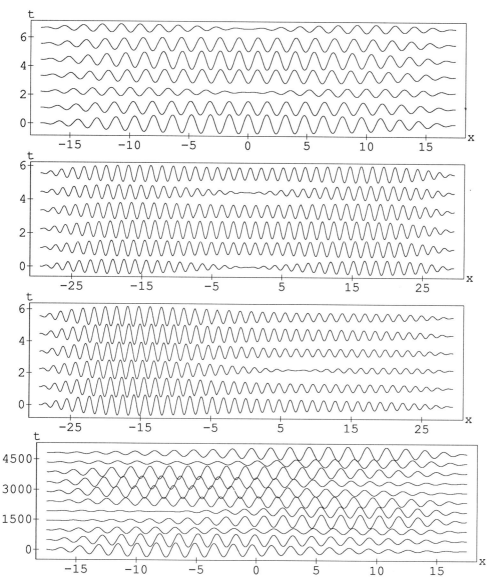

Figure 3. $W(x,t)$ for SW_0', SW_π', TW', and MW' with parameter values giving an initial bifurcation to stable spiral vortices in the perfect problem. In this case MW' describes alternating spirals: $k_c = 3.87$, $\omega_c = 0.706$, $\gamma = 0.999$, $s = 2.32$, $a = -2.23 + i0.223$, $b = -1.70 + i0.170$, $\epsilon = 0.1$, and $\mu = \nu = -0.03$ (Cross, 1988). The bifurcation parameters and aspect ratios for each plot are $\lambda = \{0.0523, -0.877, 3.12, 3.25\}$ and $\Gamma = \{35, 59, 59, 35\}$ respectively.

Equations (6) have been analyzed in detail by Dangelmayr and Knobloch (1987, 1991). The analysis shows that:

(a) Pure spirals are no longer solutions.

(b) The breaking of the translation symmetry splits the multiplicity-two Hopf bifurcation into a pair of successive Hopf bifurcations, each of which gives rise to an oscillatory pattern called SW' for which $r_1 = r_2$. The SW' that first sets in is the stable one. There are no other primary branches. Both SW' are symmetric under reflection in the midplane of the system and under rotations. Because of the structure of the eigenfunctions (16) left spirals dominate in $X < 0$ while right spirals dominate in $X > 0$, whenever $s > 0$. The opposite is the case when $s < 0$. Thus the SW' have the form of a "chevron" pattern: spirals are produced at $X = 0$ and propagate in either direction away from $X = 0$ (if $s > 0$), or they annihilate in a sink at $X = 0$ (if $s < 0$). The two types of SW', called $SW'_{0,\pi}$, differ in their total phase $\phi_1 + \phi_2 = 0, \pi$. In contrast to the translation invariant idealized system, this phase no longer decouples from the equations for r_1, r_2. The phase difference $\phi_1 - \phi_2$ continues to decouple, however, because the SO(2) rotation symmetry is not broken. For SW'_0 the crests of the spirals reach their respective ends in phase; in SW'_π they are exactly out of phase.

(c) With increasing R one or other of the SW' undergoes a secondary pitchfork bifurcation producing left- and right-handed spirals. These differ in form from the TW found in the idealized system, and we refer to them as TW'. The TW' are solutions of the form $r_1 \neq r_2$, for which the corresponding frequencies $\dot\phi_1, \dot\phi_2$ are locked in a 1:1 ratio: $\dot\phi_1 = -\dot\phi_2$. For these solutions, in contrast to TW, there is no axially translating reference frame in which the solution is time-independent. Thus spiral vortices in a finite cylinder are in fact a superposition of phase-locked left- and right-handed vortices in which the amplitude of one or other spiral dominates.

(d) Depending on the aspect ratio Γ (more precisely, the quantity Γ mod $2\pi/k$) the bifurcation to TW' may be supercritical, subcritical, or be preceded by a secondary Hopf bifurcation (see Fig.2). This bifurcation produces an oscillation with frequency $\Omega = O(|d|)$ in the amplitudes r_1, r_2. Thus in half of the period $2\pi/\Omega$ a right-handed spiral dominates and the pattern propagates upwards, while in the second half a left-handed one dominates and the pattern propagates downwards. We call the resulting state MW'. Qualitatively this pattern behaves exactly like the alternating spiral vortex state observed by Andereck et al (1986).

In Figure 3, we show the resulting SW'_0, SW'_π, TW', and MW' for particular choices of the parameters.

The theory summarized above, and described in detail by Dangelmayr and Knobloch (1991) enables us to make a number of predictions for future experiments on counter-rotating cylinders near the primary Hopf bifurcation:

(i) In a finite cylinder the initial instability is always to a chevron pattern with spiral vortices either originating in the middle of the cylinder and propagating to the ends ($s > 0$), or propagating from the ends towards the middle and annihilating there ($s < 0$).

(ii) Spiral vortices filling most of the cylinder are produced in a secondary bifurcation which may be either supercritical or subcritical.

(iii) Depending on the aspect ratio (more precisely on Γ mod $2\pi/k$) the transition to cylinder filling spirals may be preceded by a secondary Hopf bifurcation giving rise to alternating spiral vortices. With increasing Reynolds number the amplitude of the oscillation of the boundary between the left and right spirals increases, as does the reversal period. This prediction is in qualitative agreement with the experiments near the transition to Taylor vortices (R. Tagg, private communication). The alternating spirals persist for only a finite range of Reynolds numbers before they are superseded by nonreversing spirals. This may happen via either a global (heteroclinic) bifurcation at which the reversal period becomes infinite or a tertiary Hopf bifurcation. Both possibilities are hysteretic (see Dangelmayr and Knobloch (1991) for details).

When the primary instability gives rise to stable ribbons instead of stable spiral vortices ($a_r > 0$, $a_r + 2b_r < 0$), the analysis of Dangelmayr and Knobloch (1991) yields the following predictions for the behavior of the system:

(e) The initial bifurcation is, again, to a stable supercritical chevron pattern (SW'_0 or SW'_π).

(f) Depending on the aspect ratio (more precisely on Γ mod $2\pi/k$), one of two things may happen. Either the initial SW' state remains stable with increasing amplitude, or it loses stability at a secondary pitchfork bifurcation to TW'. This bifurcation may be subcritical in which case the TW' acquire stability at a tertiary saddle-node bifurcation, or supercritical in which case they are initially stable. In both cases, however, the TW' lose stability with increasing amplitude to modulated travelling waves (MW'). This bifurcation is a supercritical Hopf bifurcation and the resulting MW' are best described as pulsating TW' (Fig. 4). The period of these pulsations increases with R until a global bifurcation takes place beyond which the TW' are of the reversing type shown in Fig. 3. With increasing R the reversing MW' finally lose stability at a saddle-node bifurcation. The system then makes a hysteretic transition to the other SW' state.

The above conclusions follow directly from Fig. 10 of Dangelmayr and Knobloch on changing the signs of λ and all the eigenvalues (i.e. $t \to -t$). Note that although stable TW' and MW' exist in this case also, they do so only in a limited range of values of R.

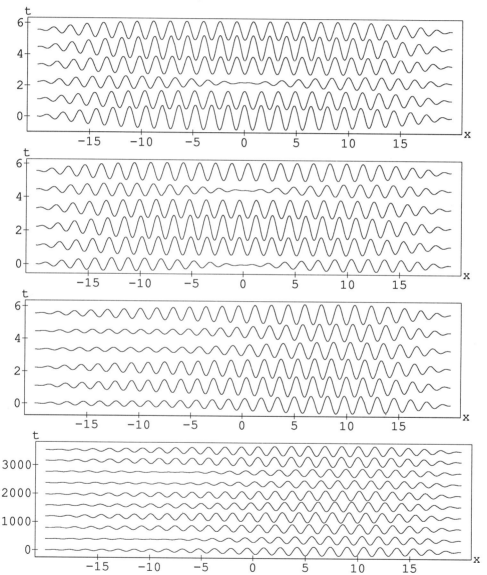

Figure 4. $W(x,t)$ for SW'_0, SW'_π, TW', and MW'. MW' appears in this case as pulsating TW', as described in the text. The parameter values are taken from Edwards (1991) and Demay and Iooss (1984), giving an initial bifurcation to stable ribbons in the perfect problem. $\Gamma = 40$, $k_c = 3.87$, $\omega_c = 0.706$,

$\gamma = 0.999$, $s = 2.32$, $a = 2.23 - i46.1$, $b = -1.70 + i16.7$, $\epsilon = 0.1$, and $\mu = \nu = -0.03$ (Cross, 1988). The bifurcation parameter used for each plot is $\lambda = \{0.844, -0.256, -0.1063, -0.05629\}$.

Acknowledgement

We are indebted to Stuart Edwards, Harry Swinney, and Randy Tagg for illuminating discussions. This work was supported by NSF grant DMS-8814702.

References

Andereck, C.D., Liu, S.S., Swinney, H.L. 1986 *J. Fluid Mech.* **164** 155.

Crawford, J.D., Knobloch, E. 1991 *Ann. Rev. Fluid Mech.* **23** 341.

Cross, M.C. 1988 *Phys. Rev. A* **38** 3593.

Dangelmayr, G., Knobloch, E. 1987 in The Physics of Structure Formation: Theory and Simulation, W. Güttinger and G. Dangelmayr, eds. (Berlin: Springer) p. 387.

Dangelmayr, G., Knobloch, E. 1990 in Nonlinear Evolution of Spatio-Temporal Structures in Dissipative Continuous Systems, F.H. Busse and L. Kramer, eds. (New York: Plenum) p. 399.

Dangelmayr, G., Knobloch, E. 1991 *Nonlinearity* **4** 399.

Demay, Y., Iooss, G. 1984 *J. Méc. théo. appl., Numero spécial* p. 193.

Edwards, W.S. 1990 Ph. D. Thesis, Univ. of Texas, Austin.

Golubitsky, M., Stewart, I. 1985 *Arch. Rat. Mech. Anal.* **87** 107.

Knobloch, E. 1986 *Phys. Rev. A* **34** 1538.

Knobloch, E. and De Luca, J. 1990 *Nonlinearity* **3** 975.

Tagg, R., Edwards, W.S., Swinney, H.L. 1990 *Phys. Rev. A* **42** 831.

Tagg, R., Edwards, W.S., Swinney, H.L., Marcus, P.S. 1989 *Phys. Rev. A* **39** 3734.

A MODEL OF THE DISAPPEARANCE OF TIME-DEPENDENCE IN THE FLOW PATTERN IN THE TAYLOR-DEAN SYSTEM

Laurent Fourtune[1], Innocent Mutabazi, C. David Andereck

Department of Physics, the Ohio State University,
174 W 18th Avenue, Columbus, OH 43210, USA

INTRODUCTION

The transition to chaos in diverse physical systems far from thermodynamic equilibrium is one of the challenging problems of modern physics. The most commonly studied models in hydrodynamics are the Rayleigh-Bénard thermal convection system and the Taylor-Couette instability[1, 2], both of which have been intensively investigated during the last two decades. The transition to chaos has been characterized and various scenarios have been discovered experimentally and in numerical simulations of those systems[3]. These two systems possess several symmetries, the breaking of which gives rise to new patterns. However, the real world is far from these simple cases and an effort is underway to study more complicated systems such as thermal convection in superposed layers of immiscible fluids[4], the horizontal Taylor-Couette system with a partially filled gap[5], the flow in curved channel[6] or the boundary layer flow over a concave wall[2].

The flow between two horizontal cylinders with a partially filled gap also known as the Taylor-Dean system, exhibits a rich variety of patterns of stationary and traveling rolls[5]. In the case when the outer cylinder is fixed, the initial instability occurs in the form of traveling inclined rolls, and upon increasing the control parameter, the roll pattern exhibits a spatio-temporal modulation[7]. In the case when the inner cylinder is fixed, the transition occurs as traveling inclined rolls which undergo a subcritical bifurcation to stationary axisymmetric rolls[8]. In this paper, we will describe in detail this transition and will give a tentative explanation using the Ginzburg-Landau equation model.

The experimental system consists of two horizontal coaxial cylinders, a stationary inner cylinder made of black Delrin plastic with radius a = 4.486 cm, and an outer cylinder made of Duran glass with radius b = 5.08 cm, which rotates with angular velocity Ω. Teflon rings are attached to the inner surface of the outer cylinder a distance L = 53.40 cm apart, giving an aspect ratio Γ = L/d = 90. The working fluid is water with 1% Kalliroscope AQ 1000 added for visualization, and its kinematic viscosity is 0.98 cstokes at the temperature T = 21C. The filling level fraction ν = 0.75 weakly influences the instability threshold[9].

The control parameter of the system is the Taylor number Ta = $(\Omega Rd/\nu)(d/R)^{1/2}$ defined with respect to the outer cylinder parameters. The flow-pattern wavelengths are

---

[1]Present address : Elève de l'Ecole Normale Supérieure, 24 rue Lhomond, F-75231 Paris Cedex 05, France

scaled by the characteristic length d (gap size), the phase velocity is scaled by ν/d and the frequencies are scaled by the inverse of the radial diffusion time $d^2/\nu = 36$ sec. We have chosen the ramping rate (experimental variation of the Taylor number) r = dTa/dt < 3 in order to achieve the quasistatic condition.

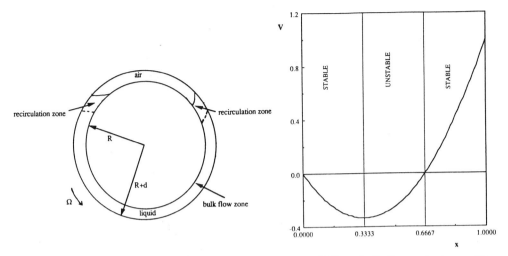

Figure 1. Flow configuration : a) Experimental geometry   b) Base bulk flow velocity profile

Flow frequencies are measured from the power spectra of single point series obtained with laser light reflected off the Kalliroscope flakes onto a photodiode detector. Spatial dependence data are obtained using a 28-85 mm variable focal length lens to form an image of the visualized flow on a 1024 pixel charge coupled device (CCD) linear array interfaced through CAMAC to a computer. The line of 1024 pixels is oriented parallel to the cylinder axis. The output consists of intensity maxima and minima which correspond to the centers and boundaries of the rolls. Space-time diagrams are then produced by displaying intensity versus axial position plots at regular time intervals. Analysis of these plots yields the roll size and the dynamics of the pattern in time and space. The spatial periodicity of the flow pattern is characterized by the wavenumber $q = 2\pi\, d/\lambda$ and its time-dependence by the phase velocity, both quantities measured from the space-time diagrams.

EXPERIMENTAL RESULTS

Rotating cylinder ends, in our case Teflon rings attached to the outer cylinder, induce Ekman cells even for low values of Ta. They become visible at Ta = 25 near both ends of the system and have an axial extension of 0.95 cm. At larger Ta, smaller stationary small rolls of average size $\lambda_E = 0.48$ cm are formed adjacent to the Ekman end cells. The maximum number of those rolls is 6 when Ta = 80. As they are localized in space and their spatial extent does not increase, this part of the system is referred to as the Ekman region in the following. For Ta = 91, cells in the Ekman region become oscillatory with a frequency f = 0.143 Hz, and at the same time, 3 to 5 traveling inclined rolls moving toward the middle of the system are generated close to each Ekman region. These rolls have a wavelength λ = 1.584 or wavenumber q = 3.967.

The phase velocity of the traveling rolls depends on their axial position : close to an Ekman region, it is v = 0.180 cm/sec. The Fourier spectrum of reflected light intensity gives a frequency peak around f = 0.136 Hz (4.90 in scaled units). The traveling rolls frequency is nearly identical to that of the oscillating rolls in the Ekman region. There are no rolls observed in the middle of the system. The mixed state of traveling inclined rolls (perturbations of finite amplitude A ≠ 0) and of laminar base flow (zero amplitude state) is stable, having been observed over many hours without noticeable changes. The transition

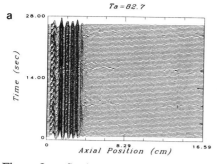
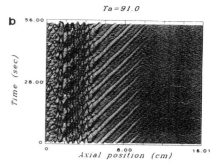

Figure 2-a. Stationary Ekman region (Ta=82.7)   Figure 2-b. Oscillatory Ekman region and emission of traveling rolls (Ta=91)

has a hysteresis of about 13% in Ta, and is a subcritical Hopf bifurcation. Increasing Ta, the number of traveling inclined rolls increases on each side and the laminar base flow extension decreases. At Ta = 110, the flow pattern becomes stationary and axisymmetric after a transient period during which the pattern relaxes by adjusting its amplitude and phase (wavelength). The spatial evolution of the transition to the stationary state for a fixed value of the control parameter shows that there exists a front between the time-dependent and stationary states, moving with a finite velocity $v_f = 0.65$ cm/sec, larger than that of the traveling rolls themselves. The phase velocity of the traveling rolls has a large discontinuity at the transition. Furthermore, the transition exhibits a large hysteresis, about 14% in Ta. The rolls become axisymmetric and the wavelength changes to $\lambda = 1.80$, corresponding to the wavenumber $q = 3.53$, together with a two-roll modulation wavelength $\Lambda = 2.81$ or wavenumber $Q = 2.24$.

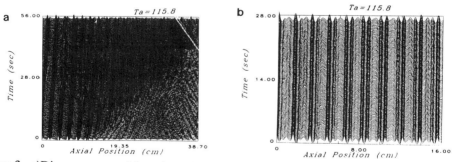

Figure 3. a) Disappearance of time-dependence and b) stationary Dean rolls with spatial modulation

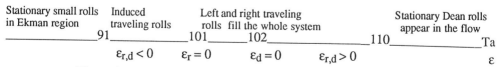

Figure 4. Transition sequence for the control parameter (Ta or ε)

DISCUSSION OF RESULTS

In order to understand the previous results, we analyze the structure of the base flow when only the outer cylinder rotates. The base flow consists of the azimuthal bulk flow, of the flow in the recirculation zone and of the flow in the Ekman region. The flow in the recirculation zone has an extension comparable to the gap size d and chosing the small gap approximation, we are able to reduce its influence. The Ekman region is small compared to the whole length of the flow extension (l_E = 0.05 L). The velocity profile of the azimuthal bulk base flow (in the small gap approximation) is given by $V(x) = 3x^2 - 2x$ (Fig.1-a) and as deduced from Rayleigh's stability criterion, it has one unstable layer sandwiched between two stable layers. The unstable layer is subject to the Dean instability as it belongs to the Poiseuille part of the profile. The linear stability theory applied to this velocity profile $V(x)$ gives a stationary critical state with the following characteristics : q_c = 2.875, Ta_c = 89.74. The stationary state appears at Ta = 110 with the wavenumber q = 3.53, therefore we consider that the transition to stationary rolls is the true Dean instability, the difference in critical parameters and the prexisting observed states are due to the boundary effects, mainly the flow in the Ekman region. The time-dependent roll pattern observed close to the Ekman region is a metastable phase due to the excitation by the Ekman region into a propagating medium (flow in our case). When the small rolls in the Ekman region become oscillating, the Ekman region can be considered as an oscillatory localized source (with frequency ω) which emits traveling rolls in the flow. The time dependent structure is inclined for symmetry arguments and has a wavenumber which is evidently selected by the frequency in the Ekman region.

<u>Amplitude equation and emission of rolls in subcritical regime ($\varepsilon < 0$)</u>

In order to understand these results, we represent the perturbative velocity in the time-dependent flow pattern (right traveling) in the separable form :

$$v'(t,x,y,z) = A(t,z)\, F(x)\, \exp\{i(\omega_c t - q_c z - p_c y)\}$$

where $F(x)$ is the structure function and the envelope $A(t,z)$ satisfies the one dimensional complex Ginzburg-Landau amplitude equation :

$$\frac{\partial A}{\partial t} - v_g \frac{\partial A}{\partial z} = \varepsilon(1+ic_0)A + \xi_0^2(1+ic_1)\frac{\partial^2 A}{\partial z^2} - g(1+ic_2)|A|^2 A \qquad (1)$$

where $\varepsilon = (Ta-Ta_c)/Ta_c$ is the relative distance from the onset of instability,
 v_g is the group velocity of the flow pattern,
 ξ_0 is the coherence length of the perturbations,
 c_0, c_1 and c_2 are respectively the corrections to the frequency due to the control parameter, the wavelength and amplitude variations, and
 g is the Landau constant of the nonlinear saturation.

We consider the Ekman region as a localized source emitting at the same frequency ω_c at z = 0 and z = L where L is the whole length of the flow extension. Far from the Ekman region, the amplitude of one induced traveling roll pattern is decreasing to zero. We therefore impose the boundary condition on the Ginzburg-Landau equation at one end, for example z = 0, without loss of generality : $A(t, z=0) = A_0 \exp\{i(\omega-\omega_c)t\}$, where A_0 represents the strength of the source. The spatial dependence of the envelope is chosen as follows $A(t,z) \sim \exp(ikz)$ where the wavenumber k is a complex number whose real part k_r is the correction to the wavenumber q_c and the imaginary part k_i gives the spatial damping length of the perturbation induced by the localized source. A similar problem formulation has been successfully applied to open flows such as wakes and jets[10]. The problem stated in this way contains many parameters : ε, v_g, c_1, c_2 and A_0, while ξ_0 and g can be removed by scaling of the axial coordinate z and of the amplitude A. The group velocity cannot be removed since the system does not possess the Galilean invariance because of the boundaries. For simplicity of the analysis, we consider $c_1 = 0 = c_2$; numerical simulations with different values have shown no qualitative difference with the results given here. For $\varepsilon < 0$, all perturbations are damped, but for small $|\varepsilon|$, the damping is slow and some strong external excitation (in our case oscillatory Ekman rolls) may lead to observable stable solutions of the CGL equation. The results are then :

1. When $\omega = 0$, $v_g = 0$ and we obtain the trivial solution $A(t,z) = 0$, which means that the stationary Ekman region does not induce a propagating flow pattern.

2. When $\omega \neq 0$, $v_g \neq 0$, $A \neq 0$: any external time-dependent excitation can propagate in the system with a $v_g = (\omega - \omega_c)/(q_c + k_c)$. The amplitude of the perturbative envelope A exhibits two successive regimes in z-dependence : nonlinear $A \sim z^{-1/2}$ and linear where $A \sim e^{(\varepsilon/v_g)z}$ (Figure 5-a shows A(z) for different values of ε). For fixed A_0 and v_g, the spatial damping coefficient k_i is estimated from our numerical calculations (from distance at which the excitation damps by a factor of 10) : it is a linear function of ε except for small values when the dependence becomes nonlinear.

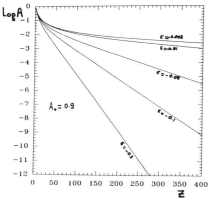

Figure 5-a . Spatial amplitude variation for different values of ε

Figure 5-b . Spatial damping coefficient k_i as function ε

<u>Hysteresis as result of modes interaction</u>

In the flow regime between Ta = 101 and Ta = 110, both traveling inclined and Dean rolls interact, therefore the flow pattern may be described by 3 amplitudes : A_r for right traveling inclined rolls, A_l for left traveling inclined rolls and A_d for Dean rolls. Without loss of generality and for simplicity, we analyze the interaction between the right traveling A_r and Dean rolls A_d. Experimentally, the bifurcation to traveling rolls is supercritical and that to Dean rolls is subcritical, therefore we have the Ginzburg-Landau equations at 5th order, neglecting the spatial dependence :

$$\frac{\partial A_r}{\partial t} = \varepsilon_r A_r - (g_r|A_r|^2 + g_{rd}|A_d|^2)A_r - h_r|A_r|^4 A_r - k_{rd}|A_d|^4 A_r - f_{rd}|A_d|^2|A_r|^2 A_r \quad (2\text{-}a)$$

$$\frac{\partial A_d}{\partial t} = \varepsilon_d A_d - (g_{dr}|A_r|^2 + g_d|A_d|^2)A_d - h_d|A_d|^4 A_d - k_{rd}|A_r|^4 A_d - f_{rd}|A_r|^2|A_d|^2 A_d \quad (2\text{-}b)$$

where $g_r > 0$, $g_d < 0$, $d_r > 0$, $d_d > 0$ after the nature of the bifurcations, and the coupling constants g_{rd}, g_{rd}, k_{rd}, $f_{rd} < 0$, as the interaction between modes is only destructive. By assumption $g_r \ll (g_d, g_{rd}, g_{dr})$, i.e., the right traveling rolls are dominated by linear growth as maintained by the external source.

Ramping up, the amplitude $|A_d|$ of the Dean rolls is very small so that all terms in equation (2-a) containing $|A_d|^{n>2}$ are negligible, and therefore the right traveling rolls have amplitude $|A_r|$ given by :

$$|A_r|^2 = -\frac{\varepsilon_r}{g_r} \quad \text{for small } \varepsilon_r$$

and

$$|A_r|^2 = -\frac{g_r}{h_r} + \left(\frac{\varepsilon_r}{h_r}\right)^{1/2}\left(1 + \frac{g_r^2}{4\varepsilon_r h_r}\right)^{1/2} \quad \text{for large } \varepsilon_r$$

The equation for Dean rolls can be written in the form, where we have neglected the spatial variations :

$$\frac{\partial A_d}{\partial t} = \varepsilon_d^{\text{eff}} A_d - g_d^{\text{eff}} |A_d|^2 A_d - h_d |A_d|^4 A_d \quad (3)$$

with $\quad \varepsilon_d^{\text{eff}} = \varepsilon_d - g_{dr}|A_r|^2 - k_{dr}|A_r|^4 \quad g_d^{\text{eff}} = g_d + f_{dr}|A_r|^2$

For small ε_r, $\varepsilon_d^{\text{eff}} < 0$, the Dean rolls are damped and the flow is dominated by the right traveling rolls. For large ε_r, the quantity $\varepsilon_d^{\text{eff}}$ can pass through zero toward positive values and therefore, the Dean rolls can grow in the flow. The onset of Dean rolls is given by the condition $\varepsilon_d^{\text{eff}} = 0$, this condition corresponds to $Ta_c = 110$. Once the Dean rolls have appeared in the flow, the linear growth and the nonlinear interaction coefficient become more important because of the coupling with the right traveling rolls. If for large ε_r, $g_d^{\text{eff}} = 0$, the transition to Dean rolls should change from subcritical to supercritical. This would explain the long time needed for the transition from the time-dependent to stationary Dean rolls.

The right traveling rolls equation becomes :

$$\frac{\partial A_r}{\partial t} = \varepsilon_r^{\text{eff}} A_r - g_r^{\text{eff}} |A_r|^2 A_r - d_r |A_r|^4 A_r \quad (4)$$

where

$$\varepsilon_r^{\text{eff}} = \varepsilon_r A_r - g_{rd}|A_d|^2 - d_{rd}||A_d|^4 \quad \text{and} \quad g_r^{\text{eff}} = g_r + f_{rd}|A_d|^2$$

With the growth of the Dean rolls, $\varepsilon_r^{\text{eff}}$ can pass through zero to negative values and as the quantity $g_r \ll g_d$, the traveling rolls are overturned by the Dean rolls through nonlinear interaction. So we have shown that the transition to Dean rolls is highly hysteretic because of the interaction with the right traveling rolls.

CONCLUSION

The flow between two horizontal coaxial cylinders with a partially filled gap, when only the outer cylinder rotates, exhibits an unusual transition from time-dependent patterns induced by the Ekman regions to a stationary axisymmetric pattern flow. This transition and its hysteretic nature is understood using the Ginzburg-Landau equation, with an external source. There is still an open problem concerning the spatial modulation of the stationary rolls.

ACKNOWLEDGMENT

We would like to thank J.E. Wesfreid for suggesting the approach taken here and P. Laure for the fruitful discussions. L.F. acknowledges the support from the Ecole Normale

Supérieure (France) and the hospitality of the Physics Department of the Ohio State University. This project is supported by ONR and NATO.

REFERENCES

1. S. Chandrasekhar, "Hydrodynamic and Hydromagnetic Stability", Oxford University Press, (1961)
2. P.G. Drazin and W.H. Reid, "Hydrodynamic Stability", Cambridge University Pres (1981)
3. P. Manneville, "Dissipative Structures and Weak Turbulence", Academic Press, N. Y. (1990)
4. S. Rasenat, Busse and I.Rehberg, A theoretical and experimental study of double-layer convection, J. Fluid Mech.199 : 519 (1989)
5. I. Mutabazi, J.J. Hegseth, C.D. Andereck and J.E. Wesfreid, Pattern Formation in the flow between two horizontal coaxial cylinders with a partially filled gap, Phys. Rev. A 38: 4752 (1988)
6. P.M. Ligrani and R. D. Niver, Flow visualization of Dean votices in a curved channel with 40 to 1 aspect ratio, Phys. Fluids 31: 3605 (1988)
7. I. Mutabazi, J.J. Hegseth, C.D. Andereck and J.E. Wesfreid, Spatio-temporal pattern modulation in the Taylor-Dean system, Phys.Rev. Lett. 64: 1729 (1990)
8. I. Mutabazi and C.D. Andereck, Transition from time-dependent to stationary flow patterns in the Taylor-Dean system, Phys. Rev.A 44: 6169 (1991)
9. I. Mutabazi, Etude théorique et expérimentale de l'instabilité centrifuge de Taylor-Dean, thèse de doctorat en Physique, Université de Paris 7 (1990)
10. P. Huerre and P. Monkevitz, Local and global instabilities in spatially developing flows, Ann. Rev. Fluid Mech. 22: 473 (1990)

END CIRCULATION IN NON AXISYMMETRICAL FLOWS

C. Normand[†], I. Mutabazi[†,‡,*], J. E. Wesfreid[‡]

[†]Service de Physique Théorique, C. E. Saclay
F-91191 Gif-sur-Yvette Cedex
[‡]E. S. P. C. I. 10 Rue Vauquelin F-75231 Paris

INTRODUCTION

The flow in the gap between two coaxial rotating cylinders has been widely studied in the Taylor-Couette axisymmetrical geometry while the Taylor-Dean flow which arises when the gap between the two cylinders is partially filled has received much less attention. In a pioneer work[1], it was assumed that the Taylor-Dean flow results from the rotation of the inner cylinder associated with the action of an external pressure gradient.

The recent interest in Taylor-Dean flow begun after experiments[2] demonstrated that by rotating the inner and the outer cylinders at different angular velocities and without any external pressure gradient the flow exhibits a rich variety of structures at the onset of centrifugal instability. One of the major difficulties to modelize these experiments is the breaking of the rotational symmetry due to the existence of two distinct air-liquid interfaces confining the fluid in the azimuthal direction.

The aim of the present communication is to characterize the behaviour of the flow near these two free surfaces by taking advantage of a large filling. Far from the free surfaces, in the core region, the flow is parallel along the azimuthal direction, with a quadratic radial profile when the small gap approximation is used. As the free surfaces are approached, the radial velocity component becomes of the same order of magnitude as the azimuthal component and we are interested in how the deviations from parallel core flow take place in these end regions. The way to handle this problem closely follows matching procedures which have been previously used in other circumstances which are reviewed below.

---

* Present address: Department of Physics, Ohio State University, Columbus, USA

Flows confined in bidimensional rectangular cavities and driven by either mechanical or thermal boundary conditions have received a special attention when their extent in one direction largely exceeds the extent in the orthogonal direction. In that case, connection between a well-known parallel core flow solution and a more complex flow in end circulation regions is established through matched asymptotic expansions. For instance, convection in a shallow rectangular cavity with differentially heated vertical walls has been investigated when the horizontal extent of the cavity, L, is much larger than its height H. A solution valid at all orders in the aspect ratio H/L was found for the core region, while the first terms of the asymptotic expansion are obtained for the end regions near the vertical walls[3]. The main feature of interest in the flow field is the existence of closed streamlines in the end regions that we shall call recirculation eddies. In this case, due to the central symmetry of the problem, the two end regions behaves in the same way. But, when the central symmetry is broken, the two end circulations are totally different. This happens for the flow induced by a constant surface stress acting along the top surface of a liquid enclosed in a rectangular basin[4]. It was found that near the upwind end the flow has a boundary layer behaviour, while near the downwind end the flow consists in a serie of damped eddies. The same behaviour occurs when the stress on the top surface is now due to surface tension variation coming from the differentially heated vertical walls of the cavity[5].

Our interest in the end circulation of Taylor-Dean flow has been drawn after the results of the stability analysis of purely azimuthal base flow revealed some discrepancies with experimental observations, in particular concerning the occurrence of time dependent instability[6]. We suspect the circulation in the end regions to have its own time dependence which possibly competes with that of the core flow instability. Our analysis of the end regions follows Bye's analysis[4] which has been improved to take into account the temporal behaviour of the flow.

EQUATIONS VALID IN THE END REGIONS

The length of the cylinders being assumed infinite, the base flow is considered as independent of the axial coordinate. The equations to be solved are the Navier-Stokes equations expressed in cylindrical polar coordinates (r, φ). The radius of the inner cylinder being R and the gap width d, we define the parameter $\delta = d/R$. The angular velocities of the inner and the outer cylinders are respectively Ω and $\mu\Omega$. A sketch of the geometry can be found in[6]. Since in the end regions, the azimuthal and radial variations of the velocity are of the same order, it is convenient to introduce non dimensional coordinates x and y defined by

$$r = R(1 + \delta x), \quad \varphi = \delta y$$

The expression for the azimuthal velocity $V_0(x)$ valid in the core region has been derived in a previous work[6] and is given by

$$V_0 = 3(1 + \mu)x^2 - 2(2 + \mu)x + 1,$$

when ΩR is chosen as the velocity scale. The method to solve the fully nonlinear Navier-Stokes equations is to use an asymptotic expansion to the core flow solution which is known as Oseen approximation. The radial and azimuthal components of the velocity are written respectively as

$$u = U_1(x, y, t), \quad v = V_0(x) + V_1(x, y, t),$$

where U_1 and V_1 are small perturbations to the core flow solution. After substituting the above expressions into the Navier-Stokes equations, the quadratic terms in U_1 and V_1 are neglected and the small gap approximation is used ($\delta \to 0$). The incompressibility condition is satisfied by introducing the streamfunction Ψ such that

$$U_1 = \frac{\partial \Psi}{\partial y}, \quad V_1 = -\frac{\partial \Psi}{\partial x}.$$

Therefore Ψ satisfies the following equation

$$\left(\frac{\partial}{\partial t} - \Delta\right) \Delta \Psi + \mathrm{Re}\left(V_0 \frac{\partial \Delta \Psi}{\partial y} - D^2 V_0 \frac{\partial \Psi}{\partial y}\right) = 0,$$

where $D \equiv \partial/\partial x$ and Re is the Reynolds number: $\mathrm{Re} = \Omega R d / \nu$. The associated boundary conditions on the cylindrical walls are: $\Psi = D\Psi = 0$ at $x = 0, 1$. Since we are only interested in the asymptotic behaviour of Ψ as the free surfaces are approched, the boundary conditions on the free sufaces limiting the fluid in the azimuthal direction are not reported here. Thus, near one of these free surfaces, arbitrarily taken as the origin of the azimuthal coordinate, a solution for Ψ that matches with the core solution when $y \to \pm\infty$, is sought under the form

$$\Psi(x, y, t) = \exp(i\omega t + \alpha y) F(x). \tag{1}$$

In the expression above (1), an oscillatory time dependence is expected by allowing ω to be a real number, while $\alpha = a + ib$, is a complex number whose real part gives the spatial decay towards the core flow solution, with $a < 0$ (respectively $a > 0$) corresponding to the end region $0 < y < \infty$ (respectively $-\infty < y < 0$). According to the standard terminology, expression (1) described spatial modes in contrast with temporal modes for which ω is complex and α purely imaginary. The imaginary part of α is related to the azimuthal periodicity of the flow ; existence of a finite value of b means that the matching with the core flow occurs via a sequence of damped eddies. Thus, the function $F(x)$ satisfies a complex Orr-Sommerfeld equation (OSE)

$$(D^2 + \alpha^2 - i\omega)(D^2 + \alpha^2) F = -\alpha \, \mathrm{Re} \, [D^2 V_0 - V_0(D^2 + \alpha^2)] F, \tag{2}$$

which differs from the standard OSE by the nature of the parameters α and ω. The boundary conditions at $x = 0, 1$ are : $F = DF = 0$. For stationary ($\omega = 0$) as well as oscillatory flows ($\omega \neq 0$) the values of α, functions of Re and μ, are obtained by numerical integration of equation (2). In fact, we get an infinite spectrum of values $\{\alpha_n\}$, from which we select the one with the smallest real part corresponding to the least damped mode. This gives us a rough estimate of the spatial structure of the flow near the end regions when the Reynolds number and rotation ratio are varied. Before presenting the results for $\alpha_1(\mu, \text{Re})$ in the general case, we want to emphasize some special cases corresponding to limiting values of Re.

Stokes regime: when Re$\to$0, eq. (2) reduces to the biharmonic equation

$$(D^2 + \alpha^2)^2 F = 0,$$

whose spectrum corresponding to the boundary conditions mentioned above is well known, the first eigenvalue $\alpha_1 = 4.21 \pm 2.25i$, being the starting point of the numerical integration of eq. (2) when Re is then increased. It should be noted that the opposite value $-\alpha_1$ is also solution, which is typical of centro-symmetrical problems.

Rayleigh regime: When Re$\to\infty$ and α takes finite values, eq. (2) reduces to

$$[D^2 V_o - (V_o + ic)(D^2 + \alpha^2)] F = 0,$$

where the phase velocity c is related to the frequency by the relation: $c = \omega/\alpha$ Re.

Boundary layer regime: When Re$\to\infty$ and $\alpha \to 0$, but the product Re$^* = \alpha$Re remains finite, eq. (2) reduces to the boundary layer flow equation

$$D^4 F = -\text{Re}^* (F D^2 V_o - V_o D^2 F).$$

RESULTS

Numerical integration of eq. (2) has been done respectively for the stationary ($\omega=0$) and the oscillatory ($\omega \neq 0$) modes and the results are presented in separate sections.

Stationary modes

The value of α corresponding to the least spatially damped mode has been obtained for different values of the rotation ratio μ, as a function of the Reynolds number. The values of the real and the imaginary parts, respectively a and b, are reported in Fig. 1 as functions of Re, for two values of the rotation ratio: $\mu = \pm 0.4$. For each case, the eigenvalue α with a positive real part (a>0) corresponding to the end region $-\infty < y < 0$ and the eigenvalue α with a negative real part (a<0) corresponding to the end region $0 < y < \infty$ have been considered.

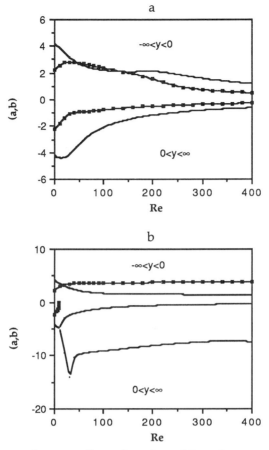

Fig. 1. eigenvalue $\alpha = a+ib$, as function of Re. a) $\mu=-0.4$, b) $\mu=0.4$.
Real part: —— ; imaginary part: –•–•–•–.

The case $\mu=-0.4$ is representative of a typical behaviour, called the quasi-symmetrical case, when the two end regions have a rather similar spatial structure along the azimuthal direction. It is shown in Fig. 1a that the eigenvalue α with either a positive or a negative real part, remains a complex number on the whole range of Reynolds numbers considered. The fact that the imaginary part always takes a non zero value b, means that the matching with the core solution occurs through a sequence of damped eddies in each of the two end regions. The term symmetrical used above is relative to the presence of eddies in the two end regions even if the wavelength of these eddies are not exactly the same near the two free surfaces.

The case $\mu=0.4$ is representative of a nonsymmetrical behaviour. It is shown in Fig. 1b that the eigenvalue α with a positive real part remains a complex number on the whole range of Reynolds numbers considered, while it is no longer true for the eigenvalue with a negative real part. On the contrary, the associated imaginary part quickly decreases and vanishes at a low value of the Reynolds number (Re=9.8). Simultaneously, for the same value of the

Reynolds number, the negative real part splits in two branches which have asymptotic limits when Re → ∞, corresponding to the boundary layer regime mentioned above. Thus, the matching with the core solution now occurs in a different way near the two end regions: damped eddies exist only on one side and on the other side a boundary layer regime is prevailing.

The value of the Reynolds number at which the imaginary part of α vanishes is a function of the rotation ratio as shown in Fig. 2. The transition between the symmetrical and the nonsymmetrical behaviour occurs across the curve drawn in Fig. 2. For μ>-0.32, the imaginary part of α (a<0) vanishes above a particular value of Re, giving rise to a boundary layer regime in the corresponding end region. For μ<-0.32, the value of α is complex, whatever the sign of a is positive or negative, and recirculation eddies exist in the two end regions.

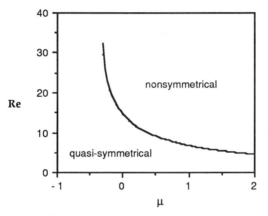

Fig. 2. Reynolds number versus μ, corresponding to the transition between symmetrical and nonsymmetrical end circulations.

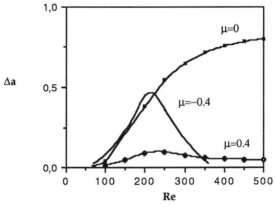

Fig. 3. Difference in the spatial damping rate between the stationary and the oscillatory modes.

Some values of μ have a special interest. When $\mu=-1$, the core velocity profile reduces to : $V_o(x)=1-2x$, which satisfies : $V_o(1-x)=-V_o(x)$. As a consequence, if α is solution of eq. (2) then $-\alpha$ is also solution and the flow is totally symmetrical in the two end regions. Direct numerical integration of the Navier-Stokes equations in the end regions have been performed[7] when the first correction to the small gap approximation is taken into account and for the particular value $\mu=0$. The results of Chen et al.[7] agree with those presented in Fig. 2 and confirm that the two end circulations are different. When $\mu=0$, the core velocity profile : $V_o(x)=(1-x)(1-3x)$, also coincides with the one examined by Bye[4] in another context. Moreover, the case $\mu=\infty$, corresponding to the inner cylinder at rest, is equivalent to the case $\mu=0$. This can be shown with the transformation: $x \to 1-x$ and $Re\,\mu \to Re$ in eq. (2).

Oscillatory modes

Spatial modes of the form (1) which oscillate in time and are damped or amplified in space have been used until now essentially in the theory of instability of parallel flows. Here, we are looking for values of α corresponding to a damping toward the core solution and we want to know whether the least damped mode corresponds to a stationary ($\omega=0$) or to an oscillatory ($\omega \neq 0$) state, when the Reynolds number is varied. Our analysis is restricted to the end region in which recirculation eddies are always present ($a>0$). It is first observed that for a set of parameters (μ, Re, ω) there are two values of $\alpha=a_t+ib_t$ satisfying eq. (2) which take their origin from the values $\alpha_0=a_0+ib_0$ and $\alpha_0^*=a_0-ib_0$ corresponding to $Re=0$. If the sign of $\Delta a = a_0 - a_t$ is positive this means that the least damped mode oscillates in time.

In general for a given point (μ, Re) the relation $\Delta a(\omega)>0$ is verified on a finite range of frequency $\Delta\omega$. Thus the oscillating solution is given by a Fourier integral

$$\Psi(x, y, t) = \exp(a_0 y) \int_{\Delta\omega} \exp\Delta a(\omega) y \, \exp[i(\omega t + b\, y)] \, F_\omega(x) \, d\omega$$

In most cases, the difference Δa reaches a maximum for a particular value of the frequency $\omega=\omega_m$, and the integral can be approximated by the contribution coming from ω_m. The maximum values of Δa have been reported in Fig. 3 as functions of Re, for three values of the rotation ratio $\mu=\pm 0.4$ and $\mu=0$. Results obtained[8] for other values of μ confirm that the most important differences between the stationary and the oscillatory value of the spatial damping occur around $\mu=0$.

CONCLUSION

Using an asymptotic matching procedure, we have characterized the spatial and temporal behaviour of Taylor-Dean flow in the end circulation regions. The simplicity of the method allows to make calculations for a large number of values of the rotation ratio µ, improving the results of Chen et al. which were limited to stationary flows and to the value µ=0. We have found that for µ<-0.32, the two end circulations are rather similar and the matching with the core flow occurs through a sequence of damped eddies. On the contrary for µ>-0.32, the two end circulations are different with a boundary layer regime on one side and damped eddies on the other side. We wonder if the transition at µ=-0.32, between a symmetrical and a non symmetrical base flow may affect the threshold of centrifugal instability. Experiments[2] have shown that within the range -0.28<µ<0.5, the preferred mode of instability is in the form of inclined travelling rolls though it was not predicted by the theory when a parallel base flow is used. Moreover, the existence of oscillatory recirculation eddies may also induce oscillatory instability modes. The stability of a nonparallel and oscillatory base flow is postponed to future work.

REFERENCES

1. D. B. Brewster, P. Grosberg, A. H. Nissan, The stability of viscous flow between horizontal concentric cylinders, Proc. R. Soc. Lond. A251 : 76 (1959).
2. I. Mutabazi, J. J. Hegseth, C. D. Andereck, J. E. Wesfreid, Pattern formation in the flow between horizontal coaxial cylinders with a partially filled gap, Phys. Rev. A38 : 4752 (1988).
3. D. E. Cormack, L. G. Leal, J. Imberger, Natural convection in a shallow cavity with differentially heated end walls. Part 1, J. Fluid Mech. 65 : 209 (1974).
4. J. A. T. Bye, Numerical solutions of the steady-state vorticity equation in rectangular basins, J. Fluid Mech. 26 : 577 (1966).
5. P. Laure, B. Roux, H. Ben Hadid, Nonlinear study of the flow in a long rectangular cavity subjected to thermocapillary effect, Phys. Fluids A2 : 516 (1990).
6. I. Mutabazi, C. Normand, H. Peerhossaini, J. E. Wesfreid, Oscillatory modes in the flow between two horizontal corotating cylinders with a partially filled gap, Phys. Rev. A39 : 763 (1989).
7. K. S. Chen, A. C. Ku, T. M. Chan, S. Z. Yang, Flow in the half-filled annulus between horizontal concentric cylinders in relative rotation, J. Fluid Mech. 213 : 149 (1990).
8. C. Normand, I. Mutabazi, J. E. Wesfreid, Recirculation eddies in the flow between two horizontal coaxial cylinders with a partially filled gap, Eur. J. Mech. B4 (1991).

BIFURCATION PHENOMENA IN TAYLOR-COUETTE FLOW SUBJECT TO A CORIOLIS FORCE

P. W. Hammer[a], R. J. Wiener[b], and R. J. Donnelly

University of Oregon, Eugene, 97403

INTRODUCTION

We have been studying the effects of an external Coriolis force on the dynamics and bifurcation behavior of Taylor-Couette flow. Our work had three motivations. The first motivation was the body of work that has been carried out investigating the effects of an external field on fluid flows.[1,2] Second, in our system the axisymmetry of the base flow is broken, whereas in systems studied previously, the external field does not break the fundamental symmetry of the system.[1-4] Finally, we wanted to introduce a second control parameter[5] to the Taylor-Couette system, in the form of an external field, to explore how the dynamics are affected. Our experiments have revealed an unexpected array of bifurcation phenomena.[6] This paper will explain the techniques used to discover the bifurcations in the flow, present a bifurcation map of the Taylor-Couette system subject to a Coriolis force, and finally, discuss some of the more interesting features revealed by the map.

Magnetic and rotational fields interact with flows analogously, and this analogy has generated both historical and contemporary interest in the effects of external fields on hydrodynamical systems. For example, Taylor-Couette and Rayleigh-Bénard flows of a conducting fluid have been studied in the presence of an external magnetic field, and the Rayleigh-Bénard system has been rotated such that any convective motion would interact with the rotational field, giving rise to a Coriolis force.[2] Until our work however, the external fields have been applied so that the symmetry of the above systems' base flow state has not been broken. In our application of the Coriolis force to the Taylor-Couette system, the axisymmetry of the base flow is broken, giving rise to interesting dynamical effects that are not seen in ordinary Couette flow, (where $\vec{\Omega}$ is zero).[4,6,7]

The Taylor-Couette system subject to an external Coriolis force can also be thought of in terms of a two control parameter system, in which the two parameters set up competing modes. It is known that systems with two competing control parameters can yield a wide array of spatiotemporal dynamics. Indeed, our system has provided insight into not only the dynamical effects of a Coriolis force, but also the effects of mode competition in a two control parameter system in which the additional parameter breaks the symmetry of the ordinary system.

Our experiments have shown that the introduction of a Coriolis force alters the sequence of bifurcations that leads to turbulence in ordinary Taylor-Couette flow. Specifically, the Coriolis force stabilizes both the base flow,[4,8,9] and the transition to the first time-dependent flow state.[3] At low Reynolds numbers there is re-emergent order from noisy flow,[6] and a direct, non-hysteretic bifurcation to turbulence.[3] This transition to turbulence occurs at a Reynolds number that is an order of magnitude lower than that for the ordinary case.[6] In addition, the re-emergent order and the direct transition occur at relatively small values of the Coriolis force so that they might be understood by means of numerical simulation or a weakly nonlinear theory.[6]

THE EXPERIMENT

An external Coriolis force is applied to the ordinary Taylor-Couette system by mounting the cylinders on a rotating turntable with the rotation axis of the cylinders orthogonal to the rotation axis of the turntable (see Appendix A for a discussion of the orientation of the cylinders). The outer-cylinder is held fixed with respect to the turntable. We find that to first-order, the Coriolis force superimposes on the ordinary base flow state (where the fluid velocity is only in the azimuthal, *ie.* θ, direction) an axial flow whose magnitude and direction vary as an $m=1$ sinusoid about the azimuth. The axisymmetry of the ordinary Taylor-Couette system is thus broken.[3,4,9] This first order correction is orthogonal to the azimuthal velocity; it is the interaction between these two modes which induces time-independent *tilted* Taylor vortex flow, TTVF, in which the Taylor vortices are tilted out of the plane normal to the cylinders' common axis. Tilted wavy vortex flow, TWVF, results from a travelling azimuthal wave superimposed on tilted Taylor vortex flow. The first order axial ($\sim \sin\theta$) mode can also be thought of as competing with the tilted wavy vortex mode in the sense that the two flows will interact and interfere. The re-emergent order bifurcation is likely a result of mode competition, as are the other dynamical flow states that are unique to this system.[6]

The fluid used in the experiments was an aqueous suspension of glycerol and Kalliroscope, with bacteriostatic stabilizer added to extend the life of the sample. Typically the fluid had kinematic viscosity, $\nu = 0.033$ Stokes. Using statistical and spectral analysis of time series of the Kalliroscope reflectance, we have semiquantitatively identified the spectral characteristics of the dynamical flow regimes and located the bifurcation boundaries for this new system.

The inner-cylinder angular velocity is scaled by the reduced Reynolds number, $\Delta = (\text{Re}/\text{Re}_{c0}) - 1$, where $\text{Re} = \omega R_1 d / \nu$ (ω is the inner-cylinder angular frequency, R_1 is the inner-cylinder radius, and d is the annular gap width). Re_{c0} is the critical Reynolds number at which the primary instability occurs for ordinary Taylor-Couette flow at the radius ratio, η, used ($\eta = 0.88$: $\text{Re}_{c0} = 120.5$). The strength of the Coriolis force depends on the value of Ω, a dimensionless rotation rate, where $\Omega = \Omega_D d^2 / \nu$ (Ω_D is the angular frequency of the rotating turntable).

To experimentally locate the bifurcation boundaries, we slowly and incrementally increased one of the parameters while holding the other fixed. At each incremental point, a time series of the Kalliroscope reflectance was measured.

The first two instabilities that occur in the ordinary Taylor-Couette system are non-hysteretic, for a sufficiently small acceleration of the Reynolds number. Using the criteria of Park *et al*,[10] and Park and Jeong,[11] we slowly increased the Reynolds number at a dimensionless acceleration $a^* \approx 0.66$, where from Ref. 10, $a^* = (R_1 d^3 \Gamma / \nu^2)(\delta\omega/\delta t)$. The flow seems to be insensitive to the acceleration rate of Ω.[8] We chose a rate of change in Ω of 0.00196 s^{-1}, and did not find any resulting hysteresis in the bifurcations.

DATA ANALYSIS

At each increment of Ω and Δ a 2048-point time series was measured (occasionally, 8192 measurements of the reflectance were measured). The data acquisition frequency was $50 \times \omega$. From the time series we determined the four statistical measures that we used to assess the flow dynamics: average relative reflectance, variance of the reflectance, spectral number distribution, and spectral degrees of freedom. The average of the reflectance was used to measure the bifurcation from base flow to secondary flow. The variance of the reflectance was used to distinguish between time-independent and time-dependent flows; the spectral number distribution and the spectral degrees of freedom revealed bifurcations between ordered states and disordered states.

Reflectance and Variance

The average relative reflectance, $<v>$, for n measurements of the reflectance v is

$$<v> = \frac{1}{n}\sum_{i=1}^{n} v_i. \qquad (1)$$

A sharp drop in $<v>$ is taken to indicate the bifurcation to secondary flow,[8] but within the secondary flow regime there exist several dynamical flow states. These states are either time-independent (ie., TTVF) or time-dependent.[6] The relative reflectance is not a sufficient measure to distinguish between time-independent flow and time-dependent flow. A reasonable measure of the fluctuations of a time series about its average is the variance of the reflectance, σ^2.[3,6] The variance is measured using the formula

$$\sigma^2 = \frac{1}{n}\sum_{i=1}^{n}(v_i - <v>)^2. \qquad (2)$$

A sharp and systematic rise in the variance indicates that there has been a transition to time-dependent flow. However, within the time-dependent flow regime there are both ordered and disordered flows, and the variance is not sufficient to discriminate between these states. Therefore, additional information is needed to determine the nature of the time-dependent flow. This information is derived from calculating the Fourier transform of the reflectance time series.

The Spectral Number Distribution

This subsection and the next discuss our attempts to resolve the problem of determining the noise level of a Fourier spectrum. The result of this search led to the use of two statistical measures of a Fourier spectrum: the first was to examine the spectral number distribution, SND, which was developed as a variation of the spectral distribution function used by Ditto et al.;[12] the second was to measure the spectral degrees of freedom, SDF.[13,14]

The SND is found by counting the number of components, $N(p)$, in the Fourier power spectrum that have power greater than p. Note that $p = 10\log_{10}(P)$, where P is the sum of the squares of the Fourier amplitudes at a given frequency.

The spectral distribution function was developed as a means for measuring scaling properties of Fourier spectra representing different dynamics.[12] The main difference between the spectral distribution function and the SND is that the spectral distribution function only counts local maxima (ie. peaks) in the spectrum, while the SND counts every component. The SND thus utilizes more information in the spectrum. In particular, the SND will take into account spectral peaks that have a width that spans more than one channel in the spectrum so that broadening of spectral peaks will be apparent in the distribution. The spectral distribution function treated peaks as delta functions, thus some information might have been lost about the broadness of peaks.[15]

The purpose of the SND was its semiquantitative use in determining relative levels of the spectral noise. The relative spectral noise was chosen to be the power at which $N(p)$ drops below 25% of the total number of components in the spectrum, n.

The Spectral Degrees of Freedom

The SND revealed an extensive amount of detail about the flow and its bifurcations, but there remained the problem of confidence in the method. Was the spectral noise a good statistic, in the sense that it was shedding reliable insight into the flow dynamics? As a check on the conclusions drawn from the SND, an alternative statistic was chosen. This alternative, the spectral degrees of freedom, was introduced in a slightly different form by Crutchfield et al.[13]

The spectral degrees of freedom is defined as[6]

$$D = \left(\sum_{i=1}^{n} P_i\right)^2 / \sum_{i=1}^{n} P_i^2, \quad (3)$$

where $P_i = (a_i^2 + b_i^2)$ is the power of the ith component of an n-component Fourier power spectrum. a_i and b_i are the Fourier coefficients. The *spectral* degrees of freedom is equal to n times the degrees of freedom, as defined by Crutchfield et al.[13] Note that for a sine wave $D = 1$, and for white noise $D \cong n$.

The degrees of freedom was put forth as an attempt to distinguish sharp spectral components from broad spectral features, but Crutchfield et al. arrived at inconclusive results for the computational model they were analyzing.[13] Park and Donnelly successfully used the degrees of freedom to measure the transition from a disordered wavy vortex flow to an ordered wavy vortex flow.[14] In their study, Park and Donnelly were only concerned with a marked change in the relative number of degrees of freedom, whereas Cruthchfield et al. only hinted at this application.[13,14] Subsequent to the work reported in Hammer et al.,[6] Babcock, Ahlers and Cannell used a similar statistic to characterize the absolute instability boundary in Taylor-Couette flow with a superimposed axial through-flow.[16] Again, as in Hammer et al.,[6] the bifurcation boundary was determined by locating a sharp change in the statistic. We found that the SDF confirmed the information that we learned from the spectral noise about the location of the boundaries.

THE BIFURCATION MAP

To determine where a bifurcation occurred in $\Omega - \Delta$ space, we plotted each statistic as a function of the parameter being increased and then looked for a sharp, systematic change in the behavior of the statistic. A bifurcation map of the Taylor-Couette system can be constructed by measuring bifurcations in the appropriate statistic. The bifurcation map shown in Figure 1 displays the points which were determined experimentally. The curves are a best fit to these points using a nonlinear least-squares fit to the function

$$\Delta(x) = a\Omega^6 + b\Omega^4 + c\Omega^2 + d. \quad (4)$$

The fits are provided to guide the eye through the data. Odd-order terms are not included in eq. [4] because theory predicts that the Coriolis force should interact symmetrically with the system.[3,4] Also note that the coefficients are different for each set of points that are fit.

The squares and the solid curve represent the bifurcation boundary separating the base flow state from the secondary flow states. This is the boundary at which the relative reflectance dropped sharply. Similarly, the circles and the broken curve indicate the bifurcation separating the time-dependent flow states from the time-independent flow state (ie. TTVF) within the secondary flow. This boundary represents where the variance has risen significantly. The triangles and the dash-dot-dot curve indicate where the flow bifurcates to a disordered flow regime, either noisy or turbulent. This bifurcation boundary was found by locating a the rise in the spectral noise from the SND. Finally, the diamonds and their accompanying dash-dot-dot curve indicate the other boundary separating ordered flow from disordered flow. This boundary was also measured using the spectral noise.

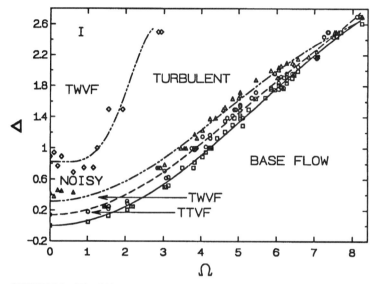

FIGURE 1. The bifurcation map for Taylor-Couette flow subject to an external Coriolis force. $\eta = 0.880$.

The error bar in the upper left corner of Figure 1 is an estimate of the error in determining the point at which the bifurcation occurs. This uncertainty is primarily due to the uncertainty in choosing where sharp breaks occur in the data. These choices were made by looking at several data sets at once, so that the sequence of bifurcations would be consistent. As a result, the choices were subject to varying degrees of judgement.

Notice that on the bifurcation map a distinction has been made between noisy flow and turbulent flow. The distinction between these two states appears to be largely qualitative, because the dynamics of the states have not yet been fully characterized. However, a distinction is made as a result of the marked difference between the spectral characteristics of noisy flow and turbulent flow. A comparison of Figures 2a and 2b shows a significant rise in the spectral noise level as the flow goes from TWVF (Figure 2a) to noisy flow (Figure 2b). On the other hand, if a noisy spectrum is compared to one from a turbulent state (Figure 3), it can be seen that the spectral noise rises even more markedly when the flow is turbulent. In Figure 3 there are no remnants of the sharp spectral components indicative of temporal periodicity in the flow, nor is there a spectral peak representing the inner-cylinder frequency. The inner-cylinder frequency is generally present as an experimental artifact, but in Figure 3 the noise has risen so high that the peak has been buried. The boundary that separates noisy flow and turbulence is not sharply defined; the spectral noise gradually and continuously rises as the system approaches turbulence from the noisy state.

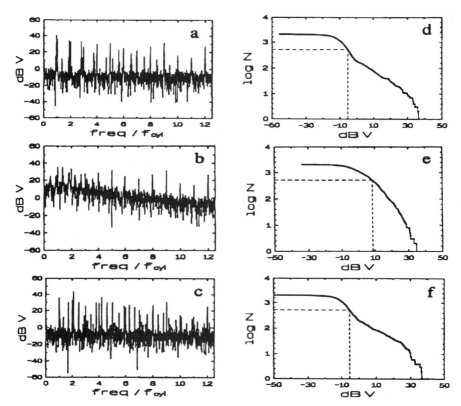

FIGURE. 2. Fourier spectra and spectral number distributions. a, b, and c are Fourier spectra showing re-emergent order. d, e, and f are the corresponding spectral number distributions. $\Delta = 0.24$ For a and d, $\Delta = 0.59$ for b and e, and $\Delta = 1.01$ for c and f. The data were taken at $\Omega = 0.62$. a, c, d, and f correspond to TWVF, while b and e correspond to noisy flow.

Figure 1 reveals that the bifurcation behavior of Taylor-Couette flow subject to an external Coriolis force is markedly different from that observed in the ordinary system. Specifically, there is evidence of re-emergent order, and a direct bifurcation to turbulence. In ordinary Taylor-Couette flow, with the outer-cylinder held fixed, there are instances of re-emergent order but they are either hysteretic,[14,17,18] or they occur at much higher Reynolds numbers.[19] There are also turbulent states in the ordinary system, but the bifurcation to turbulence occurs at $\Delta \approx 20$, and only after several bifurcations from other time dependent flows. Figure 1 shows that the turbulent state occurs at Δ as low as 1.

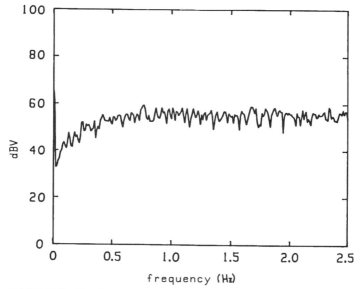

FIGURE 3. Fourier power spectrum for turbulent flow: $\Omega = 10.41$, $\Delta = 3.78$. This figure was generated using a commercial spectrum analyzer with resolution of 256 channels. Figures 2a-2c have 2048 channels, and were computed with an FFT.

Figure 1 also indicates that although the bifurcation boundaries do converge as Δ increases, the direct transition to turbulence does not occur at a sharply defined value of Ω. The absence of a sharply defined multicritical point could largely be due to a lack in experimental precision. Thus, it seems likely that given better precision a multicritical point could be located. Such a point would be of interest to theorists because it occurs for small values of both control parameters, and hence might be accessible to a weakly nonlinear theory, or even a full simulation of the Navier-Stokes equations.

Finally, note that for $\Omega = 0$, the bifurcations for increasing Δ are consistent with previous work.[18,20]

Re-emergent Order

Figure 4a shows the spectral noise for $\Omega = 0.62$. In these two figures the solid line indicates increasing Δ and the broken line indicates decreasing Δ. It is clear from this figure that the re-emergent order is nonhysteretic. This lack of hysteresis can be compared to Figure 4b which shows re-emergent order as Δ is increased at $\Omega = 0$. As Δ is decreased, however, the bifurcations do not replicate themselves and are thus hysteretic. The data in Figures 4a and 4b were taken at the same α^*, so it is clear that these two series of bifurcations are dynamically distinct. The bifurcation to a disordered state at $\Omega = 0$ (see Figure 4b) is consistent with the work of Donnelly et al.[17], Park and Donnelly[14,] and Crawford, Park, and Donnelly[18], who identified a disordered state that grows into the wavy mode. They attributed this disordered state to spatial a dislocation. This dislocation, dubbed a "turbator", was found to travel axially, and it induced a rise in the degrees of freedom. The turbator was found to mediate a transition in the axial wave number.[21] Park and Donnelly used a rise in the degrees of freedom as an indicator of the presence of turbator.[14] The Coriolis force, while inducing the disordered flow indicated in Figures 4a and 4b, also suppresses the growth of turbators.

If the spectral character of the bifurcations to and from a disordered dynamical state is examined, then it can be seen that TWVF has a well confined and low level of noise, and sharp spectral peaks (see Figure 2a). As Δ increases, the system bifurcates to noisy flow. Figure 2b shows that in the disordered state the noise level has risen significantly. However, there are still sharp spectral components. The presence of spectral components, coupled with the rise in the noise is the spectral signature of noisy flow. Recall that noisy flow is contrasted to turbulent flow, in which there is a stronger and broader rise in the spectral noise and complete absorption of all sharp spectral components (compare Figures 2b and 3). Finally, Figure 2c shows that the TWVF has re-emerged at a higher Δ. The noise has dropped back down to the level it had in Figure 2a, and the sharp components have regained their strength.

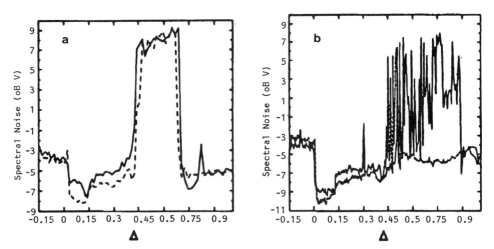

FIGURE 4. a) Spectral noise showing nonhysteretic re-emergent order at $\Omega = 0.62$. b) Spectral noise showing hysteretic re-emergent order at $\Omega = 0$. The solid curves represent Δ increasing and the broken curves are for Δ decreasing.

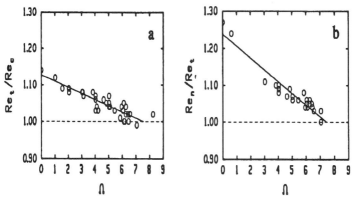

FIGURE 5. a) Re_t/Re_c vs. Ω. b) Re_n/Re_t vs. Ω.

The above three spectra can be looked at in another way using the spectral number distribution. Figures 2d, 2e, and 2f are the spectral number distributions for the power spectra of Figures 2a, 2b, and 2c, respectively. Note that $\log_{10}(N)$ is plotted vs. p, and the vertical broken lines indicate the spectral noise. Observe the difference between the profile of the TWVF (Figure 2d) and the noisy flow (Figure 2e). Figure 2d resembles the SND for a periodic system, while Figure 2e more closely resembles the SND for an aperiodic one. Thus, by using the SND, periodic states can be distinguished from aperiodic states. Also notice the close similarity between the two profiles of the TWVF before and after the noisy flow (Figures 2d and 2f). The noise selected by the so-called 25% criterion changes by about 15 dBV as the system goes through the disordered state. Also, when the ordered state re-emerges in Figure 2f, the noise level returns to what it was in the initial TWVF spectrum. Clearly the SND provides a good way to make qualitative and semiquantitative distinctions among power spectra. In addition, the SND has helped to characterize the ordered and disordered states that occur as the system undergoes its series of bifurcations that display re-emergent order.

In addition to the re-emergent order for increasing Δ, Figure 1 indicates that there is re-emergent order as Ω is increased while Δ is fixed. These bifurcations are also nonhysteretic, within experimental uncertainty.

The Direct Transition to Turbulence

For a sufficiently large value of Ω, the flow bifurcates directly to turbulence from the base flow state as Δ is increased. This can be seen in the gradual convergence of the boundaries indicating the drop in relative reflectance, the rise in the variance of the reflectance, and the rise in the spectral noise (see Figure 1).

The nature of the direct transition to turbulence is evident if the ratios Re_t/Re_c and Re_n/Re_t are plotted as functions of Ω. Note that Re_c, Re_t, and Re_n are the critical Reynolds numbers for the primary bifurcation, the bifurcation to time-dependence, and the bifurcation to disordered flow (as determined from the SND), respectively. Figures 5a and 5b show, respectively, Re_t/Re_c vs. Ω, and Re_n/Re_t vs. Ω. The ratios are determined from the points used to create the bifurcation map. The solid line is a two-parameter best fit to the data, and the broken line is where the ratio equals unity. Both lines are included as aids to guide the eye.

From Figures 5a and 5b it is clear that Re_t/Re_c and Re_n/Re_t both approach unity at $\Omega \cong 7.5$. This means that the bifurcation of the base flow directly to a time-dependent state (ie. $Re_t/Re_c = 1$) is coincident with the initial time-dependent state being disordered (ie. $Re_n/Re_t = 1$). Figure 3 shows the spectral character of such a disordered state; the flow is clearly turbulent. Thus the time-independent base flow bifurcates directly to turbulence. Niemela and Donnelly[2] experimentally found the direct transition to turbulence that was predicted by Küppers and Lortz[2] for Rayleigh-Bénard convection subject to a Coriolis force. They saw a similar convergence, whereby R_t/R_c approached unity as Ω was increased (where R is the Rayleigh number). The direct transition to turbulence in Taylor-Couette flow is significant because it is induced by the Coriolis at a Reynolds number an order of magnitude smaller than would be required for the ordinary system. Finally, the direct transition does not display hysteresis.

CONCLUSION

Taylor-Couette flow subject to a nonaxisymmetric Coriolis force displays an interesting array of bifurcation phenomena as the two control parameters, Ω and Δ, are varied. We have used time series of Kalliroscope reflectance to identify bifurcations to four dynamical regimes from the time-independent base flow state. The bifurcation map we have developed reveals that for relatively low values of Ω and Δ, the system displays nonhysteretic re-emergent order and a nonhystertic direct transition to turbulence. The spectral nature of the disordered flows, and the possible role played by mode competition in this two control parameter system, suggest that the dynamics might be accessible via a weakly nonlinear theory or a full simulation of the Navier-Stokes equations.

APPENDIX A

Figure A1 shows a cross-sectional view of the Taylor-Couette system. The system parameters are

$$
\begin{aligned}
a &= \text{inner-cylinder radius}, \\
b &= \text{outer-cylinder radius}, \\
d &= \text{annular gap width} = b - a, \\
\eta &= \text{radius ratio} = a/b, \\
\omega_1 &= \text{inner-cylinder angular frequency}, \\
\omega_2 &= \text{outer-cylinder angular frequency}, \\
\mu &= \text{speed ratio} = \omega_2/\omega_1.
\end{aligned}
$$

In considering the problem of Taylor-Couette flow subject to an external Coriolis force, it is important to consider the orientation of the Couette apparatus relative to the rotating reference frame. The apparatus may be placed in two basic configurations: the cylinders' common axis of rotation can be oriented parallel to the rotation axis of the turntable, or the cylinder axes can be placed orthogonal to the table axis. As will be shown, the former orientation simply reduces to the general problem of co- or counter-rotation of the cylinders, that is $\mu \neq 0$; in the later orientation, the Coriolis force breaks the axisymmetry of the ordinary Taylor-Couette system such that the base flow is altered.[3] The new base flow cannot be solved analytically, so it must be approximated numerically.[3]

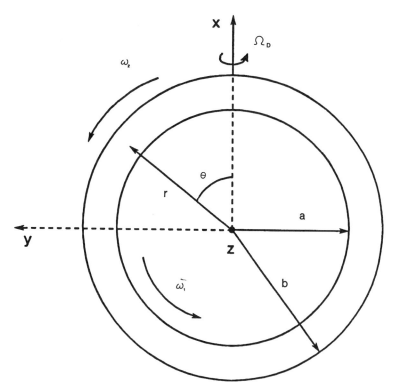

FIGURE A1. The coordinate system used for $\vec{\Omega}_D \neq 0$.

For flow in a rotating frame, where $\vec{\Omega}_D$ is the angular velocity of the system rotation, the Navier-Stokes equations become

$$\frac{\partial \vec{u}}{\partial t} = \nu \nabla^2 \vec{u} - (\vec{u} \cdot \vec{\nabla})\vec{u} - \frac{1}{\rho}\vec{\nabla}p' - \vec{\Omega}_D \times (\vec{\Omega}_D \times \vec{r}) - 2\vec{\Omega}_D \times \vec{u}. \quad (A1)$$

The double cross-product in the above equation is the centrifugal force term, and can be rewritten as the gradient of a scalar:

$$\vec{\Omega}_D \times (\vec{\Omega}_D \times \vec{r}) = -\vec{\nabla}\left(\frac{1}{2}\Omega_D^2 r'^2\right), \quad (A2)$$

where r' is the distance from the axis of rotation. If the fluid is homogeneous then the pressure term in eq. [A1] can be combined with eq. [A2] and rewritten as an effective pressure, p, such that

$$p = p' - \frac{1}{2}\rho \Omega^2 r'^2. \quad (A3)$$

This yields

$$\frac{\partial \vec{u}}{\partial t} = \nu \nabla^2 \vec{u} - (\vec{u} \cdot \vec{\nabla})\vec{u} - \frac{1}{\rho}\vec{\nabla}p - 2\vec{\Omega}_D \times \vec{u}. \quad (A4)$$

Note that subsuming the centrifugal force term into the pressure gradient term is similar to subtracting the hydrostatic pressure from a problem to eliminate gravitational effects.

In other words, the centrifugal force does not play a role in the flow dynamics. Ning, Ahlers, and Cannell[7] experimentally verified that centrifugal effects could be ignored in this system by placing their Couette apparatus at the center of the turntable and at the edge. There was no radial dependence on the dynamics they observed.

If the system has parallel orientation such that $\vec{\omega}_1 = \omega_1 \hat{z}$, $\vec{\omega}_2 = \omega_2 \hat{z}$, $\vec{\Omega}_D = \Omega_D \hat{z}$, and if one applies the customary assumptions of the ordinary Couette problem, then eq. [A4] has the azimuthal Couette flow solution, $v = Ar + B/r$. The difference is that $\omega_1 \to \omega_1 + \Omega_D$ and $\omega_2 \to \omega_2 + \Omega_D$, so that μ, A and B are changed appropriately. The pressure becomes

$$p' = \rho \int \left(\frac{v^2}{r} + \Omega_D v \right) dr + \frac{1}{2} \rho \Omega_D^2 r'^2 . \qquad (A5)$$

Again, this modification to the pressure will not have any effect on the dynamics as long as the fluid is homogeneous.

That the parallel orientation would reduce to the case of co- or counter-rotation was verified both experimentally and numerically. With $\omega_2 = 0$, Re_{c1} and Re_{c2} (the inner- and outer-cylinder critical Reynolds numbers, respectively) were calculated using $\mu = \Omega_D / (\omega_1 + \Omega_D)$. Results for the critical Reynolds numbers were obtained that were well within experimental error for the onset to Taylor vortex flow. The calculations were based on those described in Wiener, Hammer and Tagg for $\mu \neq 0$,[4] and the experiments were carried out at several different values of Ω_D. Table A1 shows the theoretical and experimental results for $\vec{\omega}_1$ parallel to $\vec{\Omega}_D$. The table indicates that within the range investigated, the difference between the theoretical and experimental critical Reynolds numbers was at most 1%. This verifies that the parallel orientation of the rotation vectors is equivalent to the general case of $\mu \neq 0$.

TABLE A1. Comparison of the theoretical and experimental results for $\vec{\omega}$ parallel to $\vec{\Omega}_D$.

Ω_D	Re_{c2}(the.)	Re_{c2}(exp.)	Re_{c1}(the.)	Re_{c1}(exp.)
.10	14.74	14.74	122.44	122.06
		14.74		121.93
.20	29.43	29.48	126.35	125.83
.30	44.67	44.22	132.38	131.01

REFERENCE

(1) S. Chandrasekhar, *Hydrodynamic and Hydromagnetic Stability*, Clarendon, Oxford, 1961.
(2) See, for example, R. J. Donnelly and M. Ozima, Proc. Roy. Soc. A **266**, 273 (1962); G. Küppers and D. Lortz, J. Fluid Mech **35**, 609 (1969); R. M. Clever and F. H. Busse, J. Fluid Mech. **94**, 609 (1979); F. H. Busse and K. E. Heikes, Science **208**, 173 (1980); and J. J. Niemela and R. J. Donnelly, Phys. Rev. Lett. **57**, 2524 (1986).
(3) R. J. Wiener, P. W. Hammer, C. E. Swanson, D. C. Samuels, and R. J. Donnelly, J. Stat. Phys., to be published.
(4) R. J. Wiener, P. W. Hammer, and R. P. Tagg, Phys. Rev. A, to be published.
(5) *Multiparameter Bifurcation Theory*, Vol. 56 of *Contemporary Mathematics* edited by M. Golubitsky and J. M. Guckenheimer, American Mathematical Society, Providence, RI, 1986.

(6) P. W. Hammer, R. J. Wiener, C. E. Swanson, and R. J. Donnelly, Bifurcations in Taylor-Couette flow subject to an external Coriolis force. Phys. Rev. Lett., submitted for publication.
(7) L. Ning, G. Ahlers, and D. L. Cannell, J. Stat Phys., to be published.
(8) R. J. Wiener, P. W. Hammer, C. E. Swanson, and R. J. Donnelly, Phys. Rev. Lett. **64**, 1115 (1990).
(9) L. Ning, M. Tveitereid, G. Ahlers, and D. S. Cannell, Phys. Rev. A., to be published.
(10) K. Park, G. L. Crawford, and R. J. Donnelly, Phys. Rev. Lett. **47**, 1448 (1981).
(11) K. Park and K. Jeong, Phys. Rev. A **31**, 3457 (1985).
(12) W. L. Ditto, M. L. Spano, H. T. Savage, S. N. Rauseo, J. Heagy, and E. Ott, Phys. Rev. Lett. **65**, 533 (1990).
(13) J. Crutchfield, D. Farmer, N. Packard, R. Shaw, G. Jones, and R. J. Donnelly, Phys. Lett. **76A**, 1 (1980). Our definition is slightly different than the one used in Eq. 2 of this paper, but the two definitions are consistent.
(14) K. Park and R. J. Donnelly, Phys. Rev. A. **24**, 2277 (1981).
(15) E. Ott, private communication.
(16) K. L. Babcock, G. Ahlers, and D. S. Cannell, Noise-sustained structure in Taylor-Couette flow with through-flow., Phys. Rev. Lett., submitted for publication.
(17) R. J. Donnelly, K. Park, R. Shaw, and R. W. Walden, Phys. Rev. Lett. **44**, 987 (1980).
(18) G. L. Crawford, K. Park, and R. J. Donnelly, Phys. Fluids **28**, 7 (1985).
(19) R. W. Walden and R. J. Donnelly, Phys. Rev. Lett. **42**, 301 (1979).
(20) R. C. Di Prima and H. L. Swinney, 1985. In *Hydrodynamics Instabilities and the Transition to Turbulence*, ed. H. L. Swinney and J. P. Gollub, 2nd ed. Berlin, Springer-Verlag.
(21) K. Park and G. L. Crawford, Phys. Rev. Lett. **50**, 343 (1983).

(a) present address: (R44), NSWC, Silver Spring, MD 20903-5000.
(b) present address: Dept. of Physics, Lewis & Clark College, Portland, OR 97219

INSTABILITY OF TAYLOR-COUETTE FLOW SUBJECTED TO A CORIOLIS FORCE

Richard J. Wiener[a] and Philip W. Hammer[b]

Department of Physics
University of Oregon
Eugene, Oregon 97403

Randall Tagg

Department of Physics
University of Colorado, Denver
Denver, Colorado 80217-3364

INTRODUCTION

Taylor-Couette flow subjected to a Coriolis force is a rich pattern-forming system.[1-5] The Coriolis force is applied by rotating a Couette apparatus so that the axis of rotation for the *system* is orthogonal to the common axis of the cylinders (see figure 1). This orientation produces a nonaxisymmetric Coriolis force through the interaction of the rotation of the system with the (primarily) azimuthal base flow. Experimentally, a small Coriolis force is found to stabilize the base flow against Taylor vortex formation. In other words, the Coriolis force increases the critical value of the Reynolds number R_c for the onset of instability in the base flow. At the transition to secondary flow, the Taylor vortices are tilted out of the plane normal to the cylinders' axis. But the flow is still time-independent. This tilted Taylor vortex flow is an example of a novel pattern which arises as a result of the nonaxisymmetric Coriolis force. Several other novel patterns also arise. At somewhat large system rotation rates, there is a direct transition from the base flow to strong spatiotemporal turbulence. At small values of Ω (where Ω is a dimensionless measure of the angular frequency of the rotating system) there is nonhysteretic re-emergent order at Reynolds numbers not far from the threshold for instability.[3] In a medium gap system, there is a Hopf bifurcation from the base flow to axially travelling tilted Taylor vortices.[4]

The change in stability behavior and variety of novel flow regimes motivated us to carry out a linear stability analysis for Taylor-Couette flow subjected to a Coriolis force.[6] Ning *et al.*[7] have performed an independent analysis. This paper will outline our stability analysis and discuss its results with an emphasis on comparison to experimental quantitative data and qualitative observations. Much of the discussion will be devoted to the limitations of the stability analysis, which suggest a number of intriguing theoretical problems that still remain to be investigated.

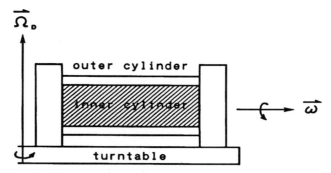

FIGURE 1. Schematic side view of the Couette apparatus mounted on a rotating turntable. Note the orthogonal orientation of the inner-cylinder rotation vector $\vec{\omega}$ with respect to the turntable rotation vector $\vec{\Omega}_D$.

STABILITY ANALYSIS

The linear stability analysis presents an interesting problem at the outset in that the Coriolis force perturbs the base flow. The dimensionless Navier-Stokes and continuity equations for viscous flow in a rotating reference frame are (see figure 2 for a defining diagram of the geometry and coordinate system and reference 6 for details on scaling)

$$\frac{\partial \vec{u}}{\partial t} = \nabla^2 \vec{u} - R(\vec{u} \cdot \vec{\nabla})\vec{u} - \vec{\nabla} p - 2\vec{\Omega} \times \vec{u}, \qquad (1)$$

$$\vec{\nabla} \cdot \vec{u} = 0, \qquad (2)$$

where the Reynolds number $R = \alpha \omega_1 (b - a)/\nu$, the dimensionless system rotation $\vec{\Omega} = \vec{\Omega}_D (b - a)^2/\nu$, and ν is the kinematic viscosity. However, the classical solution to the Navier-Stokes and continuity equations in the absence of the Coriolis force (with no-slip boundary conditions) for the Taylor-Couette base flow

$$U^{(0)} = W^{(0)} = 0,$$
$$V^{(0)} = Ar + B/r, \qquad (3)$$
$$P^{(0)} = R \int \frac{(V^{(0)})^2}{r} dr,$$

no longer obtains with the addition of the Coriolis term in eq. (1). Instead, it is necessary to determine a modified solution for the base flow, which is altered by the Coriolis force. We begin by expanding eq. (3), using Ω as a perturbation parameter:

$$U = \Omega U^{(1)} + \Omega^2 U^{(2)} + \dots,$$
$$V = V^{(0)} + \Omega V^{(1)} + \Omega^2 V^{(2)} + \dots,$$
$$W = \Omega W^{(1)} + \Omega^2 W^{(2)} + \dots, \qquad (4)$$
$$P = P^{(0)} + \Omega P^{(1)} + \Omega^2 P^{(2)} + \dots$$

This expansion is then used to derive coupled systems of differential equations for the first- and second-order corrections to the velocity and pressure fields via substitution back into eqs. (1) and (2). These systems of equations are solved by certain correction terms vanishing:

$$U^{(1)} = V^{(1)} = P^{(1)} = 0, \qquad (5)$$

$$W^{(2)} = 0. \qquad (6)$$

The remaining first-order correction term has the form

$$W^{(1)} = W_1^{(1)} e^{i\theta} + W_{-1}^{(1)} e^{-i\theta}, \qquad (7)$$

where $W_1^{(1)}$ and $W_{-1}^{(1)}$ are complex conjugates. The first-order term can be determined by using a shooting method to solve the system of ordinary differential equations which results from substitution into eqs. (1) and (2).

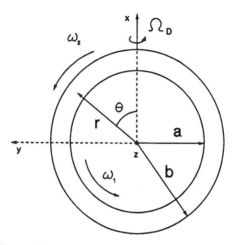

FIGURE 2. Defining diagram for an infinite cylinder Taylor-Couette system rotating with angular frequency Ω_D. The system rotation Ω_D is in the direction of the x axis, and the cylinders rotate about the z axis. The cylindrical coordinate system is co-rotating with the Taylor-Couette system.

Similarly, one can solve for the second-order correction terms for the modified base flow, which take the form

$$\begin{aligned} U^{(2)} &= U_0^{(2)}(r) + U_2^{(2)}(r) e^{i2\theta} + U_{-2}^{(2)}(r) e^{-i2\theta}, \\ V^{(2)} &= V_0^{(2)}(r) + V_2^{(2)}(r) e^{i2\theta} + V_{-2}^{(2)}(r) e^{-i2\theta}, \\ P^{(2)} &= P_0^{(2)}(r) + P_2^{(2)}(r) e^{i2\theta} + P_{-2}^{(2)}(r) e^{-i2\theta}. \end{aligned} \qquad (8)$$

To order Ω^2 the modified base flow has the form

$$U = \Omega^2 U^{(2)} + \dots,$$
$$V = V^{(0)} + \Omega^2 V^{(2)} + \dots,$$
$$W = \Omega W^{(1)} + \dots, \qquad (9)$$
$$P = P^{(0)} + \Omega^2 P^{(2)} + \dots$$

The linear stability of the modified base flow can be calculated by constructing an appropriate eigenvalue problem. We follow the standard procedure of adding a disturbance velocity $(u(\vec{r},t), v(\vec{r},t), w(\vec{r},t))$ and pressure $p(\vec{r},t)$ to the base flow solution, eq. (9), and substituting the disturbed solution into the Navier-Stokes and continuity equations. We linearize the resulting equations about the disturbance velocity $\vec{u} = 0$, and write them as a generalized linear eigenvalue problem:

$$L \begin{pmatrix} \vec{u}(\vec{r}) e^{st} \\ p(\vec{r}) e^{st} \end{pmatrix} = s M \begin{pmatrix} \vec{u}(\vec{r}) e^{st} \\ p(\vec{r}) e^{st} \end{pmatrix}, \qquad (10)$$

where L is the differential operator defined by the linearized equations and

$$M = \begin{pmatrix} 1 & 0 & 0 & 0 \\ 0 & 1 & 0 & 0 \\ 0 & 0 & 1 & 0 \\ 0 & 0 & 0 & 0 \end{pmatrix}. \qquad (11)$$

The complex eigenvalue s is the exponential growth rate of the disturbance.

We separate the operator L into its zeroth-, first-, and second-order parts and expand the eigenvalue s and the eigenfunction, which we denote by ψ, in powers of Ω:

$$(L^{(0)} + L^{(1)} \Omega + L^{(2)} \Omega^2 + \dots)(\psi^{(0)} + \psi^{(1)} \Omega + \psi^{(2)} \Omega^2 + \dots)$$
$$= (s^{(0)} + s^{(1)} \Omega + s^{(2)} \Omega^2 + \dots) M (\psi^{(0)} + \psi^{(1)} \Omega + \psi^{(2)} \Omega^2 + \dots), \qquad (12)$$

where

$$(L^{(0)} - s^{(0)} M) \psi^{(0)} = 0 \qquad (13)$$

is the zeroth-order eigenvalue problem for the stability of Taylor-Couette flow. For an explicit representation of the linear operators in eq. (12) see reference 6. The eigenvalues and eigenfunctions in eq. (12) are indexed by the azimuthal wavenumber m, and they depend on the axial wavenumber k.

To find the shift in critical Reynolds number which results from the Coriolis force, it is necessary to calculate the corrections $s^{(1)}$ and $s^{(2)}$ to the zeroth-order eigenvalue. To determine these corrections, we equate coefficients of Ω and Ω^2, respectively, in eq. (12):

$$(L^{(0)} - s^{(0)} M) \psi^{(1)} = -(L^{(1)} - s^{(1)} M) \psi^{(0)}, \qquad (14)$$
$$(L^{(0)} - s^{(0)} M) \psi^{(2)} = -(L^{(1)} - s^{(1)} M) \psi^{(1)} - (L^{(2)} - s^{(2)} M) \psi^{(0)}. \qquad (15)$$

We use the following inner-product:

$$(\tilde{\psi}^{(0)}, \psi^{(l)}) = \int_0^{2\pi/k} dz \int_0^{2\pi} d\theta \int_a^b r\, dr\, (\vec{\tilde{u}}^{(0)*} \cdot \vec{u}^{(l)} + \tilde{p}^{(0)*} p^{(l)}), \qquad (16)$$

where a tilde indicates an adjoint function and i indexes perturbative order. By projecting both sides of eq. (14) onto $\tilde{\psi}^{(0)}$, we find that the first-order eigenvalue is

$$s_m^{(1)} = \frac{(\tilde{\psi}_m^{(0)}, L^{(1)}\psi_m^{(0)})}{(\tilde{\psi}_m^{(0)}, M\psi_m^{(0)})}. \tag{17}$$

However, each matrix element in the first-order operator $L^{(1)}$ (which depends on the first-order correction $W^{(1)}$ to the base flow) has $m = \pm 1$ azimuthal dependence. When $L^{(1)}$ operates on $\psi_m^{(0)}$ it either raises of lowers each component to an $m \pm 1$ azimuthal mode. Thus by orthogonality

$$s_m^{(1)} = 0. \tag{18}$$

This result is consistent with the symmetry of the system for which the shift in the onset of instability is unaffected by the direction of $\vec{\Omega}$. Thus, one expects the growth rate s to be an even function of Ω.

To find $s^{(2)}$ we project eq. (15) onto $\tilde{\psi}^{(0)}$:

$$s_m^{(2)} = \frac{(\tilde{\psi}_m^{(0)}, L^{(1)}\psi_m^{(1)}) + (\tilde{\psi}_m^{(0)}, L^{(2)}\psi_m^{(0)})}{(\tilde{\psi}_m^{(0)}, M\psi_m^{(0)})}. \tag{19}$$

By numerically calculating the inner-products (which are written in explicit integral form in Appendix A of reference 6) on the right-hand side of eq. (19), we are able to determine the lowest-order nonvanishing eigenvalue correction. However, it is first necessary to solve for the first-order eigenfunction $\psi^{(1)}$. Combining eqs. (14) and (18) yields an equation for $\psi^{(1)}$:

$$(L^{(0)} - s^{(0)}M)\psi^{(1)} = -L^{(1)}\psi^{(0)}. \tag{20}$$

Equation (20) is an inhomogeneous version of eq. (13), the zeroth-order eigenvalue problem. In general, $\psi_m^{(1)}$ is a superposition of all azimuthal eigenmodes of $\psi^{(0)}$ other than the m eigenmode. However, the operator $L^{(1)}$ selects only $m \pm 1$ azimuthal components of $\psi^{(0)}$ when it operates on $\psi_m^{(0)}$. Thus, as a condition necessary to satisfy eq. (20), we expect $\psi_m^{(1)}$ to have the form

$$\psi_m^{(1)}(\vec{r},k) = \begin{pmatrix} (u_{m+1}^{(1)}(r,k)e^{i(m+1)\theta} + u_{m-1}^{(1)}(r,k)e^{i(m-1)\theta})e^{ikz} \\ (v_{m+1}^{(1)}(r,k)e^{i(m+1)\theta} + v_{m-1}^{(1)}(r,k)e^{i(m-1)\theta})e^{ikz} \\ (w_{m+1}^{(1)}(r,k)e^{i(m+1)\theta} + w_{m-1}^{(1)}(r,k)e^{i(m-1)\theta})e^{i(\pi/2+kz)} \\ (p_{m+1}^{(1)}(r,k)e^{i(m+1)\theta} + p_{m-1}^{(1)}(r,k)e^{i(m-1)\theta})e^{ikz} \end{pmatrix}. \tag{21}$$

Using eq. (21) and numerical solutions for the components of $\psi^{(0)}$, it is possible to generalize a numerical code for the solution of the zeroth-order problem, based on Langford et al.[8], to solve for $\psi^{(1)}$ by including the inhomogeneous terms in eq. (20) in the code. We follow similar numerical procedures used for the zeroth-order problem, and details can be found in reference 8. Once m, k, R, η, and μ are fixed, the inhomogeneous problem, eq. (20), reduces to a system of ordinary differential equations for the radial functions which occur in $\psi^{(1)}$. Equation (14) is the solvability condition which insures that $\psi^{(1)}$ can be found. We have already used this condition with $s^{(1)} = 0$ to write eq. (20). We use a shooting method with an Adams-Bashforth-Moulten predictor-corrector integrator to solve the system of differential equations.[9]

From the numerical calculation of $s^{(2)}$ it is possible to determine the marginal stability of the modified base flow. The marginal stability condition is that the real part of the total eigenvalue equals zero:

$$Re[s(R,\eta,\mu,k,m,n,\Omega^2)] = 0. \tag{22}$$

By treating m, η, and μ as fixed variables (which can be varied later), this condition implicitly defines a marginal stability surface in (R, k, Ω) space: $R = R_{marg}(k, \Omega^2)$. When R crosses this surface, at any value of k and Ω, instability occurs. However, k is not a parameter subject to direct experimental control. For a fixed Ω, which can be controlled experimentally, a critical value k_c emerges which gives the minimum critical Reynolds number. This is expressed by the condition

$$\frac{\partial R_{marg}(k, \Omega^2)}{\partial k}\bigg|_{k=k_c} = 0, \tag{23}$$

which implicitly defines both the function $k = k_c(\Omega^2)$ and the critical Reynolds number $R_c = R_{marg}(k_c(\Omega^2), \Omega^2)$. By differentiating these implicitly defined functions it is possible to determine the leading-order dependence of R_c and k_c on Ω^2 and $s^{(2)}$ (see reference 6 for details). The resulting equations are

$$\frac{k_c}{k_{c0}} - 1 = c_k \Omega^2 + O(\Omega^4) + \ldots, \tag{24}$$

where

$$c_k = \frac{1}{k_{c0}} \frac{\frac{\partial}{\partial k}\left(\frac{Re[s^{(2)}]}{\frac{\partial}{\partial k}(Re[s^{(0)}])}\right)}{\frac{\partial}{\partial k}\left(\frac{\partial R_{marg}}{\partial k}\right)}, \tag{25}$$

and

$$\frac{R_c}{R_{c0}} - 1 = c_R \Omega^2 + O(\Omega^4) + \ldots, \tag{26}$$

where

$$c_R = -\frac{1}{R_{c0}} \frac{Re[s^{(2)}]}{\frac{\partial}{\partial k}(Re[s^{(0)}])}. \tag{27}$$

The values R_{c0} and k_{c0} are the critical values in the absence of a Coriolis force.

Table 1. The results of the stability analysis for $m = 0, 1, 2$ at three different η

η	m	k_{c0}	R_{c0}	$Im[s^{(0)}]$	$Re[s^{(2)}]$	$Im[s^{(2)}]$	c_R	c_k
.753	0	3.1352	86.226	0	-0.4460	0	0.01693	-0.009888
.753	1	3.1758	88.035	12.904	-0.4568	-0.02443	0.01725	-0.009621
.753	2	3.2960	93.998	27.679	-0.4899	-0.05212	0.01830	-0.009058
.880	0	3.1294	120.50	0	-1.144	0	0.04363	-0.02395
.880	1	3.1431	121.30	8.1352	-1.159	-0.02578	0.04410	-0.02382
.880	2	3.1839	123.78	16.631	-1.202	-0.05356	0.04556	-0.02351
.950	0	3.1275	184.99	0	-3.022	0	0.1154	-0.06307
.950	1	3.1321	185.39	5.0027	-3.038	-0.02934	0.1159	-0.06302
.950	2	3.1461	186.61	10.078	-3.088	-0.05988	0.1177	-0.06291

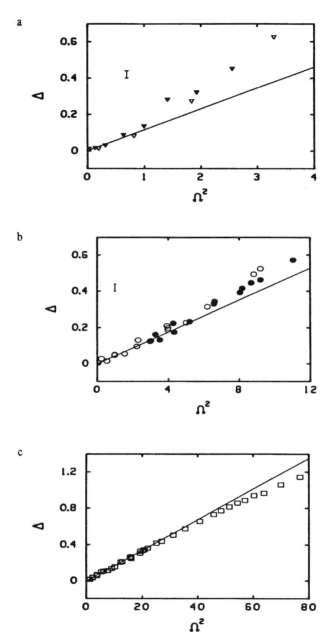

FIGURE 3. Comparison between experimental data and theoretical results. a) $\eta = .950$, b) $\eta = .880$, and c) $\eta = .753$. The slopes of the lines are given by the values of c_R for $m = 0$ in table 1.

RESULTS AND DISCUSSION

The azimuthal wavenumber m is analogous to the axial wavenumber k in the sense that its value is not subject to direct experimental control. To determine which m actually emerges, one needs to compare the marginal stability surfaces for different values of m. Table 1 shows the results of the $m = 0, 1, 2$ stability calculation for c_R and c_k at $\eta = .950$, $\eta = .880$, and $\eta = .753$. The values of η were chosen to compare to experimental results. The speed ratio μ equals zero for all results reported in table 1.

The results indicate that R_c increases as a function of Ω (i.e. the Coriolis force stabilizes the flow) and k_c decreases as Ω increases. The results also indicate that the curvature c_R increases with m for these values of m and η. Since R_{co} also increases with m, this means that to order Ω^2 there is no critical crossing of m eigenmodes such that an $m = 1$ or $m = 2$ mode becomes more unstable than an $m = 0$ mode at a critical value of Ω (for the case of $\mu = 0$). It is important that we checked for critical crossings, since they are well known to occur for counter-rotating Taylor-Couette flow in the absence of an external Coriolis force.[8] Moreover, a critical m crossing would explain the onset to the time-dependent secondary flows at larger Ω observed in experiments[1-5] (since $m \neq 0$ modes are time-dependent). For example, we suspected that the Hopf bifurcation observed by Ning et al.[4,5] for $\eta = .753$ at $\Omega > 8.5$ might be due to a critical crossing of the $m = 1$ mode with the $m = 0$ mode. However, the order Ω^2 stability analysis exhibits no such crossing.

Although the stability analysis fails to predict the direct onset to time-dependent flows at somewhat large Ω, it accurately predicts the shift in R_c for the transition to time-independent tilted Taylor vortex flow at small Ω. Figure 3 plots experimental data and our theoretical results for $m = 0$ (from table 1) for $\Delta = (R_c / R_{co}) - 1$ versus Ω^2. For $\eta = .753$ the data are from Ning et al.,[4,5] and for $\eta = .880$ and $\eta = .950$ we compare the results of our calculation with data from Wiener et al.[1,2,6] The lines in Fig. 3 are the theoretical results for c_R. For an initial range of Ω^2 there is agreement within experimental uncertainties at all three radius ratios. However, for $\eta = .880$ the value of Ω^2 at which the data diverge from the theoretical calculation is smaller than the value for $\eta = .753$. Similarly the data at $\eta = .950$ diverge from theory at smaller Ω^2 than the data at $\eta = .880$. This suggests that the range in Ω for the validity of the perturbation analysis decreases as the radius ratio increases. The theoretical results and experimental data also suggest that stability increases monotonically with η from medium to small gaps. Interestingly, the data at $\eta = .880$ and $\eta = .950$ deviate from their initial behavior with stronger than quadratic stabilization, whereas the data at $\eta = .753$ never exhibit stronger than quadratic stabilization. This suggests that the fourth-order correction for the larger η is positive, but that for $\eta = .753$ this correction is very small or negative.

The decreasing range of validity of the perturbation analysis as one approaches the small gap limit (i.e. as η increases) is a somewhat surprising result. However, the numerics reveal a simple basis for this effect. Our numerical calculations of the modified base flow show that corrections to the base flow are larger for larger η. As the gap narrows the Coriolis force increasingly perturbs the base flow. The stability analysis assumes an approximate solution to order Ω^2 for the base flow, which is valid for a smaller range of Ω as η increases, due to the increasing size of the correction terms. Thus, the stability analysis exhibits an analogous decreasing range of validity.

CONCLUSION

The linear stability analysis of the onset of instability in Taylor-Couette flow subjected to a Coriolis force agrees quantitatively with experimental measurements of increased stability for an initial range of Ω. The range of agreement depends on the radius ratio and decreases with increasing η. Calculations to order Ω^2 for several m modes do not yield any critical crossing in the stability curves such that at a critical Ω time-dependent, nonzero m modes become unstable prior to the $m = 0$ eigenmode. Thus, the onset of time-dependent flows, at values of Ω greater than those for which the analysis agrees with experiment, is still not understood. This might be fruitful ground for a weakly nonlinear theory. Another aspect of the system which merits investigation is nonzero values of μ. Counter-rotation beyond a critical value of μ causes a direct transition to axially travelling spiral vortices. How would the axial flow induced by the Coriolis force interact with spiral vortices? Would the Coriolis force alter or destroy the existence of multicritical points in counter-rotating flow? We believe that our analysis sets the stage for such further investigation of a system which has so far proven to be very intriguing.

Acknowledgments: We wish to thank Professor Russell Donnelly for discussions on this research. We gratefully acknowledge Professor John Hart for suggesting our approach to the problem, Ning Li and Morten Tveitereid for very useful discussions of the results of our parallel investigations, and Greg Bauer for discussions on computations. This research is supported through NSF Grant No. DMR-8815803.

REFERENCES

a. present address: Dept. of Physics, Lewis & Clark College, Portland, OR 97219.
b. present address: (R44), NSWC, Silver Spring, MD 20903-5000.
1. R. J. Wiener, P. W. Hammer, C. E. Swanson, and R. J. Donnelly, Phys. Rev. Lett. **64**, 1115 (1990).
2. R. J. Wiener, P. W. Hammer, C. E. Swanson, D. C. Samuels, and R. J. Donnelly, "Effect of a Coriolis Force on Taylor-Couette Flow." J. Stat. Phys., accepted for publication (1991).
3. P. W. Hammer, R. J. Wiener, and R. J. Donnelly, "Bifurcation Phenomena in Taylor-Couette Flow Subject to a Coriolis Force," this conference.
4. L. Ning, G. Ahlers, and D. S. Cannell, "Novel States in Taylor-Couette Flow Subjected to a Coriolis Force." J. Stat. Phys., accepted for publication (1991).
5. L. Ning, G. Ahlers, D. S. Cannell, and M. Tveitereid, Phys. Rev. Lett. **66**, 1575 (1991).
6. R. J. Wiener, P. W. Hammer, and R. Tagg, "Perturbation Analysis of the Primary Instability in Taylor-Couette Flow Subjected to a Coriolis Force." Phys. Rev. A15, accepted for publication (1991).
7. L. Ning, M. Tveitereid, G. Ahlers, and D. S. Cannell, "Taylor-Couette Flow Subjected to External Rotation." Phys. Rev. A15, accepted for publication (1991).
8. W. F. Langford, R. Tagg, E. J. Kostelich, H. L. Swinney, and M. Golubitsky, Phys. Fluids **31**, 776 (1988).
9. W. H. Press, B. P. Flannery, S. A. Teukolsky, and W. T. Vetterling, *Numerical Recipes* (Cambridge University Press, Cambridge, 1986).

BIFURCATIONS TO DYNAMIC STATES IN TAYLOR-COUETTE FLOW WITH EXTERNAL ROTATION

Li Ning, Guenter Ahlers, and David S. Cannell

Department of Physics and Center for Nonlinear Science
University of California
Santa Barbara, CA 93106, U.S.A.

1. INTRODUCTION

Taylor-vortex flow[1] (TVF) has been widely studied,[2,3] and has provided numerous examples of important bifurcation and pattern-formation phenomena. Recently, it was recognized that the addition of external rotation about an axis orthogonal to the cylinder axis may generate a richness of new phenomena which are associated with the fact that the resulting Coriolis force breaks the cylindrical symmetry of the TVF system. Several experimental[4-7] and theoretical[8-10] investigations have been carried out. In this contribution we would like to present additional experimental results pertaining to the nature of some of the time dependent states which are encountered.

Whereas the work of Wiener et al.[4,7] has been for radius ratios η of 0.883 and 0.95, our experiment was done with $\eta = 0.753$, with the option of having gently tapered outer-cylinder end-sections to provide "soft" boundary conditions to reduce the effects of the fixed ends.[6] With the radius ratio we used, TVF in the absence of external rotation has the primary bifurcation well separated from the secondary bifurcation, which is one from TVF to wavy vortex flow. So our experiment, in addition to extending the studies of Wiener et al. to smaller values of η, provided a larger portion of the control-parameter space for a detailed study near the primary bifurcation. We found some novel dynamic states, and the parameter ranges that they reside in seem to be accessible to theoretical study.

2. RESULTS

The apparatus and the experimental methods have been described elsewhere.[6] Briefly, we visualized the flow of water with 2% by volume Kalliroscope suspension. Images were obtained with a video camera using a digital frame grabber housed in the controlling micro-computer. To present our results, we have scaled time by the gap diffusion time $t_g = d^2/\nu$, where d is the width of the gap between the two cylinders and ν the kinematic viscosity of the fluid. The control parameters are the non-dimensionalized external rotation rate Ω, and the Reynolds number R of the rotating inner cylinder (R is proportional to the inner-cylinder angular speed $\tilde{\omega}$). We characterize the distance of the Reynolds number from the primary bifurcation point

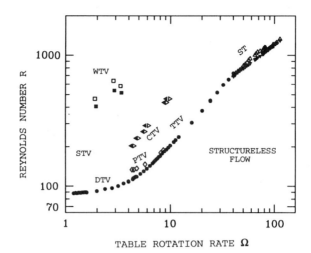

Figure 1. Experimental results for the bifurcation lines in the R-Ω plane. This figure covers the entire range of our experiments. The data are from the apparatus with two linear ramps as end sections. The solid circles are a non-hysteretic primary bifurcation which leads to very slowly drifting tilted vortices (DTV) at small Ω, and to time-periodic tilted vortices (PTV) or travelling tilted vortices (TTV) of much higher frequency for $\Omega > 8.5$. The triangles for $\Omega > 40$ are a hysteretic primary bifurcation to structureless turbulence (ST) (Open: R increasing; closed: R decreasing). The squares are a hysteretic secondary bifurcation to wavy tilted vortices (WTV) (open: R increasing; solid: R decreasing). The triangles for $4 < \Omega < 10$ are a hysteretic secondary bifurcation between a chaotic tilted vortex state (CTV) and a steady tilted vortex state (STV). The diamonds are a secondary non-hysteretic bifurcation to time-periodic tilted vortices (PTV) which meets the primary bifurcation near $\Omega = 8.5$.

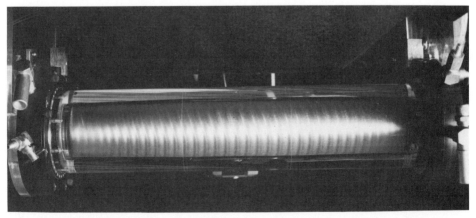

Figure 2. Photo of the steady tilted vortices (STV), $\Omega = 3.0$.

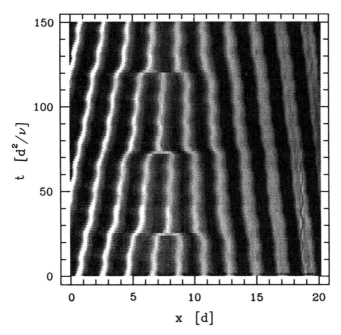

Figure 3. Space-time image of the slowly drifting tilted vortex state (DTV) at $\Omega = 4.0$, $\varepsilon = 0.066$.

$R_c(\Omega)$ with $\epsilon = R(\Omega)/R_c(\Omega) - 1$. The states we reached were prepared by quasi-statically increasing R from the uniform base state at fixed Ω. Most of the results presented were obtained from the apparatus with two symmetric linearly ramped sections attached to each end of a central homogeneous section,[6] but the same states were found to exist in the homogeneous system except for the drifting tilted vortex state, which will be explained below.

Figure 1 shows the overall experimentally determined bifurcation diagram in the $\Omega - R$ plane. The primary bifurcation was delayed as Ω increased. For small Ω, this bifurcation line was quadratic in Ω, in accordance with the symmetry that the stability of the base flow is invariant under a reversal of the direction of Ω (see for instance, Ref. 10). The transition was non-hysteretic in the homogeneous system for the range of Ω we studied, but in the system with ramped ends there was rather strong hysteresis for $40 \lesssim \Omega \lesssim 80$.

At small Ω ($\lesssim 8.5$), the first transition upon increasing R was a bifurcation from the structureless base flow to tilted vortices (STV or DTV in Fig. 1). The STV is shown in Figure 2, which is a photo taken at $\Omega = 3.0$ and $\epsilon \approx 0.3$ (the two circumferential lines which are noticeable are engraved on the outer surface of the outer cylinder and mark the joints of the ramps onto the central straight section). Consistent with the nature of the Coriolis force, the tilt direction reversed when the direction of either the inner-cylinder rotation or the table rotation was reversed, but remained the same when both rotation directions were changed.

The state of drifting tilted vortices (DTV) was characteristic only of the system with ramped ends. Immediately above the bifurcation and for $\Omega \gtrsim 2$, the tilted vortices in the ramped systems drifted at a very slow rate, with a period of order a hundred t_g, from the ramps into the straight section. In the interior of the straight section, two pairs of vortices periodically collapsed into one. This is illustrated in Fig. 3, which shows the image intensity along an axial line as a function of time in the straight section for $\Omega = 4.0$ and $\epsilon = 0.07$. Preliminary results show that the drift velocity has

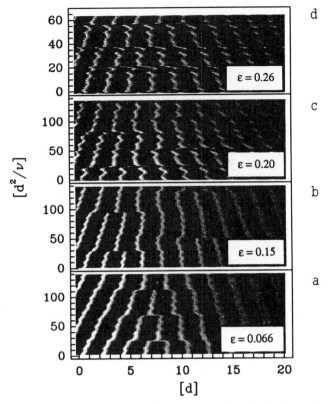

Figure 4. Sequence of space-time images of the drifting tilted vortex state (DTV) at $\Omega = 4.0$. Note the interval between phase slip events and the positions of the events in space.

a maximum as R increases, and then decreases to zero. This state was not observed in the axially uniform system. We believe that the drift is induced by the selection of an unstable state by the ramps at the two ends of the system. Although it is known that the ramps used in our apparatus select a stable state in the *absence* of rotation,[11-13] the selection of unstable states has been predicted theoretically[14] and found experimentally[15] for some cases in which both the inner and the outer cylinder radii vary axially. We presume that the rotation has altered the wavenumber selection so dramatically that the selected wavenumber is unstable close to the onset of vortex flow.

A rather intriguing consequence of the unstable wavenumber selection by the ramps is illustrated in Fig. 4, where a sequence of the space-time images of the DTV states at $\Omega = 4.0$ is recorded. At low ϵ, the collapse of two vortex pairs into one was periodic in time and the site was fixed in space (Fig.4(a)). As ϵ increased, the spatial location of the vortex-pair loss alternated between two positions, and the time interval between successive losses alternated between two values. Thus, there was both temporal and spatial period doubling (Fig.4(b)). At yet higher ϵ, the underlying state became more disordered (Fig.4(c)), until at even higher ϵ the site of the collapse event returned to one unique position again (Fig.4(d)). We believe that the phenomenon may be related to the spatio-temporal chaos induced by unstable-state selection predicted recently by Riecke and Paap.[16] Our system (aspect ratio near

20 for the straight section) might not be optimized for the observation of the whole bifurcation sequence envisioned in the theory, however.

The wavy vortex flow which occurs in TVF without external rotation was found in our apparatus at $\epsilon \gtrsim 5$ (as opposed to 0.2 in the apparatus used by Wiener et al.), and is referred to as the wavy tilted vortex (WTV) state. The ratio between the frequency ω_w of this mode and the frequency $\tilde{\omega}$ of the inner cylinder is about 0.8. Since the wavy-mode speed $(\omega_w/\tilde{\omega})/m$ is usually between 0.3 and 0.5,[17] we presume that the azimuthal mode number m is at least two and probably not 3. The space-time image of a representative state of WTV is shown in Figure 5. The meandering dark feature is the inflow region, while the outflow region represented by the other dark feature is less oscillatory.

For $4 \lesssim \Omega \lesssim 8.5$ a bifurcation from tilted vortices (DTV or STV) to a time-periodic tilted vortex state (PTV in Fig. 1) followed closely after the initial transition. The frequency of this state, when normalized by the inner-cylinder rotation-rate, is about a factor of 2 smaller than that of the previously discussed WTV state. This secondary bifurcation became the primary one at $\Omega = 8.5$. Figure 6 is a space-time plot for PTV. As can be seen, this mode differs qualitatively from the WTV shown in Fig. 5.

The primary bifurcation to the time-periodic state described above changed its nature for $\Omega \gtrsim 10$. Figure 7 shows the new state. It consists of travelling waves. It was possible in the experiment to prepare states in which the waves travelled in either direction, and sometimes a state in which travelling waves going in two directions coexisted. Figure 8 shows an example in which the waves meet in a sink near the center of the straight section of the apparatus. Initially, we believed this state to consist of tilted vortices travelling in the axial direction, and it was thus named "travelling tilted vortices" (TTV). However, closer inspection revealed that it was actually a spiral with pitch equal to one. The frequencies of the PTV and TTV states are closely related. In fact, along the bifurcation line they vary continuously from

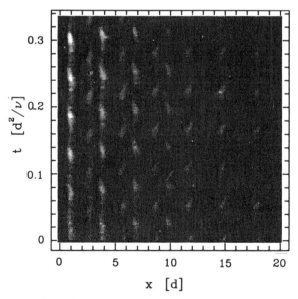

Figure 5. Space-time image of the wavy tilted vortex state (WTV) at $\Omega = 1.0$, $\varepsilon = 7.2$.

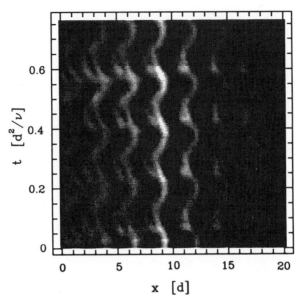

Figure 6. Space-time image of the time-periodic tilted vortex state (PTV) at $\Omega = 10.0$, $\varepsilon = 0.02$.

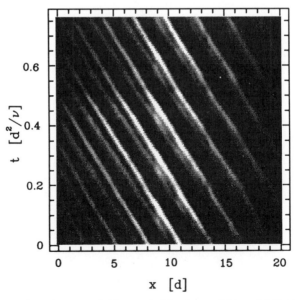

Figure 7. Space-time image of the axially travelling tilted vortex state (TTV) at $\Omega = 10.0$, $\varepsilon = 0.05$.

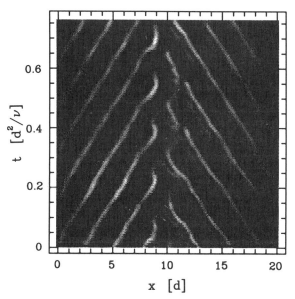

Figure 8. Space-time image of an axially travelling tilted vortex state (TTV) with waves travelling in two directions at $\Omega = 10.0$, $\varepsilon = 0.125$.

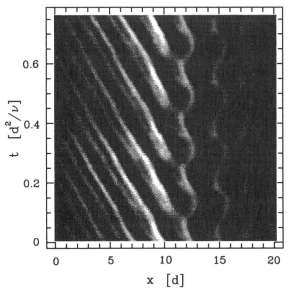

Figure 9. Space-time image showing the co-existence of PTV and TTV at $\Omega = 10.0$, $\varepsilon = 0.035$.

one state to the other.[5] Figure 9 shows that the two states can co-exist along the boundary between them.

Above PTV and TTV in the $\Omega - R$ space, the tilted vortices became chaotic and were constantly splitting and re-combining. In the ramped system, this state was observed to be detached from the end boundaries. This chaotic tilted vortex state (CTV) is depicted in Figure 10. Upon further increasing R, the flow returned to a time-independent steady tilted vortex state *via* a hysteretic transition. This dynamic state is likely to be related to the chaotic state reported by Wiener *et al.*(see these proceedings).

Upon increasing Ω to above 15, the state above the primary bifurcation became more and more irregular, and the bifurcation gradually became a direct transition to (visually) structureless turbulence (ST in Fig. 1). This turbulent state was different from those observed in Taylor-Couette flow without external rotation in that the latter still possess the Taylor-vortex structure, and the turbulence is on a smaller scale.[2] In the turbulent state observed here there was no large-scale structure, such as that of vortices, discernible. Figure 11(a) is a photo of this state, and Fig. 11(b) is a space-time image. Puffs of irregular structure were generated at the right and quickly circulated to the left. Since the images were taken in the top portion of the horizontal apparatus, it is worth clarifying that in the bottom portion the circulation was in the opposite direction. Related to this is the fact that before the onset of turbulence for $\Omega \gtrsim 20$, there already were two horizontal streaks running across the apparatus in the opposite azimuthal positions, which showed the shear boundaries of the oppositely circulating flows at the top and bottom portion of the apparatus. The initial disturbances that eventually triggered turbulence in the bulk were observed to be generated at the corners of the streak lines near the ends and propagated along the front of the streaks into the bulk of the flow.

ACKNOWLEDGEMENTS: We gratefully acknowledge discussions with M. Tveitereid. This research was supported by the National Science Foundation under Grant No. NSF DMR88-14485.

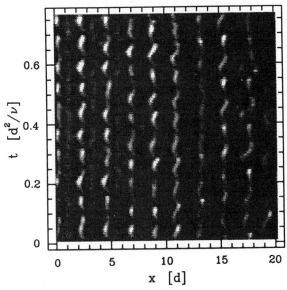

Figure 10. Space-time image of the chaotic tilted vortex state (CTV) at $\Omega = 6.0$, $\varepsilon = 0.78$.

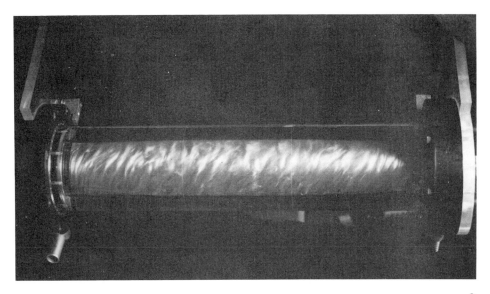

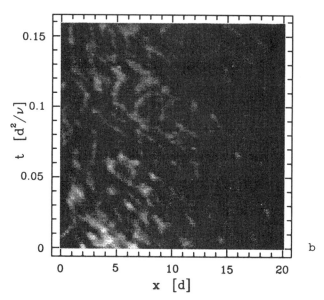

Figure 11. a A photo, and b a space-time image of the visually structureless turbulence (ST) at Ω = 30.0, ε = 0.02.

REFERENCES

1. G.I. Taylor, "Stability of a Viscous Liquid contained between Two Rotating Cylinders", *Philos. Trans. Roy. Soc. London Ser. A* **223**, 289 (1923).

2. For a recent review, see R. C. DiPrima and H. L. Swinney, in *Hydrodynamic Instabilities and Transitions to Turbulence*, edited by H. L. Swinney and J. P. Gollub (Springer, Berlin, 1981).

3. G. Ahlers, "Experiments on Bifurcations and One-Dimensional Patterns in Nonlinear Systems Far from Equilibrium", in *Complex Systems*, SFI Studies in the Sci. of complexity, ed. D. Stein. Addison-Wesley Longman, 1989.

4. R. J. Wiener, P. W. Hammer, C. E. Swanson, and R. J. Donnelly, "Stability of Taylor-Couette flow subjected to an external Coriolis force", *Phys. Rev. Lett.* **64**, 1115 (1990).

5. L. Ning, G. Ahlers, D. S. Cannell, and M. Tveitereid, "Experimental and Theoretical Results for Taylor-Couette Flow Subjected to a Coriolis Force", *Phys. Rev. Lett.* **66**, 1575 (1991).

6. L. Ning, G. Ahlers, and D. S. Cannell, "Novel Dynamic States in Taylor-Couette Flow Subjected to a Coriolis Force", *J. Stat. Phys.*, to be published.

7. R. J. Wiener, P. W. Hammer, C. E. Swanson, D. C. Samuels, and R. J. Donnelly, "The effect of a Coriolis force on Taylor-Couette flow", *J. Stat. Phys.*, to be published; and private communication.

8. Li Ning, Morten Tveitereid, Guenter Ahlers, and David S. Cannell, "Taylor-Couette Flow Subjected to External Rotation", *Phys. Rev. A*, to appear in Aug. 15, 1991 issue.

9. R. J. Wiener, P. W. Hammer, and R. Tagg, "Perturbation Analysis of the Primary Instability in Taylor-Couette Flow Subjected to a Coriolis Force", *Phys. Rev. A*, to appear.

10. M. Tveitereid, L. Ning, G. Ahlers, and D. S. Cannell, "On the Stability of Taylor-Couette Flow Subjected to an External Rotation", this proceedings.

11. D. S. Cannell, M. A. Dominguez-Lerma, and G. Ahlers, "Experiments on Wavenumber Selection in Rotating Couette-Taylor Flow', *Phys. Rev. Lett.* **50**, 1365 (1983).

12. M. A. Dominguez-Lerma, D. S. Cannell, and G. Ahlers, "Eckhaus boundary and wavenumber selection in rotating Couette-Taylor flow", *Phys. Rev. A* **34**, 4956 (1986).

13. G. Ahlers, D. S. Cannell, M. A. Dominguez-Lerma, and R. Heinrichs, "Wavenumber Selection and Eckhaus Instability in Couette-Taylor Flow", *Physica* **23D**, 202 (1986).

14. H. Riecke and H. G. Paap, "Perfect wave-number selection and drifting patterns in ramped Taylor vortex flow", *Phys. Rev. Lett.* **59**, 2570 (1987).

15. L. Ning, G. Ahlers, and D. S. Cannell, "Wave-Number Selection and Traveling Vortex Waves in Spatially Ramped Taylor-Couette Flow", *Phys. Rev. Lett* **64**, 1235 1990.

16. H. Riecke and H. G. Paap, "Spatio-Temporal Chaos through Ramp-induced Eckhaus Instability", *Europhys. Rev. Lett.* 14, 433(1991).

17. G. Ahlers, D. S. Cannell, and M. A. Dominguez-Lerma, "Possible mechanism for transitions in wavy Taylor-vortex flow", *Phys. Rev. A* **27**, 1225 (1983).

ON THE STABILITY OF TAYLOR-COUETTE FLOW SUBJECTED TO EXTERNAL ROTATION

Morten Tveitereid,* Li Ning, Guenter Ahlers, and David S. Cannell

Department of Physics and Center for Nonlinear Science
University of California
Santa Barbara, CA 93106, U.S.A.

1. INTRODUCTION

Circular Couette flow[1] is the flow between two concentric cylinders with one or both of them rotating about their geometric axis. We consider here a *modified* Couette flow, which occurs when only the inner cylinder rotates about this axis, and both cylinders rotate about a second external axis which is orthogonal to the first. In this paper we present theoretical results for the linear stability of this modified flow. This system is of particular interest because the external rotation breaks the cylindrical symmetry of Couette flow and leads to instabilities and spatio-temporal patterns which do not occur without it. Experimental studies of this system[2-5] have revealed a surprising richness of bifurcation phenomena. The first theoretical results for this problem were obtained by Ning et al.[6] and Wiener et al.,[7] who calculated the modified base flow and determined its stability for small values of the external rotation rate Ω. Using Ω as an expansion parameter, these authors carried out the analysis to $O(\Omega^2)$. The calculated values of the critical Reynolds number, the critical wavenumber, and the tilt angle of the vortices were in quantitative agreement with experiments.[6]

Measurements in our laboratory[4,5] have revealed transitions to time-periodic states at the first bifurcation for moderate values of Ω. Motivated by this result, we have extended our previous perturbation analysis to $O(\Omega^4)$, and included non-axisymmetric and time-dependent disturbances in the initial $O(1)$-perturbations. We will define the problem and outline the base flow in Section 2. In Section 3 we formulate the stability problem. The main results and a comparison with existing experimental data will be presented in Section 4.

2. GOVERNING EQUATIONS AND BASE FLOW

We consider the viscous flow of an incompressible fluid between two coaxial cylinders of infinite length. To describe the geometry and the flow, we use cylindrical coordinates (r, ϕ, z) where the z−axis coincides with the common axis of the cylinders. The inner cylinder has radius $\tilde{r}_1$ and rotates about the z−axis with angular velocity $\tilde{\omega}$, the outer cylinder has radius $\tilde{r}_2$ and no rotation about the z−axis. Moreover, both cylinders rotate with angular velocity $\tilde{\Omega}$ about the axis $\underline{\lambda} = (1, 0, 0)$ (see

---

* Permanent address: Agder College of Engineering, Grimstad, Norway.

Ordered and Turbulent Patterns in Taylor-Couette Flow
Edited by C.D. Andereck and F. Hayot, Plenum Press, New York, 1992

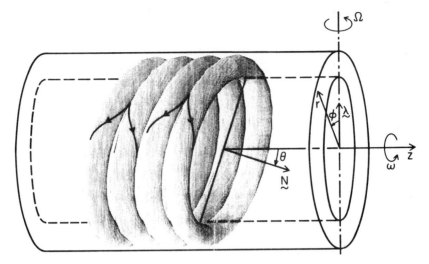

Figure 1. Illustration of the coordinate-system and the tilted Taylor-vortex flow.

Fig. 1). The variables are made dimensionless with the scales of length, time, velocity, and pressure given by $d = \tilde{r}_2 - \tilde{r}_1, d^2/\nu, \tilde{\omega}\tilde{r}_1$, and $\rho\tilde{\omega}\tilde{r}_1\nu/d$, respectively. Here ν is the kinematic viscosity and ρ the density of the fluid. The dimensionless Navier-Stokes and continuity equations become

$$\frac{\partial \underline{v}}{\partial t} + R(\underline{v} \cdot \nabla)\underline{v} = -\nabla p + \nabla^2 \underline{v} - 2\Omega \underline{\lambda} \times \underline{v} \quad , \tag{2.1}$$

$$\nabla \cdot \underline{v} = 0 \tag{2.2}$$

with boundary conditions

$$\underline{v} = (0,1,0) \quad at \quad r = r_1; \quad \underline{v} = (0,0,0) \quad at \quad r = r_1 + 1. \tag{2.3}$$

Here t is the time, $r_1 = \tilde{r}_1/d$, $\underline{v}$ is the velocity, and p is the modified pressure which includes the centrifugal and gravity terms. The two parameters are the Reynolds number $R = \tilde{\omega}\tilde{r}_1 d/\nu$, and the dimensionless angular velocity $\Omega = \tilde{\Omega}d^2/\nu$ of the rotation about $\underline{\lambda}$.

We seek a stationary z-independent solution of equations (2.1-2.3) by expanding $\underline{v}$ and p in powers of Ω:

$$(\underline{v},p) = ((U,V,W),P) = \sum_{q=0} ((U_q,V_q,W_q),P_q)\,\Omega^q \tag{2.4}$$

Here U, V and W are radial, azimuthal and axial velocity components. By equating terms of equal power in Ω we find a solution of the following form ($l = 0, 1, \ldots$):

$$\begin{aligned} V_{2l+1} &= 0, \quad V_{2l} = \sum_{m=0}^{l} a_{lm}(r)e^{i(2m)\phi} + c.c. \\ W_{2l} &= 0, \quad W_{2l+1} = \sum_{m=0}^{l} b_{lm}(r)e^{i(2m+1)\phi} + c.c. \end{aligned} \tag{2.5}$$

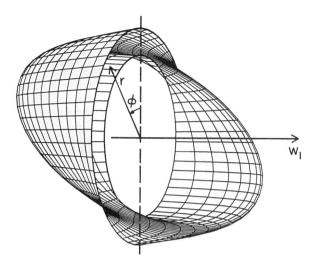

Figure 2. The leading order $O(\Omega)$ contribution $W_1(r,\phi)$ to the axial velocity of the base flow.

where $a_{lm}(r)$ and $b_{lm}(r)$ are complex functions, and c.c. denotes the complex conjugate. The solutions for U and P are given by expressions similar to that for V. Details concerning this base flow solution and a discussion of the convergence of the expansions (2.4) can be found in Ning et al.[6]

For $\Omega = 0$, the base flow solution consists of classical Couette flow[8], which is an axisymmetric steady flow purely in the azimutal direction. When $\Omega \neq 0$, the Coriolis force $(-2\underline{\Omega} \times \underline{v})$ gives rise to a modified base flow with additional velocity components in the radial and axial directions. Figure 2 displays $W_1(r,\phi)$ for $\eta = r_1/(r_1+1) = 0.75$ and $R = 85.8$.

3. FORMULATION OF THE STABILITY PROBLEM

In order to investigate the stability of the base flow, we let

$$(\underline{v}, p) = (U, V, W, P) + (\underline{v}', p') \tag{3.1}$$

where $\underline{v}' = (u', v', w')$ is the perturbation velocity, and p' is the perturbation pressure. On substituting these expressions into equations (2.1)-(2.3), subtracting the equations for the base flow, and neglecting the quadratic terms in the perturbations, we obtain linearized perturbation equations. We solve these equations by expanding the perturbations in powers of Ω and write

$$(\underline{v}', p') = e^{\sigma t} \sum_{q=0} \Omega^q (\underline{v}_q, p_q), \tag{3.2}$$

where σ is the complex growth rate. The onset of instability of the base flow is determined by the condition that $Re(\sigma) = 0$. The neutral value R_N of R, and the corresponding value σ_N of σ, which make the base flow marginally stable, depend on Ω. Thus we write R_N and σ_N as

$$(R_N, \sigma_N) = \sum_{q=0} \Omega^q (R_q, \sigma_q) \tag{3.3}$$

consistent with the expansions in (3.2). Introducing the expansions (3.2)-(3.3) into the perturbation equations and equating terms of equal power in Ω, we obtain sets of linear equations of the following form ($q = 0, 1, \ldots$):

$$(\mathbf{A} - R_0\mathbf{B} - \sigma_0)\underline{v}_q^T - (\nabla p_q)^T = \underline{h}_q^T; \quad \underline{h}_0 = 0$$
$$\nabla \cdot \underline{v}_q = 0 \qquad (3.4)$$
$$\underline{v}_q = (0,0,0) \quad \text{at} \quad r = r_1, \; r_1 + 1$$

Here $\mathbf{A}, \mathbf{B}$ are 3×3 matrix operators, and the superscript T denotes the transpose. The expressions for $\mathbf{A}, \mathbf{B}$ and the vector $\underline{h}_q$ can be deduced from the perturbation equations, and have been given elsewhere.[6]

4. RESULTS AND DISCUSSION

The $O(1)$–equations of (3.4) define an eigenvalue problem which determines the stability of Couette flow.[8,9] The solution can be written as

$$(\underline{v}_0, p_0) = (\hat{\underline{v}}_0(r), \hat{p}_0(r))e^{i(kz+n\phi)}. \qquad (4.1)$$

Here k is the axial wavenumber, and the integer n is the azimuthal wavenumber. The characteristic values of the eigenvalue problem, R_0 and σ_0, are functions of the radius ratio η and the wavenumbers. The minimum value R_{c0} of R_0 and the corresponding values σ_{c0} of σ_0 and k_{c0} of k are the critical values for Couette flow. These values are given in Table 1 for $n = 0, 1, 2$ and various values of η. For $n = 0$, they agree with previous calculations.[8–10] We notice the well known result that the lowest value of R_{c0} occurs for stationary axisymmetric perturbations ($n = 0, \sigma = 0$).

To ensure the existence of solutions of equations (3.4) for $q > 0$, the right hand side of the equations must be orthogonal to any solution of the adjoint problem,[6] i.e., $\langle \bar{\underline{v}} \cdot \underline{h}_q \rangle = 0$, where $\bar{\underline{v}}$ represents the adjoint solutions, and the bracket denotes integration over the entire fluid volume. On applying this solvability condition, we determined R_q and σ_q as functions of k, n and η. It turns out that R_q and σ_q are zero for q odd. This follows also from symmetry considerations, because the stability of the flow should not be affected by a reversal in the rotation of the frame.

A direct consequence of the Coriolis force is that additional azimuthal modes are generated at higher orders of Ω. For example, at order Ω the right hand side of Eq. (3.4) consists of terms of the form $exp(\pm i\phi)\underline{v}_0$, and the solution becomes

$$\underline{v}_1 = \hat{\underline{v}}_{10}(r)e^{i(kz+n\phi-\phi)} + \hat{\underline{v}}_{11}(r)e^{i(kz+n\phi+\phi)}. \qquad (4.2)$$

In general, the solution at $O(\Omega^q)$ can be expressed as

$$(\underline{v}_q, p_q) = \sum_{m=0}^{q} (\hat{\underline{v}}_{qm}(r), \hat{p}_{qm}(r))e^{i(kz+N_m\phi)}$$

where N_m takes on the values $(n-1, n+1)$ for $q = 1$, $(n-2, n, n+2)$ for $q = 2$ $(n-3, n-1, n+1, n+3)$ for $q = 3$, etc. The axisymmetric disturbances ($n = 0$) at $O(1)$ become non-axisymmetric when coupled to the Coriolis force at higher orders of Ω.

In order to find an expression for the critical Reynolds number R_c, we expand R_N around the critical wavenumber k_{c0} for $\Omega = 0$. This gives

$$R_N = R_{c0} + \frac{1}{2}R_{c0}''(k - k_{c0})^2 + \ldots + \Omega^2[R_{20} + R_{20}'(k - k_{c0}) + \ldots] \qquad (4.3)$$
$$+ \Omega^4[R_{40} + R_{40}'(k - k_{c0}) + \ldots] + O(\Omega^6)$$

where $R_{c0} = R_0(k_{c0})$, $R_{c0}'' = (\partial^2 R_0/\partial k^2)_{k=k_{c0}}$, etc. By solving $\partial R_N/\partial k = 0$, we find

$$R_c = R_{c0} + \Omega^2 R_{c2} + \Omega^4 R_{c4} + O(\Omega^6) \tag{4.4}$$

at

$$k_c = k_{c0} + \Omega^2 k_{c2} + \Omega^4 k_{c4} + O(\Omega^6) \tag{4.5}$$

Here $R_{c2} = R_{20}$, $R_{c4} = R_{40} - \frac{1}{2}(R_{20}')^2/R_{c0}''$ and $k_{c2} = -R_{20}'/R_{c0}''$. Since $R_{c0}' = 0$, we only need k_c to $O(\Omega^2)$ to determine R_c to $O(\Omega^4)$, so we have not calculated k_{c4}. The corresponding critical value of the complex growth rate is

$$\sigma_c = \sigma_{c0} + \Omega^2 \sigma_{c2} + O(\Omega^4) \tag{4.6}$$

where $\sigma_{c2} = \sigma_{20} - R_{20}'\sigma_{c0}'/R_{c0}''$. The values of the coefficients R_{c2}, k_{c2}, etc. are given in Table 1. From the values of the coefficients we find that R_c and $i\sigma_c$ increase and k_c decreases with increasing Ω. Moreover, we observe that $R_{c,0} < R_{c,1} < R_{c,2}$, showing that the disturbances at the onset of instability are stationary with azimuthal wavenumber n=0 (here $R_{c,n}$ denotes the critical Reynolds number with azimuthal wavenumber n).

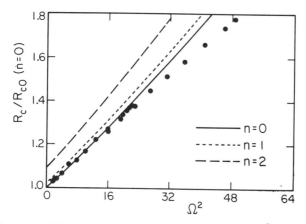

Figure 3. The critical Reynolds number as a function of Ω^2 for $\eta = 0.75$. The solid circles are experimental results from Ning et al., and the lines are the theoretical results for $n = 0, 1$, and 2.

For small values of Ω, it is found in experiments that R_c depends nearly quadratically upon Ω. This is illustrated[5,11] in Fig. 3 for $\eta = 0.75$ and Fig. 4 for $\eta = 0.88$. The figures display $R_c/R_{c0}(n=0)$ versus Ω^2: the solid circles are the experimental values, and the solid, dotted and broken curves are the results of our calculations for n equal to 0, 1 and 2 respectively all to $O(\Omega^4)$. The figures demonstrate clearly that stationary disturbances with $n = 0$ are the most unstable disturbances. The discrepancy between our calculations and the experimental values of R_c, when $\Omega^2 \gtrsim 40$ for $\eta = 0.75$ and $\Omega^2 \gtrsim 10$ for $\eta = 0.88$, can be due to many factors. For example, the radius of convergence for the perturbation expansion is approached in the calculation.

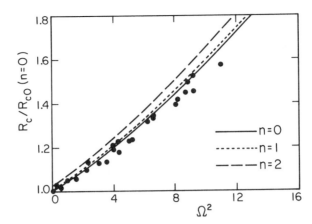

Figure 4. As in Figure 3, but $\eta = 0.88$. The experimental results are from Wiener *et al.*.

To leading order in Ω, the vortices are elliptical tori. The normal vector $\underline{N}$ of a torus makes an angle θ with respect to the cylinders' axis (see Fig. 1), and the major axis of the torus is given by $\phi = \delta$. It is clear that the tilt of the vortices is determined by θ and δ. Details about the calculation of these angles can be found in Ning *et al.*[6] We have found that θ is linear in Ω, and that δ is independent of Ω for small Ω. The coefficients for θ_1 and δ_0 in the expansions

$$\theta = \theta_1 \Omega + O(\Omega^3), \quad \delta = \delta_0 + O(\Omega^2) \tag{4.7}$$

are tabulated in Table 1.

An interesting aspect of the experiments[5] with $\eta = 0.75$ and $4 \lesssim \Omega \lesssim 8$ is a secondary time-periodic bifurcation with frequency ω_s very close to $\omega_1 = i\sigma_1 = i\sigma_c(n = 1)$. For $8 \lesssim \Omega \lesssim 14$, this bifurcation becomes the primary one. We can model this bifurcation by the following approximation of the $O(\Omega^0)$ perturbation velocity:

$$\underline{v} = A(t)\hat{v}_A(r)e^{ikz} + B(t)\hat{v}_B(r)e^{i(kz+\phi+\omega_1 t)} + D(t)\hat{v}_D(r)e^{i(kz-\phi-\omega_1 t)} + c.c. \tag{4.8}$$

Here B and D are small compared to A. Considering weakly nonlinear theory, the Landau equations for A, B and D are[12]

$$\begin{aligned} \frac{dA}{dt} &= \epsilon A - |A|^2 A \\ \frac{dB}{dt} &= (\epsilon - \epsilon_1)B - \alpha |A|^2 B - \beta A^2 \bar{D} \\ \frac{dD}{dt} &= (\epsilon - \epsilon_1)D - \bar{\alpha} |A|^2 D - \bar{\beta} A^2 \bar{B} \end{aligned} \tag{4.9}$$

Here $\epsilon = (R - R_{c,0})/R_{c,0}$, $\epsilon_1 = (R_{c,1} - R_{c,0})/R_{c,0}$; $\alpha = \alpha_r + i\alpha_i$ and $\beta = \beta_r + i\beta_i$ are complex, and the bar denotes the complex conjugate. The system of equations (4.9) gives a stationary primary bifurcation for $\epsilon = 0$, and a secondary time-periodic

TABLE 1. The expansion coefficients for critical Rayleigh number R_c, critical wavenumber k_c, critical frequency $i\sigma_c$, and the tilt-angles θ and δ (in radians) for various values of the radius ratio η and the azimutal wavenumber n (see text for definitions).

η	n	R_{c0}	R_{c2}	R_{c4}	k_{c0}	k_{c2}	$i\sigma_{c0}$	$i\sigma_{c2}$	θ_1	$-\delta_0$
	0	71.7157	0.5185	0.0004	3.14834	0.0158	0	0	0.086	0.44
0.60	1	75.8457	0.5599	0.0002	3.25054	0.0153	19.1403	0.0911		
	2	92.7029	0.7573	-0.0015	3.55969	0.0188	47.3738	0.3091		
	0	79.4907	0.9891	0.0016	3.13887	0.0239	0	0	0.102	0.56
0.70	1	81.9114	1.0387	0.0016	3.19624	0.0234	14.9228	0.1284		
	2	90.2809	1.2216	0.0017	3.36643	0.0227	33.0989	0.3384		
	0	85.7765	1.4263	0.0047	3.13541	0.0305	0	0	0.112	0.63
0.75	1	87.6160	1.4824	0.0050	3.17685	0.0298	13.0165	0.1556		
	2	93.6940	1.6763	0.0058	3.29960	0.0296	27.9643	0.3792		
	0	94.7336	2.1689	0.0153	3.13264	0.0406	0	0	0.123	0.70
0.80	1	96.1093	2.2342	0.0157	3.16142	0.0416	11.1640	0.1926		
	2	100.510	2.4498	0.0188	3.24646	0.0400	23.4255	0.4429		
	0	121.979	5.4835	0.1544	3.12928	0.0772	0	0	0.143	0.86
0.88	1	122.764	5.5762	0.1594	3.14252	0.0772	8.0155	0.3016		
	2	125.186	5.8680	0.1753	3.18200	0.0773	16.3733	0.6453		
	0	184.986	21.339	2.7201	3.12748	0.1973	0	0	0.166	1.09
0.95	1	185.390	21.493	2.7577	3.13224	0.1975	5.0027	0.5421		
	2	186.614	21.964	2.8737	3.14600	0.1978	10.0775	1.1122		

bifurcation for $\epsilon = \epsilon_2 = \epsilon_1/(1-\alpha_r-\beta_r)$ with frequency $\omega_s = \omega_1+O(\epsilon)$. When $\Omega = 0$, it is known[12] that $0 < (1-\alpha_r-\beta_r) \ll 1$, giving $\epsilon_2 \gg \epsilon_1$. For non-vanishing Ω, however, there are additional destabilizing effects which may reduce the distance between the bifurcations. It is possible that ϵ_2 may becomes less than ϵ_1 for large values of Ω. One of these effects is due to two inflection points in the azimutal distribution of the base flow(see Fig. 2). From Rayleigh-Fjørtoft's criterion for instability[9] it follows that disturbances with azimutal wavenumber $n = 1$ can be created.

5. SUMMARY

We have analyzed the linear stability of the flow between two concentric cylinders with the inner one rotating about their common axis. The two cylinders also rotate with dimensionless rotation rate Ω about an axis perpendicular to their axis. The main results of our calculations are given in Table 1. We have determined the critical values of the Reynolds number, the wavenumber, and the frequency for different values of the radius ratio. We have found, by taking into account all terms up to $O(\Omega^4)$, that the marginal disturbances of the base flow are stationary and non-axisymmetric. This is in agreement with experiments for moderate values of Ω. We have also determined the tilt of the vortices which are formed for supercritical Reynolds numbers.

ACKNOWLEDGEMENTS: We are grateful to P.W. Hammer and R.J. Wiener for sending us their experimental data. This research was supported by the National Science Foundation under Grant No. NSF DMR88-14485. M. Tveitereid gratefully acknowledges support from the Royal Norwegian Council for Scientific and Industrial Research.

REFERENCES

1. A. Mallock, "Determination of the viscosity of water", *Proc. R. Soc.* **Dec. 13**, 126 (1888); "Experiments on fluid viscosity", *Philos. Trans. R. Soc. London Ser. A.* **187**, 41 (1895); M.M. Couette, "Sur un nouvel appareil pour l'etude du frottement des fluides", *Comptes Rendus* **107**, 388 (1888); "Etudes sur le frottement des liquides", *Ann. Chem. Phys. Ser. VI* **21**, 433 (1890).

2. R. J. Wiener, P. W. Hammer, C. E. Swanson, and R. J. Donnelly, "Stability of Taylor-Couette flow subjected to an external Coriolis force", *Phys. Rev. Lett.* **64**, 1115 (1990).

3. R. J. Wiener, P. W. Hammer, C. E. Swanson, D. C. Samuels, and R. J. Donnelly, "The effect of a Coriolis force on Taylor-Couette flow", *J. Stat. Phys.*, to be published; and private communication.

4. L. Ning, G. Ahlers, and D. S. Cannell, "Novel Dynamic States in Taylor-Couette Flow Subjected to a Coriolis Force", *J. Stat. Phys.*, to be published.

5. L. Ning, G. Ahlers, D.S. Cannell, and M. Tveitereid, "Experimental and Theoretical Results for Taylor-Couette Flow Subjected to a Coriolis Force", *Phys. Rev. Lett.* **66**, 1575 (1991).

6. Li Ning, Morten Tveitereid, Guenter Ahlers, and David S. Cannell, "Taylor-Couette Flow Subjected to External Rotation", *Phys. Rev. A*, to appear in Aug. 15, 1991 issue.

7. R. J. Wiener, P. W. Hammer, and R. Tagg, "Perturbation Analysis of the Primary Instability in Taylor-Couette Flow Subjected to a Coriolis Force", *Phys. Rev. A*, to appear.

8. S. Chandrasekhar, *Hydrodynamic and Hydromagnetic Stability*, (Oxford University Press, London, 1961).

9. P. G. Drazin and W. H. Reid, *Hydrodynamic Stability*, (Cambridge University Press, Cambridge, 1981).

10. M. A. Dominguez-Lerma, G. Ahlers, and D. S. Cannell, "Marginal stability curve and linear growth rate for rotating Couette-Taylor flow and Rayleigh-Bénard convection", *Phys. Fluids* **27**, 856 (1984).

11. P. W. Hammer and R.J. Wiener, private communication.

12. P. M. Eagles, "On stability of Taylor vortices by fifth-order amplitude expansions", *J. Fluid Mech.* 49, 529-550 (1971).

NUMERICAL SIMULATION OF TURBULENT TAYLOR COUETTE FLOW

S.Hirschberg

Institut für Fluiddynamik
ETH - Zentrum
8092 Zürich
Switzerland

1. Introduction

The transition to turbulence in the Taylor Couette experiment has been extensively studied in the past by several groups. This flow with its simple geometry produces a large number of different flow patterns, depending on the gap radius ratio, the Reynolds numbers of the rotation of both cylinders and the flow history[1]. In the case with the outer cylinder at rest, several transitions precede the appearance of turbulence[2,3,4,5,6] and even the fully turbulent flow changes its characteristics up to high Reynolds numbers[7]. The large coherent Taylor vortices remain present deep within the turbulent regime. Secondary centrifugal instabilities of the flow within the Taylor vortices have been proposed as the mechanism creating both the wavyness of the vortices at lower Reynolds numbers[8] and an ordered small scale vortical structure at both walls found in some experiments[9,10].

In this paper, the results of direct numerical simulations of the transition to turbulence and fully developed turbulence in the Taylor Couette flow are discussed. In the first part, the behaviour of the numerical model, involving periodic boundary conditions in x and θ is compared to experiments. In a second part, results of visualisations of the flow's vortical structures are presented.

2. Numerical method

A spectral numerical scheme has been used for the calculations. It is described in detail in [11]. A comprehensive treatment of spectral schemes for fluid dynamical applications is given in [12].

The incompressible Navier Stokes equations are solved on a 3 dimensional domain assuming periodic boundary conditions in circumferential and axial directions and no slip boundary conditions at the cylinder walls. Fourier functions are used for the representation of the flow field in the two periodic directions and Chebychev polynomials for the radial direction. A second order Runge Kutta scheme with three partial steps is used for the time integration except for the viscous term involving second derivatives normal to the wall. This term is treated by a set of Green's functions in spectral space. They assure the physically accurate damping of high frequency modes in time and at the same time lead to favourable stability properties. The aliasing errors arising in the calculation of the inertial term are filtered. Pressure is calculated using the continuity condition at the end of each partial time step. No pressure boundary condition is necessary. The over all count of operations for the calculation of the pressure term and the application of the Green's functions is $O(N_r^2 N_\theta N_x)$ but for the resolutions used, the FFT, taking $O(N_r N_\theta N_x log_2(N_r + N_\theta + N_x))$ requires the biggest part of the computation time.

The code has been tested by the simulation of the amplification of unstable perturbations to the Couette solution which were compared with amplification rates predicted by linear theory. Further more, the wavy vortex flow cases measured and calculated in [13,14] have been computed giving good agreement in the resulting wave velocities.

3. Simulations and comparison with experiments

All the calculations presented were done for a system with gap radius ratio $r_1/r_2 = \eta = 0.875$ for which an extensive experimental experience exists. The axial wavelength of the vortices was chosen to be $\lambda/d = 2.5$ (d denotes the gap width), well within the range of observed wavelength in experiments. The simulations were always started from an interpolated solution of an earlier run and then continued until the transients clearly had disappeared. Typically one run was continued for around 24 full turnarounds of the inner cylinder at medium Reynolds numbers. The higher the Reynolds number is chosen, the faster the transients disappear. The highest Reynolds number simulated was $Re = \omega_1 r_1 d/\nu = 6000 = 50 Re_c$ (ω_1 denotes the rotation speed of the inner cylinder). A study of the energy spectra showed, that the grid resolution was in all cases adequate to resolve all relevant scales of the flow.

The results of nine simulations are reported for which the parameters are presented in table 1. The first three of them are wavy vortex flow (WVF) cases. The number of waves m around the cylinder was chosen to be m=6 for $Re = 300$ and $Re = 600$ and m=5 for $Re = 900$. Simulations with m=4 did not converge to a steady WVF solution, being a steady wave pattern moving around the cylinder with constant speed. Similar experiences with azimutal wavelength selection are reported by [2], for an experimental investigation with an identical system except for the finite annulus length. The wave velocities calculated agree well with the ones measured in [2].

Table 1. Computed cases

Re	N_r	N_θ	N_x	m	$2\pi f_1/\omega_1 m$	$2\pi f_3/\omega_1$	Flow state
300	33	32	32	6	0.379	-	WVF
600	33	32	32	6	0.347	-	WVF
900	33	32	32	5	0.344	-	WVF
1250	33	48	64	4	0.331	0.44	MWV
1500	33	48	64	4	0.332	0.47	TMWV
2000	33	48	64	4	0.334	0.48	TMWV
2500	33	64	64	4	0.342	0.49	TMWV
3000	33	64	64	4	0.347	0.49	TMWV
6000	49	128	128	4	-	-	TTV

The simulations with modulated wavy vortices (MWV) are all m=4, k=0 cases[15], as only a quarter of the circumference was resolved. The case with $Re = 1250$ is a MWV case without turbulence. In this case, the second frequency which is usually called f_3 in the literature turned out to be exactly a third of the first frequency f_1 and therefore the velocity signal was exactly periodic. This is surprising, as in most experimental investigations the ratio of the two frequencies is found to be nonrational. However in the investigation of [15], exactly the same case is reported and the frequencies of our system agree well with their measurements for this case.

The velocity signal at some points in the flowfield as well as the amplitude and the phase angle of some fundamental modes of the flow were stored. In fig. 1, the radial velocity component measured at midgap close to the outflowing jet is plotted for some Reynolds numbers. The first case is the WVF at $Re = 300$, for which a periodic signal results. The second trace is the case with $Re = 1250$, for which an additional low frequency has appeared. Still the signal is periodic. At the next higher Reynolds number $Re = 1500$, the periodicity has disappeared and seemingly some additional small scale fluctuations are present.

For Reynolds numbers larger than $12 Re_c$, modulated wavy vortices with turbulence (TMWV) have been found in experimental investigations. The first of the present computations, for which the Reynolds number exceeds $12 Re_c$ is the case with $Re = 1500 = 12.7 Re_c$. In order to reveal whether the numerical model shows a chaotic behaviour for this case, its computation has been continued up to a sample length of around 42 full turnarounds of the inner cylinder or roughly 20 modulation cycles. The amplitude of one fundamental spectral mode versus time is a singly periodic signal in the case of MWV as only the waveform influences this integral measure and the position of the wave is contained in the phase angle. For the case with $Re = 1250$, the phase portrait of the amplitude signal is therefore a limitcycle. At $Re = 1500$, the periodicity of the amplitude signal is lost

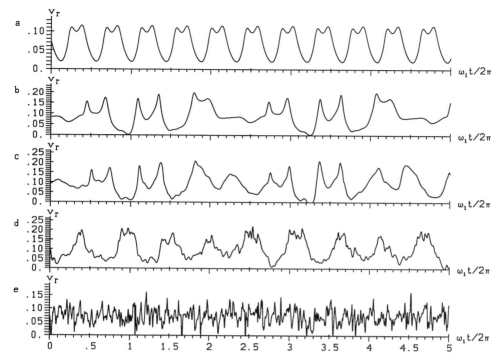

Figure 1. Traces of the radial velocity at midgap close to the outflowing jet.
a $Re = 300$, b $Re = 1250$, c $Re = 1500$, d $Re = 3000$, e $Re = 6000$

and small scale fluctuations appear at the amplification peaks (fig. 2a). The Fourier spectrum of this signal (fig. 2b) contains in addition to the peaks of the fundamental frequency some broad band noise which indicates a chaotic behaviour of this flow. It is further assumed, that all cases with higher Reynolds numbers, for which the small scale noise in the signal is larger in amplitude, show chaotic behaviour too.

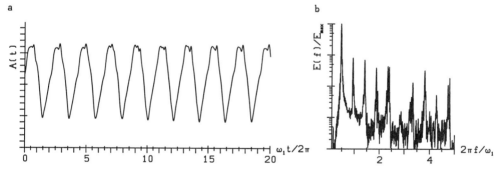

Figure 2. a Time trace of the amplitude of a fundamental spectral mode at $Re = 1500$
b Fourier spectrum

For increasing Reynolds numbers, the small scale noise in the velocity and amplitude signals continuously increased. The ratio of the two frequencies of the system changed slowly with increasing Reynolds numbers and became nonrational (see table 1). The ratio of the amplitudes of the fundamental spectral mode to modes with higher wave number in azimutal direction has been reduced at higher Reynolds numbers. At $Re = 3000$, the amplitude of the fundamental mode has been reduced so remarkably, that the waves can hardly be recognized in flow visualisations. The persistence of waves up to Reynolds number $Re = 25 Re_c$ is consistent with the experimental findings of [5], that the Reynolds number at which the waves disappear increases from $22Re_c$ to $26Re_c$ with increasing gap aspect ratio.

The calculation of the $Re = 6000$ case has been performed on a grid with $49 \times 128 \times 128$ resolution in r, θ ,x respectively. 11.5 turnarounds of the inner cylinder have been computed and statistics have been gathered over 4.5 turnarounds. At this Reynolds number, straight turbulent Taylor vortices (TTV) appeared. They remained steady over the whole simulation time. The intensity of the small scale turbulence was strongly increased compared to the lower Reynolds number cases and it was focussed mainly within the two jets. (see fig. 1e and 3a, 3b) At the walls, the angular momentum profiles rv_θ developed a logarithmic shape in the region, where the jets leave the walls. The law fitting the curve differs strongly from the one known for turbulent flows over plane walls. The law (1) was used in fig.4.

$$(v_\theta r)^+ = \frac{v_\theta r - \omega_1 r_1^2}{r_1 u_\tau} = 3.9 + \frac{1}{0.74} log r^+$$
$$r^+ = \frac{u_\tau (r - r_1)}{\nu} \qquad (1)$$

Over the rest of the wall, the distribution of mean angular momentum is nearly linear up to the momentum level of the channel center. Examples of profiles of angular momentum are shown in Figure 4.

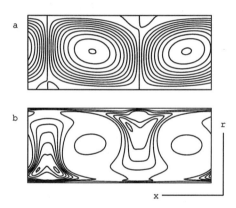

Figure 3. Turbulent flow at $Re = 6000$
a Streamlines of the mean flow
b Kinetic energy of turbulence

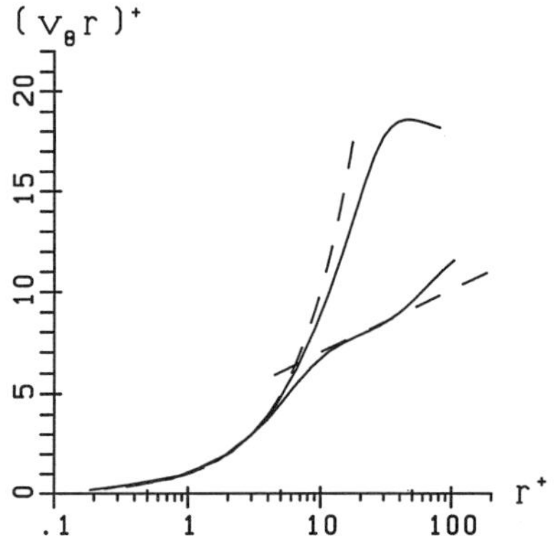

Figure 4. Mean profiles of ang. momentum: at the center of the outflowing jet with a logarithmic law and besides the jet with a nearly linear profile

To judge, whether the simulated flowfield had reached a steady state, the mean torque $r^2 \tau_{r\theta}$ has been calculated. In a steady state, this torque has to be constant over the gap. In Figure 5, The distribution of the mean torque, composed of three separate parts, is plotted. At the walls, the viscosity dominates, while in the channel center, the momentum exchange by the Taylor vortices dominates. Turbulent fluctuations contribute only a small part of the mean torque. Their importance lies in their control over the intensity of the secondary Taylor vortex motion, which is an indirect torque control mechanism. The sum of the three components is constant up to ±0.2%.

As a further test, the torques calulated from the simulations have been compared to measured ones[16]. For the exponential law (2), the exponent γ has been found to be close to 1.5 in the region of the simulations. More exactly, it was 1.56 in the WVF range, 1.5 in the MWV and TMWV range and clearly below 1.5 in the range between $Re = 3000$ and $Re = 6000$.

$$G(Re) = cRe^\gamma \qquad (2)$$

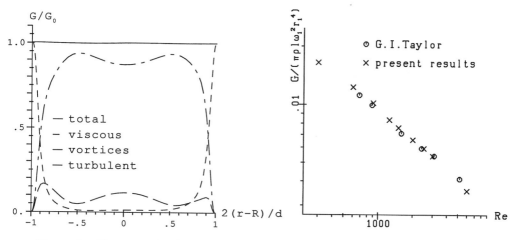

Figure 5. Components of mean torque

Figure 6. Torques

4. Visualisation of Vortices in shear flows

Several techniques for the visualisation of vortical structures in shear flows have been tested. The best results were obtained by a quantity related to the critical point theory[17,18]. As all the visualisations presented are based on that technique, it is shortly introduced here.

The streamline pattern in the close vincinity of the point investigated is studied using the theory of autonomous differential equations. A reference system moving with the point investigated is chosen in order to make the velocity at that point vanish. The flow field is linearised around the point of interest. The eigenvalues and eigenvectors of the velocity gradient tensor characterize the streamline pattern of the surrounding flowfield. If a conjugate complex pair of eigenvalues exists, the streamline pattern shows a circulation around the point investigated. The real eigenvector is the rotation axis, around which the surrounding fluid is rotating. The time, a particle would take to surround the point in the frozen flow field can be integrated up to $T = 2\pi/\lambda_i$ and the circulation frequency is therefore proportional to λ_i. A rotation vector Ω is constructed, having the direction of the real eigenvector and the magnitude of λ_i.

$$\Omega = -\lambda_i \frac{a \cdot (r \times c)}{|a \cdot (r \times c)|} \cdot \frac{a}{|a|} \qquad (3)$$

Where a denotes the real eigenvector, r and c the real and imaginary part of a complex eigenvector and λ_i the imaginary part of the corresponding eigenvalue.

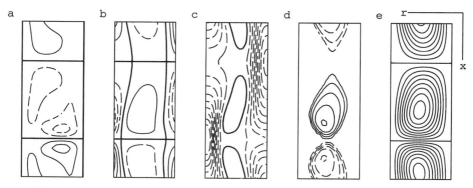

Figure 7. Axisymmetric Taylor vortex Flow. a) ω_r b) ω_θ c) ω_x
The contour levels are the same for all three components of ω
d) Ω_θ e) Streamlines of this flow.

The effectiveness of Ω in displaying vortices is demonstrated using an axisymmetric Taylor Vortex Flow example. In this case, all three vorticity components are nonzero. ω_x takes the highest values due to the shear driving the flow. In ω_θ, the shear layers at the two walls are larger in magnitude than the vorticity of the Taylor vortex. In contrast, Ω_θ focusses on the Taylor vortex itself and the two other components of Ω vanish both (see fig. 7). Ω is therefore a useful quantity for the visualisation of vortical structures. However this does not mean that the shear layers appearing in the vorticity have no relevance.

5. The observed vortical structures

In the following, the results of visualisations of the vortical structures in the simulated flowfields are presented. In all figures, the unwound gap with the azimutal length $2\pi/m$ and the axial length 5 d or 2 periods is shown. Clouds with constant magnitude of Ω are drawn which are coloured along the sign of the θ component of Ω to indicate the rotation sense of a vortex. Light coloured clouds turn with positive circumferential rotation and dark ones with negative rotation. Ω has been nondimensionalised by $|\Omega|^* = \omega_1 r_1/0.5(r_1+r_2)$

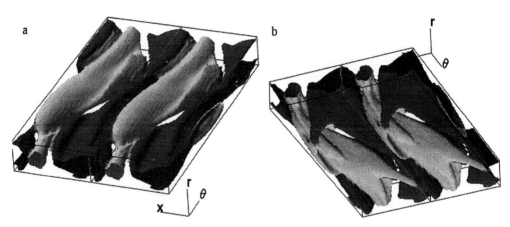

Figure 8. Vortex skeleton of WVF at $Re = 300$; a) view from the outer cylinder b) view from the inner cylinder ; $|\Omega|= 0.5$ in both cases

Fig. 8a and 8b show two views of the same WVF solution at $Re = 300$. One view is from the outer cylinder and one from the inner cylinder. The perspective projection used, makes the azimutal length of the gap appear shorter than it actually is. The vortices are divided in a bulbous and a waisted part. Ω takes higher values in the bulbous part and it is restricted to the channel center in the waisted part. One interesting point is the wrapping of parts of one vortex around its neighbour at the inner cylinder, visible in fig. 8b. Similar phenomenas are seen in all simulations with waves at the inner and sometimes also at the outer cylinder. At higher Reynolds numbers, such phenomenas appear also at a smaller scale. In this case, small parts of one Taylor vortex are torn around the neighbouring Taylor vortex in the region where the jets approach the walls. This creates arrays of thin, long vortices at the cylinder walls, oriented upstream from the outflowing jet towards the inflowing jet and turning in the opposite direction than the large Taylor vortex. The region where the jets approach the walls is the part of the flowfield, where the largest vortex streching rates are observed. Therefore it is suggested, that vortex streching is the mechanism, responsible for this phenomenon. Fig. 10a shows an example of such vortices at the outer cylinder wall.

Fig. 9a to 9c show four stages in the modulation cycle of the MWV solution at $Re = 1250$. The observer is moving with the mean wave speed in order to better recognize the changes in waveform during the modulation cycle. The cycle starts with nearly straight vortices. Within these straight vortices, Ω is by no means equally distributed

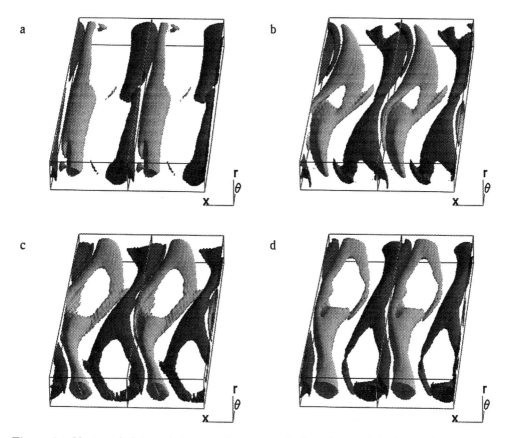

Figure 9. Vortex skeleton at 4 successive stages during the modulation cycle, $Re = 1250$ $|\Omega| = 1.8$ in all four cases

over the whole vortex but takes high values in the part of the vortex that has been waisted shortly before. It is this region with high Ω, which is in a second stage broken up in two centers and turned around at the same time. One can recognize a vortical structure within the outflowing jet following the streamlines of the flow which may be related to the instability causing the waves to grow. When the amplitude of the wave has reached a certain limit, this instability disappears. The two centers of the bulbous part become longer and less intense and at the same time, Ω starts to increase in the waisted part. As Ω takes higher and higher values in the waisted part, the wavyness begins to decrease and the modulation cycle goes towards its initial stage. Over the whole cycle, the shift and reflect symmetry is fulfilled.

At even higher Reynolds numbers, the flow becomes weakly turbulent with the modulation cycle still present. The modulation cycle actually keeps its main features but they become increasingly difficult to observe at the higher Reynolds numbers. Although the flow with $Re = 1500$ has been found to show chaotic behaviour in time, the shift and reflect symmetry holds to a high degree for its velocity fields, even in the small scale structures. The small nonsymmetric part of the flowfield is hardly amplified even after long simulation times. In the higher Reynolds number cases however, the symmetry is broken for the small scale fluctuations. Small scale vortical structures appear for example in the jets, where the rotational zones on the sides of the jets facing the bulbous part of the vortex become unstable and build smaller scale vortices. Those small vortices within the jets are oriented upstream from the inner moving wall towards the outer wall in both jets. They move slower than the wave in the inflowing jet and faster in the outflowing jet.

When the waves start to raise in amplitude, spiral subvortices within the main Taylor vortices appear. They correspond to the spiral vortex driving the instability in fig. 7b. Their orientation is downwind from the inner wall towards the outer wall within the outflowing jet and also downwind from the outflowing jet towards the inflowing jet at the outer wall. This is unusual as most disturbances take the upwind direction in both cases. As they are oriented perpendicular to the flow, they may well be caused by a secondary centrifugal instability of the flow within the Taylor vortices[8]. When the amplification of the wavyness in the modulation cycle is over, the spiral vortices disappear. Fig. 10b shows an example of such spiral subvortices.

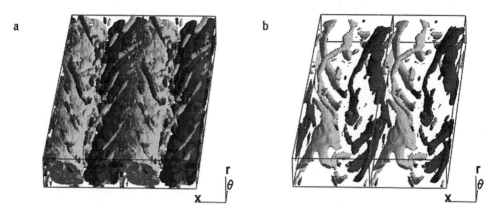

Figure 10. a) Arrays of small vortices at the outer cylinder wall. $Re = 3000$, $|\Omega| = 1$.
b) Spiral vortices within the large Taylor vortices. $Re = 2500$, $|\Omega| = 2.5$

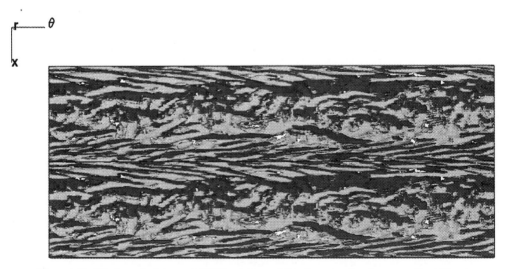

Figure 11. Vortex skeleton of TTV at $Re = 6000$ close to the inner wall, $|\Omega| = 2$.

In the highest Reynolds number simulated, the waves have completely died out. The flow is fully turbulent. Still, the large Taylor vortices are present and they are an important feature as they transport momentum from the inner cylinder towards the outer cylinder. The turbulent small scale vortices are located mainly within the jets and in the shear layers at the walls. In both jets, they are oriented upstream from the inner wall towards the outer wall. They are of different scales and they interact with each other. In the region, where the jets leave the walls, at the walls, an ordered streaky vortex pattern is visible. This streaky pattern has also been documented in experimental visualisation studies and has been referred to as the 'Herring bone streaks' in [9,10]. Those streaks are oriented mainly in the azimutal direction with a slight deviation angle ϑ which makes them point towards the center of the jets. The arrow built by those streaks points downstream at the outer cylinder and upstream at the inner cylinder. At the inner cylinder, the range of ϑ is found to be $-8° \leq \vartheta \leq 8°$ while at the outer cylinder $-15° \leq \vartheta \leq 15°$ holds. The observed wavelength of a vortex pair lie between 30 and 50 wall units in both cases.

6. Conclusions

The simulations presented show good agreement with experiments in all features, that have been compared. They are therefore believed to be a reliable tool to study the transition to turbulence and turbulence at moderate Reynolds numbers in this system.

The visualisations have revealed some new phenomenas taking place in the transition to turbulence like the wrapping of parts of one Taylor vortex around its neighbour and the appearance of spiral vortices within the Taylor vortices during the amplification of the wavyness.

In the case of the fully turbulent TTV, the flow close to the rigid walls has been found to be divided in two regions. Where the jets approach the walls, the wall layer is mainly linear while it contains some features similar to turbulent flows over flat walls in the region where the jets leave the walls.

References

1. C.D.Andereck, S.S.Liu and H.L.Swinney, J.Fluid Mech. 164, 155(1986)
2. D.Coles, J. Fluid Mech. 21, 385(1965)
3. P.R.Fenstermacher, H.L.Swinney and J.P.Gollub, J. Fluid Mech. 94, 103 (1979)
4. J.P.Gollub and H.L.Swinney, Phys. Rev. Lett. 35, 14, 927(1975)
5. R.W.Walden and R.J.Donnelly, Phys. Rev. Lett. 42, 301(1979)
6. R.C.Di Prima and H.L.Swinney, in Hydrodynamic Instabilities and the Transition to Turbulence,(Springer Verlag 1985)
7. G.P.Smith and A.A.Townsend, J. Fluid Mech. 123, 187(1982)
8. P.S.Marcus, J. Fluid Mech. 146, 65(1984)
9. A.Barcilon, J.Brindley, M.Lessen and F.R.Mobbs, J. Fluid Mech. 94, 453(1979)
10. A.Barcilon and J.Brindley, J. Fluid Mech. 143, 429(1984)
11. S.Hirschberg, to be published (1991)
12. C.Canuto, M.Y.Hussaini, A.Quarteroni and T.A.Zang, Spectral Methods in Fluid Dynamics (Springer Verlag, New York, 1988)
13. G.P.King, Y.Li, W.Lee, H.L.Swinney and P.S.Marcus, J.Fluid Mech. 141, 365(1984)
14. P.S.Marcus, J. Fluid Mech. 146, 45(1984)
15. M.Gorman and H.L.Swinney, J. Fluid Mech. 117, 123(1982)
16. G.I.Taylor, Proc. Roy. Soc. A, 157, 564 (1936)
17. A.E.Perry and M.S.Chong, Ann. Rev. Fluid Mech. 19, 125 (1987)
18. H.Vollmers, H.P.Kreplin and H.U.Meier, AGARD Conference Proceedings No. 342, 141(1983)

INTERMITTENT TURBULENCE IN PLANE AND CIRCULAR COUETTE FLOW

J. Hegseth, F. Daviaud, and P. Bergé

Service de Physique de l'Etat Condensé, Centre d'Etudes
de Saclay, F-91191 Gif-sur-Yvette Cedex, France

INTRODUCTION

Intermittent turbulence, the coexistence of laminar and turbulent flow[1], is well known to exist in circular Couette flow when the two cylinders counter-rotate[2]. In particular, the turbulent and laminar regions may organize into the Spiral Turbulence pattern (alternating regions of laminar and turbulent flow which wrap around the axis of the cylinders to form a rotating spiral) or a V-shaped pattern (local spiral patterns of opposite helicity which connect to form V shaped regions). Evidence for the subcritical nature of the transition to Spiral Turbulence and V-shaped patterns is given by the large hysteresis in parameter space and by perturbation experiments[3,4]. The role of centrifugal instabilities in this state is not so clear although the interpenetrating spirals appear to initiate turbulent spots under certain circumstances. It is also thought that the boundedness in the azimuthal direction in circular Couette flow leads to the spiral or V-shaped intermittent patterns[3,4]. Because centrifugal instabilities and boundedness are absent in plane Couette flow, a comparison of intermittent turbulence in this case to circular Couette flow may help to further understand these patterns. We have recently completed a plane Couette flow apparatus and will present some preliminary results of experiments in which we have also observed intermittent turbulence.

The transition to turbulence generally occurs in one of two ways[1]. The first type of transition is characterized by the development of velocity and pressure fluctuations at approximately the same rate throughout the flow. The development of turbulence for corotating circular Couette flow is an example of this type. In the second type of transition, the turbulence first occurs in localized patches. These patches where the flow exhibits strong velocity fluctuations coexist with regions where there are little or no velocity fluctuations. This state, which we will call intermittent turbulence, is also well known to occur in open flow systems near a wall, e.g. boundary layer flow[5] or plane Poiseuille flow[6]. Almost all of the examples of flow with intermittent turbulence are subject to linear instabilities at some Reynolds number R. While turbulent spots may be generated by relatively strong localized perturbations below the onset of linear instability, the effect of these instabilities if any remains ambiguous. Plane Couette flow is believed to be linearly *stable* at all R[7] eliminating this ambiguity. Thus we expect any instabilities which occur in this flow to be subcritical. In addition to instability to finite amplitude perturbations and hysteresis, we also expect to observe the stable coexistence of laminar and turbulent states and a moving front behavior between these states[8].

We will first give a brief review of some of the results of intermittent turbulence in circular Couette flow. We will then discuss plane Couette and how it is different from and similar to circular Couette flow. This will be followed by a description of our realization of plane Couette flow and a presentation of some preliminary results.

INTERMITTENT TURBULENCE IN CIRCULAR COUETTE FLOW

Coexisting turbulent and laminar regions in circular Couette flow (intermittent turbulence) occurs most often in the circular Couette system when the cylinders counter-rotate. Spiral turbulence occurs when there is sufficient shear in the gap, i.e. the outer cylinder rotates relatively fast and the inner cylinder rotates in the opposite sense at a lower rate (large $|R_o|$ and a comparatively low R_i, where R_i (R_o) is the inner (outer) cylinder Reynolds number based on the gap between the cylinders, d, and the cylinder velocity[9]). Spiral turbulence is an unusual form of intermittent turbulence because it exhibits a well ordered pattern at large scales, i.e. it is a large scale coherent turbulent structure. Spirals of both helicities exist in a wide range of R_o and R_i at lower aspect ratios ($\Gamma \approx 30$ where Γ is the aspect ratio i.e. the ratio of the cylinder length, L, to the gap, d) at radius ratio $\eta = 0.882$ ($\eta = r_i/r_o$ where r_i = radius of inner cylinder and r_o = radius of the outer cylinder)[3,4]. This, however, was not the case for larger aspect ratios, in fact, at $\Gamma = 73$ persistent spirals occur only near $R_o = -3000$ while at higher R_o the system favored V-shape patterns. These V-shaped patterns usually consist of spirals of opposite helicity at the top and bottom of the cylinder which connect to form a V in the center of the cylinder. At $R_o \approx -8000$ large scale coherent patterns throughout the container have not been seen at any R_i at $\eta = 0.882$. At this R_o there are local spirals with the spiral helicity changing sign over an axial distance of the order of the cylinder diameters. Because these spiral-like patterns do not always connect to make V-shaped patterns, these turbulent patches formed a "broken" spiral pattern. Sometimes the turbulence would not form local spirals or V-shaped structures, just turbulent patches.

Localized perturbations have been applied to laminar Couette flow by quickly injecting a small amount of fluid through a small hole[3,4]. The azimuthal and axial expansions of the resulting turbulent spot was videotaped in the spiral rest frame and the velocities determined[3,4]. Figure 1 shows a turbulent spot ≈ 0.5 seconds after the perturbation. This spot initially expands much faster (≈ 2 times) in the azimuthal direction than in the axial direction[3]. The turbulent spot stops its azimuthal expansion as soon as it is about as wide as half of the perimeter length. This is consistent with the azimuthal extent of Spiral Turbulence which is always observed to be greater than $\frac{1}{2}$ the average cylinder circumference[10]. Although the spot quickly stops growing azimuthally it continues to grow along the cylinder axis for ~ 2 seconds. At this time, it breaks into two spots which continue to propagate axially in the rest frame of the spiral with their widths remaining approximately that of the final spiral. At later times these spots propagate both axially and azimuthally and may also undergo further splitting and growth etc. This complicated development process continues for ~ 10 minutes when the different pieces connect to form a spiral. A similar process occurs without an external perturbation when interpenetrating spirals have gained sufficient strength to trigger a turbulent spot. This process may in part be explained by noting that the Reynolds stress in the turbulence must generate an azimuthal and axial backflow. This backflow then limits the azimuthal and axial extent of the turbulence. When the spiral is developing the Reynolds stress and the backflow are irregular and complicated until the spiral pattern is constructed. After the spiral is constructed the backflow also becomes regular in both the axial and azimuthal direction as shown by the measurements of Coles and Van Atta[2]. Thus the simple spiral shape is a result of the circular geometry where the feedback of the Reynolds stress through the backflow on to itself is very regular and direct. The splitting of the spot is due to the internal structure of the turbulence which is made of progressive finite amplitude waves that propagate either up or down, on average, along the axis[3,4]. The initial spot which consists of both types of waves separates into two spots one which consist of primarily up moving waves and the other of primarily down moving waves. The resulting spiral may be one of two helicities where one helicity consists of mostly up moving waves and the other helicity consists of mostly down moving waves. In either helicity one end of the spiral is a emitter of waves and the other end an absorber of waves.

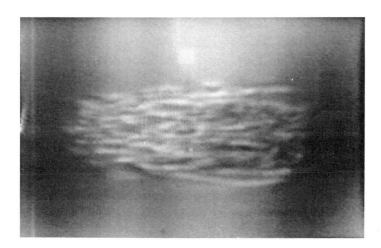

Figure 1

A growing turbulence spot in circular Couette flow ≈ 0.5 seconds after it has been triggered by a perturbation at the center of the spot. $R_o = -3000$ and $R_i = 750$. Video image courtesy of the Nonlinear Dynamics Laboratory at the Department of Physics, Ohio State University.

PLANE COUETTE FLOW

Circular Couette flow has been an attractive system for studying fluid instabilities because of its simple symmetric geometry which leads to an exact solution of the Navier-Stokes equations for low Reynolds numbers. An even simpler example of fluid flow is that of the linear velocity profile which results from the relative motion of two infinite parallel walls at given distance apart. This simple flow, known as plane Couette flow, is also an exact solution of the governing nonlinear equations. Unlike circular Couette flow which is subject to a linear centrifugal instability, this simple velocity profile is believed to be linearly stable at all R^7 ($R = \frac{Uh}{\nu}$ where $U =$ the speed of either wall, $h = d/2$ is half the gap and ν is the kinematic viscosity of the fluid). Because a small perturbation cannot destabilize this flow, this is an ideal system to study subcritical instability. In particular we have observed that this flow yields a direct transition to turbulence for finite amplitude perturbations. It has recently been reported that the turbulent state invades the laminar state with a constant velocity that increases with R after a turbulent spot has been initiated by a localized perturbation[11]. We report here the observation of an intermittent turbulent state (coexisting laminar and turbulent regions) within a range of R, $R \approx 350$ to $R \approx 450$.

Methods

Several different geometries have been tried in attempts to make plane Couette

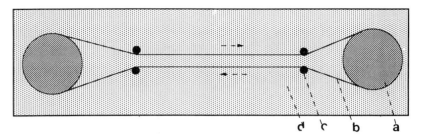

Figure 2

Schematic drawing which illustrates a horizontal cross section of the geometry of the plane Couette flow apparatus. Two large cylinders (a), two pairs of small cylinders (c) and a clear plastic (b) are all placed in a water tank (d) and an external motor drives the system.

flow[12,13]. Our system utilizes a geometry which results in no mean flow, i.e. the walls move in opposite directions with the same velocity. This has the advantages of simplifying the apparatus and increasing the time available to observe turbulent spots. In this geometry the two vertically oriented parallel walls move in a horizontal direction. We define these walls by using an endless transparent film belt (363.0cm long, 25.4cm wide, and 0.16mm thick), two pairs of small rotating plexiglass cylinders, and two large rotating plexiglass cylinders as shown schematically in Figure 2. The endless belt is driven, through friction, by one of the two large cylinders (12.00cm in diameter) which are placed at opposite ends of the system. The cylinder which drives the belt is mechanically coupled to an A. C. motor which drives the system. The two "infinite" parallel planes are defined by guiding the endless belt through two pairs of smaller cylinders 7.00mm apart all of which are parallel to the large cylinders as shown in Figure 2. Using h as a length scale, the transverse aspect ratio Γ_y (dimensionless width of the channel) is $\Gamma_y = 72.7$ and the longitudinal aspect ratio is (dimensionless length of the channel) $\Gamma_x \approx 340$ (this is approximate because the length of the straight section of the channel is changed when the belt tension is changed). Near the entrance, one may expect that the velocity will require some length (or equivalently some time) to diffuse from the wall into the domain before a linear profile is established. Assuming a Blasius boundary layer growth we obtain the formula $l \approx R \frac{h}{25}$ so that 90% of the flow domain is fully developed at $R = 500$. We have chosen to use a transparent belt so that optical methods in data acquisition, such as Laser Doppler Velocimetry (LDV) and flow visualization with image processing, may be used. The entire assembly is placed in a glass tank which is filled with water, our working fluid, which has a viscosity of $\nu \approx 0.01$ stokes. We use an A. C. motor and a 100:1 planetary gear box to drive the system. This speed reduction allows us to operate at a Reynolds number between 100 and 500 while maintaining sufficient power to drive the system and a high resolution in the speed.

We visualize the laminar and turbulent regions by seeding the flow with a dilute solution of Merck Iriodin 100 Silver Pearl. Iriodin consists of thin and flat reflective mica platelets (diameter of the large reflecting side is between $\approx 10\mu m$ and $\approx 60\mu m$). These platelets or flakes align, on average, with the stream planes of the flow. They also respond very quickly to any local change in the flow field giving an almost instantaneous change in the light reflectance field whenever there is a change in the velocity field[10]. This makes Iriodin an excellent indicator of turbulence on all scales[14]. When the flow is seeded with Iriodin, turbulence in indicated by a relatively rapid fluctuation in the reflected light field whereas in the laminar flow the light reflectance is steady. The light reflectance field may then recorded using a video camera and subsequently analyzed. We have also visualized the velocity field in two dimensions by seeding the flow with a white ceramic powder (Pyroceram 7575, 100 Mesh) and illuminating the flow with a Argon laser light sheet. We have visualized both the vertical and horizontal cross sections by changing the orientation of the light sheet. The light scattered by the particles is reflected by a mirror into a camera for recording. We have verified that the velocity profile is linear by following the particles in the streamwise cross section using a video camera. This is accomplished by selecting a line of the the video image (512 pixels in length) which is parallel to the wall and digitizing it at a specified sampling rate. As a particle travels along this line, the light reflected from this particle is recorded in time and its velocity subsequently determined. Since there are usually several particles in a given line at a given time, the error of this velocity measurement may also be estimated. The specific line that is digitized may then be moved to a different cross stream position and the process repeated. In this way we have found the velocity profile and Figure 3 shows the results of these measurements.

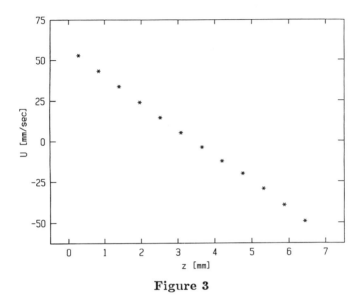

Figure 3

Plot of the streamwise velocity U as a function of the cross stream position z at $R = 200$. These measurements, which show a linear velocity profile, were made by illuminating particles with a laser light sheet in the streamwise cross section and following their displacement.

Preliminary Results

The threshold of instabily in subcritical bifurcations generally depends on the strength of the perturbation. To find the minimum Reynolds number for the transition to turbulence in these preliminary experiments, we have used a perturbation of sufficient strength to induce a turbulent spot. This is done by perturbing the flow with a turbulent jet injected into the laminar flow from the spanwise direction (from above the channel). By injecting $\approx 10 ml$ of water in ≈ 5 seconds through a hole $2mm$ in diameter, we found that we could induced turbulent spots of the order of the size of the spots observed in the intermittent state.

After a turbulent spot is initiated at a given R, as described above, the subsequent evolution is visualized by shining light through the transparent belt and observing the light reflected from the Iriodin in the flow. This procedure allows almost all of the flow across and along the belt to be visualized. As expected, we have found that the turbulent spots in this system are not advected with a mean velocity.

This allows us to easily observe a turbulent spot for long a time compared to other systems where intermittent turbulence is known to exist.

We have performed preliminary experiments where we have searched for the minimum R for the subcritical transition to turbulence. In these experiments we have induced a turbulent spot in laminar flow at a given R and observed if the flow relaxes back to laminar flow. We have seen in these preliminary observations that below $R \approx 380$ the turbulent spots generated in laminar flow relax within an observation time of 1 minutes. At $R \approx 380$ these induced spots are self sustaining for greater then 3 minutes. Figure 4a and 4b shows two images of the intermittent turbulent state at $R = 380$. We emphasize that these numbers are only the result of several trials and that we do not have yet measurements of the deviation, if any, in the minimum R or the relaxation time or the survival time of the spots near the threshold.

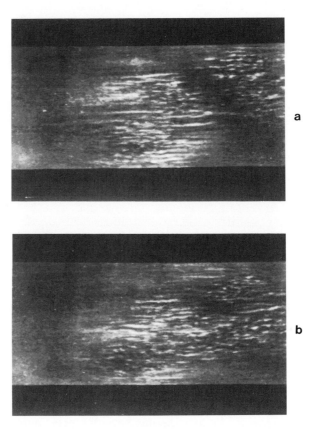

Figure 4

a) Image of the visualized flow field ≈ 1.5 minutes after the a large perturbation has been applied at $R = 380$. Spatial variations in the light reflectance can seen in the in the turbulent region. This region also exhibit reflectance fluctuations in time. b) Image of the visualized flow ≈ 0.5 minute later than the image in a. The shape and position of the separate turbulent and laminar regions have evolved slowly in time (much more slowly than the fluctuations in the turbulence).

As mentioned above, the turbulent regions in this system are not advected with a mean velocity. They do, however, have a slow spatial and temporal evolution. Above the threshold for sustained intermittent turbulence we have observed that the position and shape of the turbulent regions evolve much more slowly than the time for light reflectance fluctuations within the turbulent regions. Figure 4 illustrates how the shape and position of the turbulent regions change in time. The Figure also illustrates that the turbulent patches generally do not have straight vertical (spanwise) edges. The edges of the turbulent patches are usually either curved or straight and inclined to the horizontal. Throughout the turbulent region horizontal streaks can be seen, especially when the light source is placed nearly vertically above the flow domain (see Figure 4). We have also directly observed vortical structures by illuminating the gap in the spanwise (vertical) direction (i.e. the spanwise cross section) and watching Pyroceram particles. As the turbulent patches evolve they grow, split, and merge. A large spot may grow while smaller pieces split away from it and these smaller spots almost alway relaminarize.

At R higher than the threshold ($R \approx 380$) the state of intermittent turbulence appears to last an arbitrarily long time (as long as one would wish to turn the belt). The faction of the total space which is turbulent appears to increase as R is increased. At $R \approx 450$ nearly all of the space is turbulent. At these higher R values the turbulence may also form straight inclined patches (in both senses) across the channel, a shape similar to that of Spiral Turbulence. This configuration, however, is as temporary as any other because it soon splits in half with both halves changing position and shape often by the top half growing up and the bottom half growing down. When a turbulent patch grows up or down the eddies (the mostly horizontal streaks) also appear to travel in the same direction.

CONCLUSION

We have constructed a plane Couette flow apparatus and observed a intermittent turbulent state which occurs in a range of R (R between 380 and 450). Below this R, the laminar state is stable to finite amplitude perturbations and above this range the entire flow domain is turbulent. For $380 < R < 450$ and after a perturbation, the dynamical regime shows a fluctuating mixture of laminar and turbulent domains which is reminiscent of spatiotemporal intermittency. We have not seen any large scale coherent structure as exists in circular Couette flow. This implies that bounded geometry in circular Couette flow is influential in the Spiral and V-shaped patterns. This also implies that the intermittent state in circular Couette flow is probably not greatly influenced by the linear centrifugal instability.

The laminar flow to intermittent turbulence transition in both circular Couette flow and plane Couette flow are subcritical in nature. Subcritical bifurcations in flows with infinite domains can lead to either expanding or contracting turbulent domains in laminar flow[8]. This argument requires the existence of a potential functional in which there are at least two minima corresponding to a metastable (a local minimum) and a stable state (the global minimum, corresponding to the laminar state). The case of spiral turbulence, as seen in the laboratory frame of reference, shows that there are complicating effects in this bounded system since there is both an expanding turbulent front at the leading edge of the spiral and a contracting turbulent front at the trailing edge. The case of plane Couette flow shows that, at least in the range of R close to threshold that we have investigated above, there is a complicated mixture of laminar and turbulent flow and not a simple expanding or contracting turbulent front. Work is currently under progress to study this transition.

We wish to thank M. Bonetti, G. Balzer, and Y. Pomeau for stimulating discussions P. Hede and B. Ozenda for their technical assistance. We wish to especially thank Martial Labouise for technical assistance and helping to debug the apparatus.

REFERENCES

1. D. J. Tritton, "Physical Fluid Dynamics," Van Nostrand Reinhold, New York (1977). M. Van Dyke, "An Album of Fluid Motion," Parabolic, Stanford (1982).

2. D. Coles, Transition in circular Couette flow, J. Fluid Mech. 21:385 (1965). Also see C. W. Van Atta, Exploratory measurements in spiral turbulence, J. Fluid Mech. 25:495 (1966) and D. Coles, C. W. Van Atta, Progress report on a digital experiment in spiral turbulence, AIAA J. 4:1969 (1966).
3. J. J. Hegseth, C. D. Andereck, F. Hayot, and Y. Pomeau, Spriral Turbulence and Phase Dynamics, Phys. Rev. Lett. 62:257 (1989).
4. J. J. Hegseth, C. D. Andereck, F. Hayot, and Y. Pomeau, Spiral turbulence: development and steady state properties, Euro. J. Mech. B (Fluids) no. 2 - Suppl. 10:221 (1991).
5. M. Gad-El-Hak, R. F. Blackwelder, and J. J. Riley, On the growth of turbulent regions in laminar boundary layers, J. Fluid Mech. 110:73 (1981).
6. D. R. Carlson, S. E. Widnall, and M. F. Peeters, A flow visualization study of transition in plane Poiseuille flow, J. Fluid Mech. 121:487 (1983).
7. S. A. Orszag, and L. Kells, Transition to turbulence in plane Poiseuille and Couette flow, J. Fluid Mech. 96:159 (1980).
8. Y. Pomeau, Front motion, metastability, and subcritical bifurcations in hydrodynamics, Physica D, 23:1 (1986).
9. C. D. Andereck, S. S. Liu, and H. L. Swinney, Flow regimes in a circular Couette system with independently rotating cylinders, J. Fluid Mech. 164:155 (1986).
10. J. Hegseth, "Spatiotemporal Patterns In Flow Between Two Independently Rotating Cylinders," Ph.D. Thesis, Ohio State University, (1990).
11. N. Tillmark, and P. H. Alfredsson, Transition in plane Couette flow - the first flow visualization experiments, Bull. Am. Phys. Soc. 35:2248 (1990).
12. M. Aydin and J. Leutheusser, Novel experimental facility for the study of plane Couette flow, Rev. Sci. Instrum. 50:1362 (1979).
13. P. H. Alfredsson, and N. Tillmark, private communication (1990).
14. K. W. Schwarz, Evidence for organized small scale structure in fully developed turbulence, Phys. Rev. Lett. 64:415 (1990).

ON THE ECKHAUS AND THE BENJAMIN-FEIR INSTABILTY IN THE VICINITY OF A TRICRITICAL POINT

Helmut R. Brand

Fachbereich Physik
Universität Essen
D 4300 Essen 1, West Germany

INTRODUCTION

For pattern forming systems with spatial variations in one direction the Eckhaus instability, the instability against spatial modulations of the pattern, has been studied for a stationary forward bifurcation for a number of systems theoretically[1-5] and experimentally[6,7] and a characteristic band of wavelengths, which are linearly stable against spatial modulations, emerges in all these cases. Here we perform the analogous analysis at the tricritical point, at which the coefficient of the cubic term in the envelope equation vanishes and where we assume saturation to quintic order. It is found that the band of Eckhaus stable wavelengths is a factor of $(1.5)^{\frac{1}{2}}$ wider in this case. We also summarize the results obtained for the Eckhaus instability recently in the case of a weakly inverted stationary bifurcation[8], where the envelope equation takes the form

$$\dot{A} = \epsilon A + \gamma A_{xx} + \beta |A|^2 A - \delta |A|^4 A \tag{1}$$

and where ϵ, β, γ, and δ are real with $\beta > 0$ for a weakly inverted bifurcation.

The modulational instability in one dimension for oscillatory instabilities was discussed for forward bifurcations in particular by Benjamin and Feir[9] for the purely dispersive case and by Newell[3,10-12], who also incorporated the effects of dissipation. Benjamin, Feir, and Newell showed that spatially homogeneous solutions of the envelope equation can become unstable provided the coefficients of the envelope equation satisfy a certain inequality.

Here we give the analogous analysis for the case of a vanishing real part of the cubic coefficient in the envelope equation and summarize the results obtained near a weakly inverted Hopf bifurcation[8]. These questions are naturally becoming more important, as there has been recently a growing interest in spatial patterns near weakly inverted bifurcations[13-20].

This contribution is organized as follows. In the next section we present the analysis for the Eckhaus instability and for the Benjamin-Feir-Newell instability at the tricritical point. In the third section we give a brief summary of the results of ref.8 on the Eckhaus and the Benjamin-Feir-Newell instability for a weakly inverted stationary and Hopf bifurcation, respectively, followed in the last section by a comparison with the literature and a perspective.

THE ECKHAUS INSTABILITY AT A TRICRITICAL POINT

At the tricritical point the envelope equation for a stationary instability reads

$$\dot{A} = \epsilon A + \gamma A_{xx} - \delta |A|^4 A \qquad (2)$$

since the coefficient β of the cubic term in eq.(1) vanishes at the tricritical point. Eq.(2) has time-independent finite amplitude plane wave solutions

$$A_{st} = \left(\frac{\epsilon - \gamma k^2}{\delta}\right)^{\frac{1}{4}} e^{ikx} \qquad (3)$$

In order to test for linear stability of these solutions against compressions and dilations, we follow ref.2. Looking for solutions of the form $A = A_{st} + \hat{A}(x,t)$ we have upon linearization in $\hat{A}$

$$\frac{\partial \hat{A}}{\partial t} = \epsilon \hat{A} + \gamma \hat{A}_{xx} - 3\delta \hat{A}|A_{st}|^4 - 2\delta |A_{st}|^2 A_{st}^2 \hat{A}^* e^{2ikx} \qquad (4)$$

Eq.(4) has solutions of the form

$$\hat{A}(x,t) = \hat{A}_1 e^{\lambda t} e^{i(k+M)x} + \hat{A}_2 e^{\lambda t} e^{i(k-M)x} \qquad (5)$$

which gives

$$\lambda_{1,2} = -(\gamma(k^2 + M^2) + 2\epsilon - 3\gamma k^2) \pm 2\sqrt{\gamma^2 k^2 M^2 + (\epsilon - \gamma k^2)^2} \qquad (6)$$

having a relative maximum for

$$M^2 = k^2 - \frac{(\epsilon - \gamma k^2)^2}{\gamma^2 k^2} \qquad (7)$$

and

$$\lambda = \frac{(\epsilon - 2\gamma k^2)^2}{\gamma k^2} \qquad (8)$$

From inspection of eqs.(7,8) we read off that the finite amplitude solutions (eq.(3)) are Eckhaus stable at the tricritical point for

$$k^2 < \frac{\epsilon}{2\gamma} \qquad (9)$$

For stationary forward bifurcations with saturation to cubic order one obtains instead of eq.(9)

$$k^2 < \frac{\epsilon}{3\gamma} \qquad (10)$$

From this we conclude that the Eckhaus stable band at a tricritical point is by a factor of $(1.5)^{\frac{1}{2}}$ wider than for a forward bifurcation, a prediction one might be able to test experimentally.

Next we consider the analogue of the Benjamin-Feir-Newell instability at a tricritical point for an oscillatory instability. In general the imaginary part of the cubic coefficient will still be nonvanishing at such a point. Correspondingly the envelope equation reads

$$\dot{A} = \epsilon A + (\gamma_r + i\gamma_i)A_{xx} - i\beta_i|A|^2 A - (\delta_r + i\delta_i)|A|^4 A \qquad (11)$$

Eq.(11) allows for spatially finite amplitude equations of the form

$$A = (\epsilon/\delta_r)^{\frac{1}{4}} e^{i\omega t} \quad (12)$$

with

$$\omega = -(\epsilon/\delta_r)^{\frac{1}{2}}(\beta_i + \delta_i(\epsilon/\delta_r)^{\frac{1}{2}}) \quad (13)$$

Proceeding along the lines of refs.10-12 we find that the spatially homogeneous solutions (eqs.(12) and (13)) are linearly unstable against space-dependent perturbations if

$$\gamma_i(\beta_i + 2\delta_i(\epsilon/\delta_r)^{\frac{1}{2}}) + 2\gamma_r(\delta_r\epsilon)^{\frac{1}{2}} < 0 \quad (14)$$

Inequality (14) is the analog of the Benjamin-Feir-Newell instability criterion for a tricritical point. In the special case that the imaginary part of the cubic coefficient vanishes simultaneously with the real part, we have

$$\gamma_i \delta_i + \gamma_r \delta_r < 0 \quad (15)$$

an expression, which looks very similar to that of the forward bifurcation: one only needs to replace the coefficients of the cubic terms (β_r, β_i) by those of the quintic terms (δ_r, δ_i).

We close this section by pointing out, that in the appendix of ref. 21 the Swift-Hohenberg equation with the cubic nonlinearity $|A|^2 A$ replaced by the more general nonlinearity $|A|^{2n} A$ was studied and the Eckhaus boundary was given by restricting the stability analysis to the vicinity of onset. While focussing the analysis on the cases $n \to 0$ and $n \to \infty$ relevant for their considerations, the result by Paap and Kramer coincides with inequality (9) for $n = 2$.

THE ECKHAUS AND THE BENJAMIN-FEIR-NEWELL INSTABILITY FOR A WEAKLY INVERTED BIFURCATION

In this section we summarize the results of our detailed recent analysis[8] concerning the Eckhaus instability near a weakly inverted stationary bifurcation and the Benjamin-Feir-Newell instability near a weakly inverted oscillatory instability. For more details we refer the reader to our original article (ref.8).

The envelope equation for a weakly inverted stationary instability takes the form

$$\dot{A} = \epsilon A + \gamma A_{xx} + \beta |A|^2 A - \delta |A|^4 A \quad (16)$$

where ϵ, γ, β and δ are real and where β and δ are assumed to be positive to guarantee that the bifurcation is inverted and saturates to quintic order. The diffusion coefficient γ is assumed to be positive.

Eq.(16) allows time-independent finite amplitude, plane wave solutions of the form

$$A_{st} = A_0 e^{ikx} \quad (17)$$

with

$$|A_0|^2_{1,2} = \frac{\beta}{(2\delta)} \left(1 \pm \sqrt{1 + \frac{4(\epsilon - \gamma k^2)\delta}{\beta^2}}\right) \quad (18)$$

In eq.(18) the upper (+) sign in the bracket refers to the stable branch of solutions, which exists for a band of k values

$$\gamma k^2 < \frac{\beta^2}{4\delta} + \epsilon \quad (19)$$

whereas the lower (-) sign is associated with the unstable branch of solutions.

The finite amplitude solutions (eqs.(17, 18) with positive sign) are Eckhaus stable, provided[8]

$$k^2 < \frac{\epsilon}{2\gamma} + \frac{3\beta^2}{32\gamma\delta}\left(1 + \sqrt{1 + \frac{32\epsilon\delta}{9\beta^2}}\right) \qquad (20)$$

For a weakly inverted oscillatory instability one has for spatial variations in one dimension the envelope equation[22,23,15-19]

$$\dot{A} = \epsilon A + \gamma A_{xx} - \beta |A|^2 A - \delta |A|^4 A \qquad (21)$$

where γ, β, and δ are complex and thus of the form $\alpha = \alpha_r + i\alpha_i$ and where $\beta_r < 0$ and $\delta_r > 0$ to guarantee that the bifurcation is inverted and saturates to quintic order. Furthermore we have discarded nonlinear gradient terms in writing down eq.(21).

Eq.(21) admits spatially homogeneous solutions of finite amplitude

$$A_h = A_0 e^{i\omega t} \qquad (22)$$

where

$$|A_0|^2_{1,2} = -\frac{\beta_r}{2\delta_r}\left(1 \pm \sqrt{1 + \frac{4\epsilon\delta_r}{\beta_r^2}}\right) \qquad (23)$$

and

$$\omega = -|A_0|^2(\beta_i + \delta_i |A_0|^2) \qquad (24)$$

and where the (+) sign in the bracket of eq.(23) corresponds to the stable branch, whereas the (-) sign is associated with the unstable branch.

Our analysis[8] shows that the spatially homogenous solution (22) - (24) is linearly unstable against spatially inhomogeneous perturbations of the type considered here if

$$\gamma_i(\beta_i + 2\delta_i |A_0|^2) + \gamma_r(\beta_r + 2\delta_r |A_0|^2) < 0 \qquad (25)$$

Eq.(25) represents the analog of the Benjamin-Feir-Newell criterion for a weakly inverted bifurcation.

In ref.8 we have also shown that both criteria, namely that for the Eckhaus instability (inequality (20)) and that for the Benjanim-Feir-Newell instability (inequality (25)) can be generalized easily to the case of envelope equations for anisotropic systems with spatial variations in two dimensions. The simplicity of this generalization can be traced back to the fact that the envelope equations for uniaxially anisotropic systems contain only second order derivative terms whereas for isotropic systems fourth order derivatives also come into play. Anisotropic systems with spatial variations in two dimensions include for example Rayleigh-Bénard and electro- convection in nematic liquid crystals. The reader is referred to ref.8 for the details of the analysis.

COMPARISON WITH THE LITERATURE AND PERSPECTIVE

The Eckhaus instability near a weakly inverted stationary bifurcation seems to have been discussed first in ref.8 and we have summarized the results of this analysis in the last section. The underlying assumption in writing down the corresponding envelope equation (eq.(16)) is that the scaling for the slow spatial modulations is unchanged compared to that for a forward bifurcation, meaning that the spatial derivatives in the envelope equation still enter to second order. This assumption parallels that of having the same structure for the gradient energy when going from a second order to a weakly first order phase transition in equilibrium systems.

The Benjamin-Feir-Newell instability near a weakly inverted oscillatory instability, which has been discussed in detail in ref.8 and summarized in the last section, has also been looked at for spatial modulations in one dimension in ref.24. Close inspection of their eq.(1.5) shows that eq.(25) is contained implicitly in their result. We note, however, that due to the rescaling of the equation in ref.24 the transition from a weakly inverted to a forward oscillatory instability cannot be studied straightforwardly in their notation, but can be read off from our result (eq.(25)) immediately.

Finally we would like to point out that the vicinity of a tricritical point with vanishing real part for the cubic coefficient in the amplitude equation has been investigated in ref.25. In this study an expansion in two parameters is performed, namely a) in the distance from the critical value of the bifurcation parameter and b) in the distance between the critical wavevector and the wavevector for which the real part of the cubic coefficient vanishes. A large variety of phenomena is found, including mechanisms for strong pattern selection. The authors stress in particular the importance of the nonlinear gradient terms as they have been discussed first near a weakly inverted Hopf bifurcation in refs.22 and 23.

From the results of all these studies it appears most important to have well-controlled experiments in the vicinity of a tricritical point crossing over for example from a forward to a weakly inverted oscillatory bifurcation. From the results of such experiments and their comparison with the predictions made it should be possible to evaluate the range of applicability of the approaches suggested. One candidate for such an experimental system seems to be the electrohydrodynamic instability in nematic liquid crystals, which was found to show a weakly inverted Hopf bifurcation in thin cells very recently[26], but which is showing a forward bifurcation in thicker samples[27].

ACKNOWLEDGEMENTS

It is a pleasure to thank Guenter Ahlers, Lorenz Kramer and Alan C. Newell for stimulating discussions.

Financial support by the Deutsche Forschungsgemeinschaft is gratefully acknowledged.

REFERENCES

1. W. Eckhaus, *Studies in nonlinear stability theory* Springer, Berlin 1965
2. A.C. Newell and J.A. Whitehead, *J.Fluid Mech.* **38**, 279 (1969)
3. A.C. Newell and J.A. Whitehead, p.284 in *Proceedings of the IUTAM Symposium on instability in continuous systems*, H. Leipholz, Ed., Springer, N.Y. 1971
4. L. Kramer and W. Zimmermann, *Physica D - Nonlinear Phenomena* **16**, 221 (1985)
5. S. Kogelman and R.C. DiPrima, *Phys.Fl.* **13**, 1 (1970)
6. M. Lowe and J.P. Gollub, *Phys.Rev.Lett.* **55**, 2575 (1986)
7. G. Ahlers, D.S. Cannell, M.A. Dominguez-Lerma, and R. Heinrichs, *Physica D - Nonlinear Phenomena* **23**, 202 (1986)
8. H.R. Brand and R.J. Deissler, submitted for publication
9. T.B. Benjamin and J.E. Feir, *J.Fluid Mech.* **27**, 417 (1966)
10. A.C. Newell, p.157, in *Lectures in Appl.Math.* M. Kac., Ed. **15** (1974)
11. A.C. Newell, p.244 in *Proceedings of an international workshop on synergetics*, H. Haken, Ed., Springer, N.Y. 1979

12 A.C. Newell, *Solitons in Mathematics and Physics*, Society for Industrial and Applied Mathematics, Philadelphia, 1985
13 V. Steinberg and H.R. Brand, *Phys.Rev.* **A30**, 3366 (1984)
14 H.R. Brand and B.J.A. Zielinska, *Phys.Rev.Lett.* **57**, 3167 (1986)
15 R.J. Deissler and H.R. Brand, *Phys.Lett.* **A130**, 293 (1988)
16 O. Thual and S. Fauve, *J.Phys.(Paris)* **49**, 1829 (1988)
17 H.R. Brand and R.J. Deissler, *Phys.Rev.Lett.* **63**, 2801 (1989)
18 W. van Saarloos and P.C. Hohenberg, *Phys.Rev.Lett.* **64**, 749 (1990)
19 R.J. Deissler and H.R. Brand, *Phys.Lett.* **A146**, 252 (1990)
20 R.J. Deissler and H.R. Brand, *Phys.Rev.* **A44**, yyyy (1991)
21 H.G. Paap and L. Kramer, *J.Phys.(Paris)* **48**, 1471 (1987)
22 H.R. Brand, P.S. Lomdahl and A.C. Newell, *Phys.Lett.* **A118**, 67 (1986)
23 H.R. Brand, P.S. Lomdahl and A.C. Newell, *Physica D - Nonlinear Phenomena* **23**, 345 (1986)
24 B.A. Malomed and A.A. Nepomnyashchy, *Phys.Rev.* **A42**, 6009 (1990)
25 W. Eckhaus and G. Iooss, *Physica* **D39**, 124 (1989)
26 I. Rehberg, S. Rasenat, M. de la Torre, W. Schöpf, F. Hörner, G.Ahlers, and H.R. Brand, *Phys.Rev.Lett.* **67**, xxxx (1991)
27 I. Rehberg, S. Rasenat, J. Fineberg, M. de la Torre, and V. Steinberg, *Phys.Rev.Lett.* **61**, 2449 (1988)

PHASE VS. DEFECT TURBULENCE IN THE 1D COMPLEX GINZBURG-LANDAU EQUATION

A. Pumir[1,2], B.I. Shraiman[3], W. van Saarloos[3,4]
P.C. Hohenberg[3], H. Chaté[3,5] and M. Holen[3,6]

[1] LASSP, Cornell University, Ithaca, NY 14853
[2] LPS, Ecole Normale Supérieure, F-75231 Paris, France
[3] AT&T Bell Laboratories, Murray Hill, NJ 07974
[4] Instituut-Lorenz, University of Leiden, The Netherlands
[5] SPSRM, CEN Saclay, F-91191 Gif-sur-Yvette, France
[6] Dept. of Math., Princeton University, Princeton, NJ 08544

The understanding of the dynamical properties of spatially extended systems remains a major challenge, in spite of some recent progress. In this paper, we will restrict ourselves to the very rich variety of cellular instabilities observed in hydrodynamic systems, such as the Taylor-Couette flow at moderate Taylor numbers, as discussed in this book. This is only one of many possible situations where the space-time dynamics leads to non trivial properties[1].

One of the simplest and of the most useful model equations describing this class of phenomena is the well-known supercritical, complex Ginzburg-Landau equation in 1 dimension[2]:

$$\partial_t A = A + (1 + ic_1)\partial_{x^2} A - (1 - ic_3)|A|^2 A \tag{1}$$

where c_1, c_3 are two real coefficients, which can be computed or measured experimentally in a variety of cases, like in the case of the oscillatory instability in a low Prandtl number fluid[3]. The function A is complex and describes typically the envelope of a wave or cellular pattern. We restrict ourselves here to the 1-dimensional case. By analogy with other fields of statistical physics, one expects that space dimension will play a crucial role. Dimension one was chosen here since it is conceptually the simplest, and because of the wealth of experimental data now available on quasi-one dimensional systems[4,5,6,7].

It is easy to check that Eq.(1) has a continuous family of travelling waves (TW) solutions: $A = A_0 e^{i(qx - \omega t)}$ with $A_0 = \sqrt{1 - q^2}$ and $\omega = -c_3 + q^2(c_1 + c_3)$, provided $q^2 < 1$. It is also possible to study the linear stability of these waves. It turns out that some of these travelling wave solutions are unstable with respect to the phase perturbations. The condition for instability reads[2]:

$$q^2 \geq \frac{(1 - c_1 c_3)}{(3 - c_1 c_3 + 2c_3^2)} \tag{2}$$

It results from this analysis that when $(1 - c_1 c_3) \leq 0$, all the TW solutions are unstable with respect to an instability, similar to the Eckhaus instability, which is refered to here as the Benjamin-Feir [8] instability.

Theoretically, Kuramoto[9] predicted that in the Benjamin-Feir regime, for $(1 - c_1 c_3) \leq 0$ and $|(1 - c_1 c_3)|$ small enough the system could be described by an unstable wave, modulated in phase :

$$A(x,t) = a(x,t)e^{-i(\omega t + \phi(x,t))} \tag{3}$$

where the phase $\phi(x,t)$ is a function slowly depending on space and time, and $a(x,t)$ is a real function that remains close to 1 and follows adiabatically the vatiations of the gradient of ϕ. The phase ϕ of the system has a dynamics on its own, which is described by the following equation :

$$\partial_t \phi + \phi \partial_x \phi + \partial_{x^2} \phi + \partial_{x^4} \phi = 0 \tag{4}$$

Such a regime, describable by the phase of the system only has been called 'phase turbulence'. Only recently has this regime been observed by Sakaguchi [10]. It has also been recently shown [7] that in fact, the Benjamin-Feir instability for waves of finite wavenumbers, may lead to a supercritical bifurcation leading to a phase modulated wavetrain, the number of waves in the structure remaining constant. Such a behavior has been observed numerically for values of c_1 greater than 1 only. Only recently has this regime been observed numerically [10]. The pure phase description only holds in the absence of phase slips, or space-time dislocations, that is points where $a(x,t)$ goes through 0. When such a defect occurs in the system, the winding number, defined for a finite periodic system of size L by :

$$\nu = \frac{1}{2\pi} \int_0^L \partial_x \phi dx \tag{5}$$

jumps by an integer value. Of course, it is conceivable that other regimes may exist as well. In this paper, we present and discuss some selected results of a careful study of the parameter space, described by the two coefficients (c_1, c_3). Because we are interested in spatially extended systems, we will be considering very large systems. Ideally, we would like to understand the properties of infinitely large systems, the so called thermodynamic limit. Note that complementary results have been obtained by Bretherton and Spiegel[11], in a different part of the parameter space.

Finally, we would like to point out a possible connection between dynamical regimes and the theory of critical phenomena, in statistical mechanics. Bunimovitch and Sinai[12] have shown that for a very simplified system of coupled oscillators, the space time dynamics can be described by a 2 dimensional Ising model. It would be extremely interesting to know if this connection still exists and can be exploited for understanding more realistic models of spatio-temporal dynamics.

Numerically we have used very standard pseudo spectral methods, with typically 512 or 1024 collocation points, and either a fourth order Runge-Kutta algorithm or a leap-frog scheme (second order in time).

We report here some results in the range $c_1 \gtrsim 1$. Note that this is the regime where the Benjamin-Feir instability has been observed to lead to a supercritical bifurcation[7]. Our results can be summarized as follows :

Close to the Benjamin-Feir curve in the (c_1, c_3) plane, we found a phase regime. Very different dynamical behavior can be observed, according to the mean wavenumber in the system. At 0 (or small enough) wavenumber, the dynamics of the phase exhibits space-time fluctuations, very reminiscent of the behavior predicted by the Kuramoto-Sivashinsky equation, Eq. (4). In particular, the spatial spectrum of the fluctuations of the phase gradients is flat for small wavenumbers, has a maximum at a finite wavenumber, and decays exponentially at large k, see Fig. 1. It is hard to make very quantitative comparisons with the predictions of Eq. (4), since the Kuramoto-Sivashinsky equation is valid very close to the Benjamin-Feir limit of instability, where the dynamics is very slow, and the gradients are very small. A detailed numerical study would require enormous computer resources. When a mean wavenumber is present in the system (that is, for a system of finite length, L, at non zero winding number), we found that in general, the system tends to a modulated TW state, as in the Benjamin-Feir stable case. Starting from arbitrary conditions, one ends up with a state exhibiting a wave train, made of compressions or dilations of the structure, and propagating at a finite velocity. The 'pulses' of wavenumbers have no particular spatial order. An example is shown in Figure 2. It is rather surprising to find temporally ordered states in a region of parameter space where one would naively expect a completely disordered dynamics ! This behavior is probably related to the results of Kawahara[13], who considered an equation very close the the phase equation relevant to our case, derived in Ref. 7.

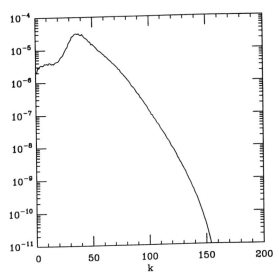

Fig.1. Spectrum of the spacial gradients of the phase. The parameters are $c_1 = 3.5; c_3 = 2.$; corresponding to a phase regime. The mean wavenumber in the system is 0.

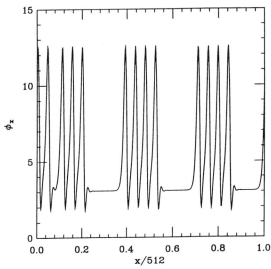

Fig.2. The wavenumber modulation of a steadily propagating solution obtained for $c_1 = 1.5$ and $c_3 = .83$, and a winding number equal to 5 ($L = 512$).

Further away from the Benjamin-Feir limit in the parameter space (c_1, c_3), a qualitative change occurs. Space time defects, where the amplitude of the function A goes through 0, appear in the system, and the phase description is no longer valid. We have tried to characterize the nature of the transition between the two regimes - phase and defect -. In the defect regime, it is possible to define a spatio-temporal density of defects, ρ. Practically, one records every space time defect in a system of size L, and for a time T; the space time density of defects is defined by the limit of the total number of defects, divided by LT. In our numerical study, we never found any significant difference between a system of length $L = 512$, and a system of length $L = 1024$. Measurements of the space time density of defects reveal that at fixed c_1, when c_3 is decreased, the defects disappear smoothly. Our data is compatible with the following formula:

$$\rho \sim (c_3 - c_3^*)^2 \qquad (6)$$

where c_3^* depends upon c_1 (see Fig. 3). For $c_1 \gtrsim 1.7$, a unique 'attractor' was found. Namely, the system evolved either to the phase or to the defect regime, depending on whether c_3 is larger or smaller than c_3^*. Interestingly, for $c_1 \lesssim 1.7$, a region of the parameter space was found where either a phase or a defect regime could be found.

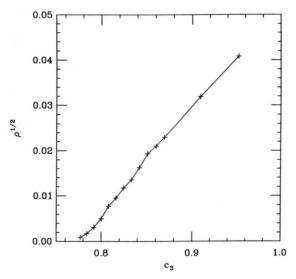

Fig.3. The square root of the density of defects, ρ, vs. c_3 for $c_1 = 3.5$

In order to study a fluctuating system, it is often fruitful to study its correlations. Here we have studied the following functions:

$$c(r) = < e^{i(\phi(r,t) - \phi(0,t))} > \qquad (7a)$$

and

$$g(r) = < \nabla\phi(r,t)\nabla\phi(0,t) > \qquad (7b)$$

where the brackets $< ... >$ refer to a time average. In the defect regime, one finds that $c(r)$ decays exponentially with r, like $c(r) \sim e^{-r/\xi}$. The correlation length, ξ strongly increases when the transition is approached, see Fig. 4. Obviously to numerically extract a diverging correlation length requires system sizes much larger than the correlation length. Finite size effects in our data can be seen, when systems of sizes 512 and 1024 are compared. Our data for $c_1 = 2.5$ and $c_1 = 3.5$ is compatible with the following expression for the growth of the correlation length near a transition $(c_3 \gtrsim c_3^*)$:

$$\xi \sim (c_3 - c_3^*)^{-1} \qquad (8)$$

with the same notation as before. However, in 1 dimension, there are compelling reasons to believe that there is no divergence of the correlation length ξ at c_3^*, but rather that ξ saturates at a large value, corresponding to the correlation length of c in the phase regime. Unfortunately, the correlation lengths one finds there are much larger than the sizes of the systems one can realistically study. We will elaborate on this rather subtle question elsewhere[14].

To conclude, we have studied two rather different types of behavior in the 1-dimensional complex Ginzburg-Landau equation : a phase regime, as predicted by Kuramoto and a defect regime. They exist in the $c_1 \gtrsim 1$ part of the parameter space. We have demonstrated the existence of a well defined transition between these two different states. In any case, our results, although still incomplete, show that the simplest (and possibly the most popular !) equation used in the theory of hydrodynamic instabilities generates a rich variety of dynamical behaviors. It would be interesting to observe these regimes in a real experimental system.

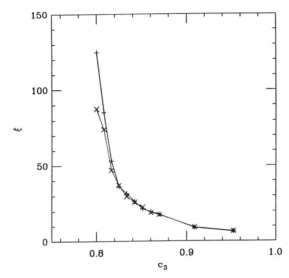

Fig.4. The correlation length ξ vs. c_3 for $c_1 = 3.5$. The $\times$ symbols corresponds to a system of size $L = 512$, the $+$ symbols to a system of size $L = 1024$. These finite size effects are not observable on the density of defects.

Acknowledgements

A. Pumir has been partially supported by a grant from the Air Force Office of Scientific Research AFOSR-91-011.

References

1. P.C. Hohenberg and B.I. Shraiman, Physica D **37**, 109 (1989)
2. Y. Kuramoto, 'Chemical oscillations, waves and turbulence', Springer Verlag, 1984
3. V. Croquette and H. Williams, Phys. Rev. A**39**, 2765 (1989)
4. P. Kolodner, D. Bensimon and C. Surko, Phys. Rev. Lett. **60**, 1723 (1988)
5. A. Simon, J. Bechhoefer, A. Libchaber, Phys. Rev. Lett. **61**, 2574 (1988)
6. M. Rabaud, S. Michalland, Y. Couder, Phys. Rev. Lett. **64**, 184 (1990)
7. B. Janiaud, A. Pumir, D. Bensimon, V. Croquette, H. Richter and L. Kramer, preprint (1990)
8. T.B. Benjamin and J.E. Feir, J. Fluid Mech. **27**, 417 (1966)
9. Y. Kuramoto, Prog. Theor. Phys. Suppl. **64**, 346 (1978)
10. H. Sakaguchi, Prog. Theor. Phys. **83**, 169 (1990) and **84**, 792 (1990)
11. C.S. Bretherton and E. A. Spiegel, Phys. Lett. **96A**, 152 (1983)
12. L. A. Bunimovitch and Ya. G. Sinai, Nonlinearity **1**, 491, (1988)
13. T. Kawahara, Phys. Rev. Lett. **51**, 381 (1983)
14. B.I. Shraiman, A. Pumir, W. van Saarloos, P.C. Hohenberg, H. Chaté and M. Holen , in preparation (1991)

DOUBLE EIGENVALUES AND THE FORMATION OF FLOW PATTERNS

Rita Meyer-Spasche

MPI für Plasmaphysik, D-W8046 Garching bei München, Germany

Abstract Numerically obtained solution diagrams for Taylor vortex flows look quite asymmetric. In this review of the nonlinear interaction of stationary 2-vortex solutions with stationary 4-vortex solutions it is emphasized that these asymmetries are due to the broken symmetry of the Taylor apparatus itself. To get global results on the mathematical structure of the solution set, more symmetric equations (Bénard problem, model problem) are also considered.

1. Introduction

Numerically obtained solution diagrams for Taylor vortex flows with periodic boundary conditions look quite asymmetric when they show all bifurcating branches ([11, 14, 25, 26, 27, 29, 33, 39, ...] and *Fig.1*). Asymmetry is not so obvious in the (measured or computed) plots of torque versus Reynolds number, in which the two simultaneously bifurcating flows, the $\lambda/2$-*shifted twins*, are projected onto each other. The asymmetries become most obvious in those parts of the solution diagrams where flows are unstable. To understand mathematically the structure of the solution set, it seems necessary to be interested in *all* stationary solutions in a certain parameter range, no matter wether they are stable or not. Stability is thus ignored here. It is treated by other authors; see [17, 33] and the references therein.

Because wavelength-halving bifurcations had been observed in deep-water waves [10], P. Saffman suggested that we compute solutions as a function of the wavelength λ, and we actually found such wavelength-halving bifurcations [27]. As explained below, they could be traced back to an intersection point of neutral curves. The fact that such doubly singular points can lead to secondary bifurcations and thus to the formation of new patterns has been observed in other problems as well (e.g. [2, 21], but [40]), and the structures of their solution sets are often quite similar to those described here, e.g. [13, 3, Fig.4.7 in 22]. Also, there are some other flows that look quite similar to unstable ones found by us, and some of them are even stable [4, 18, 31, 32, 35, 42].

In section 2 investigations on secondary bifurcations of Taylor vortex flows are reviewed, in section 3 the simpler, more symmetric narrow-gap limit (Bénard problem) is considered, and in section 4 we discuss a global, analytic result for model equations derived from the Bénard problem.

2. Secondary bifurcations of Taylor vortex solutions

The curve of neutral stability of Couette flow in the (Re, λ)-plane with minimum $(Re_{cr}, \lambda_{cr} \approx 2)$ intersects with its periodic repetition [27] with minimum $(Re_{cr}, 2\lambda_{cr} \approx 4)$ in a point (Re_{24}, λ_{24}); see Fig.2. On the first curve 2-vortex solutions bifurcate from Couette flow; on its periodic repetition 4-vortex flows bifurcate. On both neutral curves the Jacobian of the system is singular, and in (Re_{24}, λ_{24}) it is singular of order 2. Re_{24} is thus a double eigenvalue of the linearised system for $\lambda = \lambda_{24}$. It is called the $(2,4)$ *double eigenvalue* in the following.

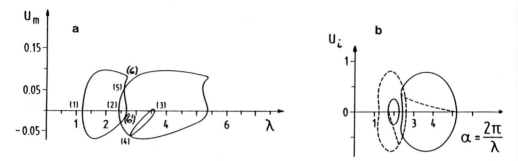

Fig.1. a) Variation of u_m, the radial velocity at the midpoint, with the period λ for $Re = \sqrt{1.5} Re_{cr}$, $\eta = .727$, $\mu = 0$. Points (1), (2), (3): Bifurcation of 2-vortex and 4-vortex flows from Couette flow ($u_m \equiv 0$). Points (4), (5): Wavelength-halving bifurcation points. (6) is a point on the basic (2,4)-fold, while (6') is the corresponding point on the branch of solutions shifted by $\lambda/2$. b) First (——) and second (- - -) harmonic in the midpoint versus wave number α for the same branches as a). The computations were performed with M = 31 inner radial grid points and N = 8 Fourier components [37].

In [27] we studied the structure of families of solutions with respect to continuous variation of both parameters λ and Re for $\eta = 0.727$, $\mu = 0$. Varying λ with fixed $Re = \sqrt{1.5} Re_{cr}$ (or $T = 1.5 T_{cr}$), we found that the 2-vortex solution bifurcating from Couette flow on the low-λ side of the neutral curve (point (1) in *Fig.1a*) continuously changes with increasing λ into a 4-vortex solution which passes a fold point (6) at $\lambda \approx 2.68$ and then undergoes a period-halving bifurcation at point (5), $\lambda \approx 2.59$. When we started with maximum λ at the other side of the neutral curve in point (3), we found essentially the same thing, but without fold. The period-halving bifurcation occurs at point (4), $\lambda \approx 2.8$. The wavelength-halving bifurcations were shown to be secondary bifurcations generated by the $(2,4)$ double eigenvalue, (Re_{24}, λ_{24}). Using perturbation methods, Andreichikov [1] discussed qualitatively what can happen in a neighborhood of (Re_{24}, λ_{24}) for arbitrary η and μ. In the case $\eta = 0.5$, $\mu = 0$ he gave a detailed analysis by numerically evaluating the formulas previously derived for the general case. With Fig.7 of [14] we repeated Fig.4 of [1] with adapted nomenclature. Meanwhile, Andreichikov's analysis has been confirmed by Tavener&Cliffe [39] for $\eta = 0.615$, $\mu = 0$. These authors give much more details about the various stages of the interaction. In Fig.1 of [37] we summarized their analysis by sketching the solution surface in a neighborhood of (Re_{24}, λ_{24}). Note that Schaeffer's analysis [34] is not applicable to the wide gap case for other than 'technical' reasons.

The fold (6), (6') mentioned before was missing in Andreichikov's analysis. The reason seems to be that it emanates from the curve of secondary bifurcations for a Reynolds umber slightly larger than Re_{24}; see [29] and the references therein. This fold seems to exist at least in the range $0.5 \leq \eta < 1$, $-0.2 \leq \mu \leq 0$ [29]. It is called the *basic* $(2,4)$-*fold* or the *basic* $(n, 2n)$-*fold* since there exist infinitely many periodic repetitions of it. Additionally, many more folds of the solution manifold exist. There is the *subcritical fold* which was detected for $\eta = 0.5$ [1, 14] and $\eta = 0.615$ [39] near (Re_{24}, λ_{24}), but which was not found for $\eta = 0.727$ and does not exist for the self-adjoint case $\eta \to 1$. A single solution branch with more than 20 fold points was found ($Re = 3.65 Re_{cr}$; $\eta = 0.727$ fixed, λ varying [28, Fig.3]). We also found harmonic bifurcations (period doubling, period tripling, ...) other than the periodic repetitions of those generated by the (2,4)-interaction [37]. Some of the additionally found secondary bifurcations could be traced back to other double eigenvalues, e.g. (Re_{26}, λ_{26}) [37, Fig.8]. Furthermore, there is strong evidence that tertiary branches are generated through coalescence of two different points of secondary bifurcation, etc [37, Fig.11]. Coalescence of three or more points of secondary bifurcation has to be expected, and this will give rise to new patterns of solutions. A glimpse of the zoo to be expected (free stagnation points, separatrices, triangular vortices, ...) is afforded by [37, Figs. 5, 7, 11, 12]. The neighborhoods of the $(2,6)$ and of the $(4,6)$ double eigenvalues were systematically investigated in [39].

Excellent agreement was found for those parameter values for which results were obtained with both codes. For instance, we found transcritical bifurcations with a very small shift from a pitchfork (bifurcation in $\lambda \approx 2.65$, fold point in $\lambda \approx 2.63$) [37, 39]. Tavener&Cliffe used a finite-element discretization, while we used Fourier sums/finite differences. The discretization errors are thus of a different nature, and the unfoldings caused by numerical approximation are thus likely to be different. In [37] we reviewed previous investigations on the discretization error of the TAYPERIO code and added new investigations which focus on the discretization error near fold points.

Where do the lines of secondary bifurcation go which start in the (n, m) double

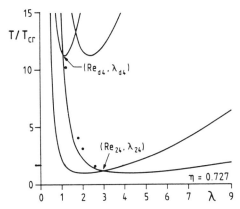

Fig.2. Bifurcation curves of the 2-vortex and double-vortex flows in the $(Re^2/Re_{cr}^2, \lambda)$ plane, together with their first periodic repetitions. • : harmonic bifurcation as described in the text.

eigenvalues? In [29] we speculated that one line of secondary bifurcations connects the (2,4) double eigenvalue with (Re_{d4}, λ_{d4}), i.e. with the intersection point of the marginal curve for 4-vortex flows with the marginal curve for *double vortex flows* (2 vortices in the r-direction though the outer cylinder is at rest), see the flow in *Fig.3* with $T = 11.24\ T_{cr}$ ($Re = 3.35\ Re_{cr}$) and $\lambda = 1.1$. This is supported by the computations reported in [29] and illustrated in *Fig.3*. The point ($T = 10.13\ T_{cr}$, $\lambda = 1.2$) produced one of the points marked by '•' in *Fig.2*; point (5) in *Fig.1a* produced another one. The conjecture that the curve of secondary bifurcations passes through the point (Re_{d4}, λ_{d4}) is supported by the investigations on the Bénard model problem reported below.

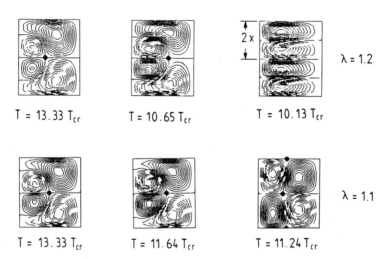

Fig.3. Streamline plots of flows computed with fixed wavelength λ and varying Reynolds number Re. While there is harmonic bifurcation for $\lambda = 1.2$, there is a continuous change into a double-vortex flow and bifurcation to Couette flow for $\lambda = 1.1$. • : free stagnation point. M = 31, N = 12 [29].

3. The symmetric limiting case

The branches shown in *Fig.1a* are quite asymmetric with respect to the axis $u_m \equiv 0$. This is due to the fact that the Taylor experiment itself has a broken symmetry. As already described, there is a change of structure on both 2-vortex branches bifurcating from Couette flow in points (1) and (3) of *Fig.1a*. Both branches end in a wavelength-halving bifurcation on a branch of the primary 4-vortex solutions. The change of structure takes place continuously, thus indicating that the fold points (6), (6') originate from perturbed bifurcations. As has been pointed out by Busse [6, 7], the unperturbed bifurcation cannot be found in the Taylor problem, but is provided by the limiting case $\eta \to 1$, $\mu \to 1$. In this limit our equations become mathematically equivalent to the Boussinesq approximation describing the 2D Bénard problem for $Pr = 1$. This was already conjectured by G.I. Taylor. It is now well-known [16, 9, 12, 8, 38, 5, 30]. To actually perform the limit, η and μ have to be changed simultaneously, according to eqs (161), (162) and (191) in Chandrasekhar's book [24, 8].

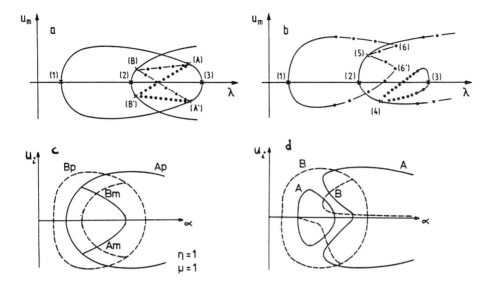

Fig.4. Breaking of the midplane symmetry changes the bifurcation diagram.
a), b): Radial velocity u_m versus wavelength λ. *a)* Symmetric case (Bénard problem). Isola (1) − (3) − (1): pure-mode 2-roll solutions bifurcating from the trivial solution in points (1) and (3) on the neutral curve. The upper and the lower branches contain solutions shifted by $\lambda/2$. The periodic repetition of this loop bifurcates in point (2). Isola $(B) - (A) - (B') - (A') - (B)$: transition solutions. These are mixed-mode solutions with 2 large and 2 small rolls, interpolating between 2-roll and 4-roll solutions. *b)* Asymmetric case (Taylor problem). Bifurcation points $(A), (A')$ become perturbed and generate the fold points $(6), (6')$. Because of the asymmetry with respect to the midplane, 'inward transition solutions' (— • — , [37, Fig.6c, d]) and 'outward transition solutions' (• • • , [37, Fig.7b]) are clearly different.
c), d): Sketch of first (——) and second (- - -) Fourier components of the solution branches shown in *a)* and *b)* versus wave number $\alpha = 2\pi/\lambda$, after Busse [7]. *c)* Bénard problem; the pure-mode solutions A_p (2-roll) and B_p (4-roll) bifurcate from the trivial solution. The mixed-mode solutions exist in the α-interval where A_m and B_m are both $\neq 0$. In the bifurcation points $(B), (B')$ $A_m = 0$ and $B_m = B_p$. In the bifurcation points $(A), (A')$ $B_m = 0$ and $A_m = A_p$. *d)* Perturbed Bénard problem. The similarity between *d)* and *Fig.1b)* is clearly seen. Since our computations treat a wide-gap problem, the perturbations are quite strong.

We thus now consider convection layers between two infinitely extended parallel plates (periodic and stress-free boundary conditions). The analysis of the bifurcation pattern in a neighborhood of the corresponding intersection of the neutral curves for 2-roll and 4-roll solutions ([6, 7, 20, 24] and references therein) yields, briefly: There are a pure-mode 2-roll solution and its periodic repetition, a pure-mode 4-roll solution, and then two antisymmetric mixed-mode solutions called *transition solutions*, and also non-antisymmetric *mean component solutions*. The two branches of transition solutions and their $\lambda/2$-*shifted twins* bifurcate from the 2-roll solution and from the 4-roll solution and thus connect them as a closed loop. The situation is very similar to the one described in [2, Fig.4] : There are two branches of primary solutions, orthogonal to each other, and a closed loop of secondary solutions which interpolate between them, having growing and decaying components in both directions. In *Fig.4a* we sketch this situation.

The basic convection rolls are symmetric with respect to the midplane between the heated plates [7]. This symmetry is lost in the Taylor problem because of the curvature of the two cylinders and because of their different speeds. In *Fig.4b* we sketch how the solution diagram is perturbed for $\eta, \mu \neq 1$. This sketch shows the change of structure where we typically found it in our computations (*never* in the fold point). *Fig.4c* (Bénard problem) and *d* (perturbed Bénard problem) demonstrate this in a different projection. They are adapted from Busse [7]. *Fig.1a* and *b* result from computations on the Taylor problem [37] and confirm [6, 7]. This shows that the asymmetries of the solution branches shown in *Figs.1a, 4b* with respect to the λ-axis are mostly caused by the perturbation of the second Fourier component.

Why do the bifurcations on the 2-roll branch become perturbed, while those on the 4-roll branch persist? The answer can be found from symmetry considerations as presented in [39, 41] : In the Bénard problem, all four bifurcations $(A), (A'), (B), (B')$ are symmetry-breaking and thus of the pitchfork type. In the Taylor problem, the 2-vortex solution lost its symmetry because of the asymmetries of the apparatus. It is invariant with respect to the identity operator only. (The mirror symmetry with repect to $z \equiv 0$ is a property of all interacting solutions and thus does not count here.) The codimension-zero singularity is thus a quadratic turning point – bifurcations $(A), (A')$ turn into the fold points $(6), (6')$. The 4-vortex solution is still invariant with respect to translation by $\lambda/2$. The loss of this invariance is equivalent to loss of reflectional symmetry at a distance $\lambda/4$ from the midplane $z \equiv 0$. The codimension-zero singularity is thus a symmetry-breaking pitchfork bifurcation: Bifurcation points $(B), (B')$ turn into bifurcation points (4) and (5). Note the distortion in the parameter space.

4. A global result on secondary bifurcations

As explained before, we found [29] strong evidence that a line of secondary bifurcations on the 4-vortex solutions connects two different double eigenvalues, i.e. two different intersection points of primary bifurcation curves in the (Re, λ)-plane for fixed η and μ; see *Figs.2, 3*. It seemed desirable to confirm this by analytic investigation of the Taylor problem, but this was found to be too complicated. The most severe difficulty was: The two interacting and pattern-forming double eigenvalues seem to stay distinct for all parameter values. The perturbation method [1] is thus not applicable. Furthermore, the many perturbed and transcritical bifurcations [37] make it difficult to understand bifurcation patterns. As we have seen in section 3, some of the fold points on Taylor vortex branches correspond to bifurcation points on convection roll branches. An analytic treatment of the Bénard problem is thus less complicated and promises more insight than the corresponding investigations on the less perfect Taylor problem.

There are several investigations of secondary bifurcations in the Bénard problem [36, 20, 6]. Essentially, these authors use a Fourier decomposition and an amplitude expansion to 3rd order in $(\alpha - \alpha_0)^{1/2}$ and $(Ray - Ray_0)^{1/2}$, where α_0 and Ray_0 are those values of the wave number and Rayleigh number, respectively, for which the critical curves of two different numbers of rolls intersect. The results are thus restricted to a neighborhood of (Ray_0, α_0), i.e. to a neighborhood of one of the interacting double eigenvalues. The two interacting double eigenvalues cannot merge for any parameter value in the problem. This can easily be seen from the simple formulas for the intersecting curves [20, 24].

To get analytic results which are valid in a domain containing both pattern-forming double eigenvalues, we [24] constructed a model problem by chopping off the Fourier expansions of the stream function and temperature deviation after the second mode (Galerkin method). Perturbation arguments are then used to decide which other terms of the equations can be neglected. This process turns the PDEs of the Boussinesq approximation into a nonlinear algebraic system of 7 equations which can be solved explicitly for the pure-mode solutions. Furthermore, an implicit formula can be derived for the loci of secondary bifurcations on the 4-roll solutions. In the limiting case $Pr \to \infty$, this formula can easily be evaluated, and it clearly shows that the computed curve meets both double eigenvalues. The conjecture [29] is thus confirmed for the model problem.

Is this result applicable to the Bénard problem? A comparison of the equations solved in [24] with those treated in [20] for their case $k = 1$ shows that we lost the terms relating to the third horizontal, second vertical mode. This is likely to produce a wrong curvature of the center manifold at the double eigenvalue and thus to give a wrong inclination of the curve of secondary bifurcations in the point corresponding to (Re_{24}, λ_{24}) [19].

5. References

1. I.P. Andreichikov (1977): Branching of secondary modes in the flow between rotating cylinders. Fluid dynamics, translated from Izv. Akad. Nauk SSSR, Mekh.Zhidk. Gaza, No. 1, 47 – 53
2. L. Bauer, H.B. Keller, E.L. Reiss (1975): Multiple eigenvalues lead to secondary bifurcation. SIAM Review **17**, 101 - 122
3. M.J. Bennett, R.A. Brown, L.H. Ungar (1987) Nonlinear interactions of interface structures of differing wavelength in directional solidification. In [15]
4. P.J. Blennerhassett, P. Hall (1979) Centrifugal instabilities of circumferential flows in finite cylinders: linear theory. Proc. R. Soc. Lond. A **365**, 191 – 207
5. K. Bühler (1984): Der Einfluß einer Grundströmung auf das Einsetzen thermischer Instabilitäten in horizontalen Fluidschichten und die Analogie zum Taylor-Problem. Strömungsmechanik und Strömungsmaschinen **34**, 67 – 76
6. F.H. Busse, A.C. Or (1986): Subharmonic and asymmetric convection rolls. ZAMP **37**, 608–623
7. F.H. Busse (1987): Transition to asymmetric convection rolls. [23], 18 – 26
8. F.H. Busse, private communications (1978, 1989)
9. S. Chandrasekhar (1961): **Hydrodynamic and Hydromagnetic Stability.** Oxford University Press
10. B. Chen, P. Saffman (1979) Steady gravity-capillary waves on deep water, I, II. Stud. Appl. Math. **60** (1979) 183 – 210; **62** (1980) 95 – 111
11. N. Dinar, H.B. Keller (1989): Computations of Taylor vortex flows using multigrid continuation methods. Lecture Notes in Engineering **43**, Springer Verlag Berlin
12. P.G. Drazin, W.H. Reid (1981): **Hydrodynamic Stability.** Cambridge University Press, Cambridge
13. J.C. Eilbeck (1987) Numerical study of bifurcation in a reaction-diffusion model using pseudo-spectral and path-following methods. In [23]
14. G. Frank, R. Meyer-Spasche (1981) Computation of transitions in Taylor vortex flows. ZAMP **32**, 710 – 720
15. W. Güttinger, G. Dangelmayr, eds. (1987) **The Physics of Structure Formation,** Proc. Tübingen 1986, Springer Series in Synergetics
16. H. Jeffreys (1928): Some cases of instability in fluid motion. Proc. Roy. Soc. A **118**, 195 – 208. Also **Collected Papers** (1975), vol. 4, 469 – 84. London: Gordon & Breach

17. C.A. Jones (1981) Nonlinear Taylor vortices and their stability. JFM **102**, 249 – 261
18. H.S. Kheshgi, L.E. Scriven (1985): Viscous flow through a rotating square channel. Phys Fluids **28**, 2968 – 2979
19. E. Knobloch, private communication (1991)
20. E. Knobloch, J. Guckenheimer (1983): Convective transitions induced by a varying aspect ratio. Phys. Rev. A **27**, 408 – 417
21. G.A. Kriegsmann, E.L. Reiss (1978): New magnetohydrodynamic equilibria by secondary bifurcation. Phys. Fluids **21**, 258 – 264
22. M. Kubiček, M. Marek (1983): **Computational Methods in Bifurcation Theory and Dissipative Structures.** Springer Verlag, New York, Berlin, Heidelberg, Tokyo
23. Küpper, Seydel, Troger, eds. (1987) **Bifurcation: Analysis, Algorithms, Applications** Proc. Dortmund 1986, ISNM 79, Birkhäuser Verlag
24. D. Lortz, R. Meyer-Spasche, P. Petroff: A global analysis of secondary bifurcations in the Bénard problem and the relationship between the Bénard and Taylor problems. Proc. Oberwolfach 1989, Methoden und Verfahren der mathematischen Physik **37**, Verlag Peter Lang, Frankfurt a.M., 1991
25. K.A. Meyer (1967) Time-dependent numerical study of Taylor-vortex flow. Phys Fluids **10**, 1874 – 1879
26. R. Meyer-Spasche, H.B. Keller (1980): Computations of the axisymmetric flow between rotating cylinders. J. Comp. Phys. **35**, 100 – 109
27. R. Meyer-Spasche, H.B. Keller (1985): Some bifurcation diagrams for Taylor vortex flows. Phys. Fluids **28**, 1248 – 1252
28. R. Meyer-Spasche, M. Wagner (1987): Steady axisymmetric Taylor vortex flows with free stagnation points of the poloidal flow. [23] 213 – 221
29. R. Meyer-Spasche, M. Wagner (1987): The basic (n, 2n)-fold of steady axisymmetric Taylor vortex flows. [15], 166 – 178
30. M. Nagata (1986): Bifurcations in Couette flow between almost corrotating cylinders. J. Fluid Mech. **169**, 229 – 250
31. M.D. Neary, K.D. Stepanoff (1987) Shear-layer-driven transition in a rectangular cavity. Phys. Fluids **30**, 2936
32. M. Neveling, D. Lang, P. Haug, W. Güttinger, G. Dangelmayr (1987) Interaction of stationary modes in systems with 2 and 3 spatial degrees of freedom. [15], 153 - 165
33. H. Riecke, H.-G. Paap (1986) Stability and wave vector restriction of axisymmetric Taylor vortex flows. Phys. Rev. A **33**, 547
34. D. Schaeffer: Qualitative analysis of a model for boundary effects in the Taylor problem, Math. Proc. Camb. Phil. Soc. **87**, 307 – 337 (1980)
35. G. Schrauf (1986) The first instability in spherical Taylor-Couette flow. J. Fluid Mech **166**, 287 - 303
36. L. A. Segel (1962): The non-linear interaction of two disturbances in the thermal convection problem. J. Fluid Mech. **14**, 97 – 114
37. H. Specht, M. Wagner, R. Meyer-Spasche (1989): Interactions of secondary branches of Taylor vortex solutions. ZAMM **69**, 339 – 352 (FU-Preprint Nr. A-88-15, 1988)
38. J.T. Stuart (1986): Taylor-vortex flow: a dynamical system. SIAM Rev. **28**, p. 315 – 342
39. S.J. Tavener, K.A. Cliffe: Primary flow exchange mechanisms in Taylor-Couette flow applying non-flux boundary conditions. (1987) to be published
40. L. Turyn (1986): Bifurcation without mixed mode solutions. Contemporary Mathematics **56**, 335 – 341
41. B. Werner (1984) Regular systems for bifurcation points with underlying symmetries. Proc. Dortmund 1983, (Küpper, Mittelmann, Weber, eds.), ISNM 70, Birkhäuser Verlag.
42. M. Wimmer (1988): Viscous flows and instabilities near rotating bodies. Prog. Aerospace Sci **25**, 43 – 103

THE EFFECT OF TRHOUGHFLOW ON RAYLEIGH BERNARD CONVECTIVE ROLLS

H. W. Müller[1], M. Lücke[1], and M. Kamps[2]

[1]Institut für Theoretische Physik, Universität des Saarlandes
D–6600 Saarbrücken, F.R.G.
[2]Stabsstelle Supercomputing, Forschungszentrum Jülich
D–5170 Jülich, F.R.G.

We investigate Rayleigh–Benard convection in the presence of a horizontal Poiseuille flow transversal to the convective roll chain. Using a 1d amplitude equation and a 2d numerical simulation of the basic field equations, we study how different boundary conditions at the inlet and outlet of the channel affect nonlinear convection. If convection is suppressed near the cell apertures, spatially localized traveling wave states appear with a uniquely selected bulk wavelength. For convectively unstable parameters the system becomes very sensitive to perturbations and noise driven convection occurs. Phase pinning boundary conditions lead to stationary roll patterns with a space dependent wavelength decreasing downstream. Strengthening the throughflow causes local Eckhaus instabilities which finally generate a transition to propagating rolls.

I. INTRODUCTION

From the theoretical point of view the properties of nonlinear convection in heated shear flows are hardly examined, so that the knowledge is mainly based on experiments[1-7] and 2–dimensional numerical simulations[8-13] of the hydrodynamic field equations. Convection in broad channels heated from below usually appears in the form of stationary rolls aligned parallel to the throughflow (longitudinal rolls). Traveling transversal rolls (axes perpendicular to the shear flow) have been detected in narrow channels, however, for small flow rates only. This is a result of two competing mechanisms: Perturbations in the form of longitudinal rolls are preferred by the throughflow, whereas traveling transversal disturbances are favoured by the influence of the lateral channel sidewalls[14,15]. Luijkx et al.[4] reported transversal traveling roll patterns which did not fill the whole length of the channel. Such spatially localized structures have also been observed in numerical simulations[11-13]. Similar patterns also appear in the Taylor–Couette system with an axial throughflow[16-18].

Here we investigate transversal convective structures and in particular the question how the boundary conditions (b. c.) at the inlet and outlet of the channel affect the nonlinear convection structure. Our investigation is based on 2d computer simulations of the full field equations as well as a 1d amplitude equation. A summary of the results has been published earlier[12].

II. METHODS OF INVESTIGATION

We consider a horizontal layer of an incompressible Boussinesq fluid between two rigid perfectly heat conducting plates at $z=0$ and $z=1$. The fluid is heated from below and a lateral pressure gradient drives the throughflow in x–direction. In the absence of lateral boundaries the basic conductive state is described by a linear temperature profile $T_{cond}=T_0+Ra(1-z)$ and a plane Poiseuille velocity $U(z)=6\sigma Re\ z(1-z)e_x$.

Here $Ra = \alpha g d^3 \Delta T/(\kappa \nu)$ is the Rayleigh number given in terms of the thermal expansion coefficient α, gravitational constant g, layer thickness d, thermal diffusivity κ, kinematic viscosity ν, and the temperature difference between the plates ΔT. We scaled lengths by d, times by d^2/κ and temperature by $\kappa\nu/(\alpha g d^3)$. The second control parameter is the Reynolds number $Re = \bar{U}d/\nu$ being proportional to the vertically averaged flow velocity $\bar{U}$. The Prandtl number $\sigma = \nu/\kappa$ is a material parameter of the working fluid. The vertical component w of the hydrodynamic velocity field is used as order parameter of the system.

In a laterally unbounded layer linear stability analysis predicts[19] stationary longitudinal convection rolls to grow first above threshold. However, in ducts with a small aspect ratio in y direction sidewall forcing dominates[20,21] and the instability occurs in form of traveling transversal rolls if the flow rate Re is weak enough[14,15]. We adopt here a 2d description in the x–z plane (perpendicular to the roll axes) taking the convective fields of temperature θ and velocity $\mathbf{u} = (u,0,w)$ to depend on x,z,t only. Since the reflexion symmetry $x \to -x$ of the system is conserved under simultaneous reversal of the flow direction (Re $\to$ −Re) the critical Rayleigh number Ra_c as well as the wave number k_c are even functions of the Reynolds number. In Fig. 1 we show the stability threshold $\epsilon_c(Re)$ of the reduced Rayleigh number

$$\epsilon = Ra/Ra_c^0 - 1 \tag{2.1}$$

where $Ra_c^0 = 1707.76$ is the critical Rayleigh number without flow. For the same symmetry reasons the oscillation frequency at onset, ω_c, is an odd function of the flow rate growing in lowest order proportional to Re.

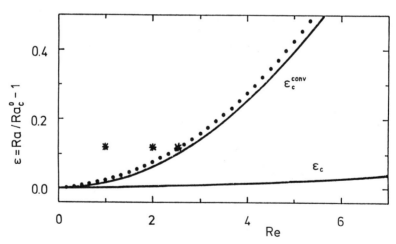

Fig. 1. The basic conductive state is stable below the threshold of linear stability ϵ_c, convectively unstable for $\epsilon_c < \epsilon < \epsilon_c^{conv}$ and absolutely unstable for $\epsilon > \epsilon_c^{conv}$. For convection suppressing inlet and outlet boundary conditions and in the absence of a continuous source of perturbations permanent convection is only possible in the absolutely unstable region. Stars indicate parameter combinations of Fig. 2. If ϵ_c^{conv} is approached form the left the streamwise growth length ℓ of the roll pattern diverges. At the dotted line $\ell = 10$.

A. Amplitude equation

Close above threshold the hydrodynamic fields $F = (u,w,\theta)$ have the form of a traveling wave

$$F(x,z,t) = A(x,t) \hat{F}(z) e^{i(k_c x - \omega_c t)} + \text{c.c.} \qquad (2.2)$$

where $\hat{F} = (\hat{u}, \hat{w}, \hat{\theta})$ is the complex eigenvector of the linear stability problem. The amplitude A controls saturation and dynamics of the order parameter. It is governed by the complex Ginzburg–Landau equation

$$\tau_0(\partial_t + v_g \partial_x)A = [\mu(1+ic_0) + \xi_0^2(1+ic_1)\partial_x^2 - \gamma(1+ic_2)|A|^2]A . \qquad (2.3)$$

The bifurcation parameter

$$\mu = \frac{Ra}{Ra_c(Re)} - 1 = \frac{\epsilon - \epsilon_c(Re)}{1 + \epsilon_c(Re)} \qquad (2.4)$$

measures the distance between Rayleigh number and bifurcation threshold $Ra_c(Re)$. The coefficients τ_0, ξ_0^2, γ contain corrections $\sim Re^2$ to their $Re=0$ values. The group velocity v_g and the imaginary parts c_i are odd functions in Re, they increase in lowest order proportional to the flow rate.

In experimental channels of finite length the instability of the basic conductive state results from localized rather than extended perturbations. Applying the concept of absolute and convective instability[22] to eq. (2.3) one finds that the conductive state, $A \equiv 0$, is convectively "unstable" if $\epsilon_c < \epsilon < \epsilon_c^{conv}$ and absolutely unstable if $\epsilon > \epsilon_c^{conv}$. The borderline ϵ_c^{conv} (Fig. 1) between the two subregions is given by

$$\epsilon_c^{conv} = \epsilon_c + (1+\epsilon_c)\tau_0^2 v_g^2 / 4\xi_0^2(1+c_1^2). \qquad (2.5)$$

If the control parameters ϵ and Re are such that the basic state, $A \equiv 0$, is convectively unstable localized initial disturbances are carried away by the flow and rolls cannot grow globally. If, however, the conductive state is absolutely unstable perturbations grow and expand also upstream until nonlinear saturation occurs.

B. Numerical simulation

We compare results obtained by the amplitude equation (2.3) with 2 d numerical solutions of the hydrodynamic field equations. We use a finite difference MAC algorithm[23] which expresses spatial derivatives by central differences and applies a forward Euler–step for the time integration. At each time step pressure and velocity fields are iteratively adapted to each other with a variant[24] of the SOLA–code[25]. A main objective of our study was the influence of different boundary conditions at the inlet and outlet of the channel on the nonlinear convective structure. In order to investigate this question we simulated convection in a channel of length $\Gamma=25$. Since information is transported downstream by the shear flow the entrance b.c. is of special importance for the convective behavior within the bulk.

III. NONLINEAR CONVECTION

A. Convection suppressing boundary conditions

As a first set of b.c. the basic conductive field profiles $T_{cond}(z)$ and $U(z)$ have been enforced at the apertures of the channel. For the amplitude equation the corresponding boundary conditions are $A = 0$ at $x = 0, \Gamma$. These b.c. are motivated by experiments of Luijkx et al.[4] where a porous plug at both ends of the duct has been used. In Fig. 2 we show representative examples of our results. Thin lines are numerical solutions of the full hydrodynamic equations for the vertical velocity field $w(x,z=0.5)$. The thick envelope is the final–state solution of the amplitude equation which has the form $A(x,t) = B(x)e^{i\omega t}$. The stationary envelope $|B(x)|$ increases over a distance ℓ to half of its saturation value while the roll pattern propagates downstream. Within the bulk of this localized convective state the roll size is spatially uniform as indicated by the solid squares in Fig. 2. The selected wave number is a unique function of ϵ, Re, and the Prandtl number σ. It is independent of the channel length and history of the system. The upstream front of the envelope is pushed more and more towards the outlet of the channel if the flow rate is increased. This behavior is in qualitative agreement with earlier results[4,11]. A quantitative investigation of the length $\ell(\text{Re}, \epsilon)$ and the selected wave number $k(\text{Re}, \epsilon)$ can be found in refs. 12,13. Note that the (ϵ,Re) control parameter combinations used in Fig. 2 are within the absolutely unstable subregion (cf. stars in Fig. 1). There the left intensity front of the localized convective structure would move upstream in a laterally infinite system. The convection suppressing b. c. at $x = 0$, however, pins the front. Entering into the convectively unstable subregion at ϵ_c^{conv} by increasing Re or decreasing ϵ an initially present convective structure is "blown" out of the system – left and right fronts move downstream. Permanent convection therefore is not possible for $\epsilon_c < \epsilon < \epsilon_c^{conv}$.

B. Noise sustained convection structures

Our foregoing discussion is based on the assumption that the undisturbed basic conductive state is realized at the channel entrance. Experiments, however, always show a certain level of perturbations that, at least in principle, have to be incorporated into the hydrodynamic equations for the convective fields. One might think of inlet turbulence but also other noise sources – spatially localized or extended – are

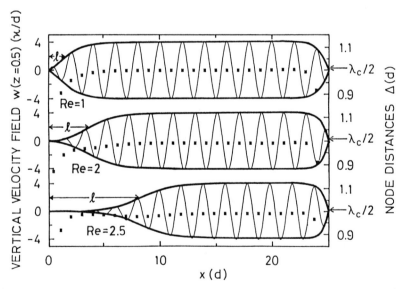

Fig. 2. Snapshots of the fully developed velocity field under convection suppressing inlet/outlet b.c. The control parameters ($\epsilon = 0.114$, Re as indicated) are such that the conductive state is absolutely unstable (c.f. stars in Fig. 1). Thin lines show the vertical velocity field $w(x,z=0.5)$ taken from computer simulations of the full hydrodynamic equations. Local node distances $\Delta(x)$ (squares) are smaller than half the critical wavelength λ_c. The structures propagate to the right (= downstream) under stationary envelopes. Thick lines are numerical solutions of the amplitude equation with $A = 0$ at inlet/outlet. The envelopes grow over a length ℓ to half of the bulk value.

conceiveable. Such a permanent source of disturbances turns out to be crucial if the system is convectively unstable.

Here we investigate a special variant of this problem by using time–dependent, uncorrelated random numbers of vanishing mean for the vertical velocity field $w(x = 0,z,t)$ at the inlet. We performed computer simulations with average noise amplitudes between 1 and 100% of the downstream convective saturation amplitude of w. For absolutely unstable parameters we found that the convection behavior is hardly affected in comparison to that one shown in Fig. 2. Only in a short entrance region of a few roll diameters stochastic variations of the fields reflect the random inlet b.c. With increasing distance from inlet the influence of the noise quickly decays and convection behaves as described in the previous section.

For convectively unstable parameters, however, the situation is completely different. Fig. 3 shows the simulation results for $\epsilon=0.114$ and Re=3. The intensity

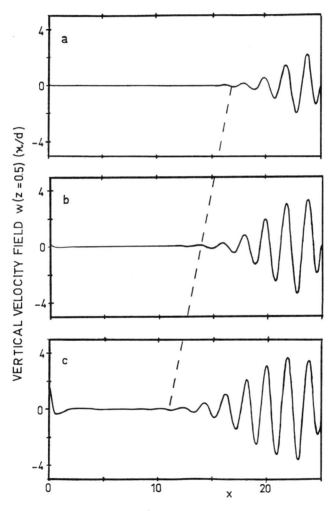

Fig. 3. Snapshots of the vertical velocity field $w(x, z = 0.5)$ under the influence of noise at the inlet (x = 0) of the channel. Parameters, $\epsilon = 0.114$ and Re = 3, are such that the basic conductive state is convectively unstable (c.f. Fig. 1). Noise amplitude is 1 % (a), 10 % (b), 100 % (c) of the saturation value. The roll pattern propagates downstream while its envelope is almost stationary except for small fluctuations. If the inlet noise source is removed convection dies out.

envelope of the propagating rolls is no longer stationary but fluctuates slightly around a mean profile and also the phase of the convective structure fluctuates. The temporal average of the growth length ℓ is a monotonously decreasing function of the noise strength (a: 1%; b: 10%; c: 100%). According to (2.3) a small inlet–noise amplitude, $A(x=0)$, of the critical convection mode grows initially exponentially

$$A(x) = A(x=0)\, e^{\kappa x} \tag{3.1a}$$

$$\kappa = \tau_0 v_g/(2\xi_0^2) - \sqrt{[\tau_0 v_g/(2\xi_0^2)]^2 - \mu/\xi_0^2} \tag{3.1b}$$

if one ignores the small imaginary coefficients in (2.3). Our numerical solutions of the full hydrodynamic field equations for noise sustained convective states do show such an exponential growth with a common exponent ($\kappa \simeq 0.57$ for the parameters of Fig. 3): The straight dashed line in Fig. 3 connects downstream positions of the same small field amplitude resulting from noise that increases linearly on a log– scale from Fig. 3a to 3c. Thus the system acts like an amplifier of disturbances with temporal and spatial delay. This property might be a useful tool for quantitative measurements of experimental noise. As soon as the noise source is removed, all convection structures die out after some time. Numerical simulations with the amplitude equation also show this dynamics if the homogeneous inlet b.c. is replaced by a noise term. This result is in agreement with earlier work of Deissler[26] who has investigated the influence of noise on convectively unstable systems.

C. Phase pinning boundary conditions

While we fixed in Sec. III. A the amplitude of the convective state to be zero at inlet/outlet we consider here a b. c. that pins the phase of the convective structure. Such a situation occurred in experiments of Pocheau et al[27] with azimuthally opposite flow in two halves of an annular container. Then a stationary deformed pattern appears at small Re while roll propagation sets in beyond a critical flow rate. Here we derive from the amplitude equation (2.3) a phase equation to describe this behavior in comparison with numerical simulations.

1. Derivation of the phase equation   First we introduce

$$\tilde{A}(x,t) = A(x,t)e^{-i\omega_c t} = r(x,t)e^{i\psi(x,t)} \tag{3.2}$$

to compensate the critical time dependence. The evolution equation for this amplitude directly follows from (2.3). Since deformed patterns only occur for very weak flow rates, we expand

$$r(x,t) = r^{(0)} + r^{(1)}(x,t)\mathrm{Re} + \mathcal{O}(\mathrm{Re}^2) \tag{3.3a}$$

$$\psi(x,t) = \psi^{(0)}(x) + \psi^{(1)}(x,t)\mathrm{Re} + \mathcal{O}(\mathrm{Re}^2). \tag{3.3b}$$

For Re = 0 the coefficients $v_g, c_0, c_1, c_2, \omega_c$ vanish so that

$$r^{(0)} = \sqrt{(\epsilon - \xi_0^2 q^2)/\gamma} \quad ; \quad \psi^{(0)} = qx. \tag{3.4}$$

This solution describes stationary rolls with homogeneous wave number $\bar{k} = k_c + q$. The first order in Re yields after adiabatic elimination of $\partial_x r^{(1)}/r^{(0)}$ the inhomogeneous phase diffusion equation

$$\partial_t \psi = D_\parallel \partial_x^2 \psi - (\omega_c + \Omega) \tag{3.5a}$$

with the diffusion constant

$$D_\parallel = \frac{\xi_0^2}{\tau_0} \frac{\epsilon - 3\xi_0^2 q^2}{\epsilon - \xi_0^2 q^2} \qquad (3.5b)$$

coming from the zeroth-order equation. The inhomogeneity

$$(\omega_c + \Omega) = \omega_c + v_g q + \frac{\xi_0^2}{\tau_0} q^2(c_1 - c_2) - \frac{\epsilon}{\tau_0}(c_0 - c_2) \qquad (3.5c)$$

increases in lowest order proportional to the flow rate Re. In deriving (3.5) we used $\mu = \epsilon + \mathcal{O}(\text{Re}^2)$. Obviously, the phase dynamics depends via $D_\parallel$ and Ω on the wave number $\bar{k} = k_c + q$ of the undisturbed roll pattern. The stationary solution of (3.5)

$$\psi(x) = \frac{1}{2} \frac{(\omega_c + \Omega)}{D_\parallel}(x - \Gamma)x + q x \qquad (3.6a)$$

describes a standing deformed roll structure. Its wave number

$$k(x) = k_c + q + \frac{(\omega_c + \Omega)}{D_\parallel}(x - \frac{\Gamma}{2}). \qquad (3.6b)$$

increases linearly in flow direction with a gradient $(\omega_c + \Omega)/D_\parallel$ proportional to Re. The integration constant in (3.5) has been chosen such that the undisturbed wave number $\bar{k} = k_c + q$ occurs at $x = \Gamma/2$ in the middle between inlet and outlet as in the experiments[27].

2. *Comparison with simulations*   In order to incorporate phase pinning into our simulations we used as inlet/outlet boundary conditions

$$\partial_x w = \partial_x T = 0 \quad \text{at } x = 0 \quad \text{and} \quad x = \Gamma \qquad (3.7)$$

by which maximal convective upflow or downflow is fixed at the apertures. The flow rate Re was increased step by step keeping $\epsilon = 0.215$ constant. The time interval between two adjacent steps in Re was sufficently long for the pattern to relax to the new steady structure. The initial pattern at Re = 0 consisted of 27 convection rolls. Its uniform wave number $\bar{k} = 27\pi/\Gamma = 3.393$ differed from $k_c = 3.116$ by $q = \bar{k} - k_c = 0.277$. After switching on the throughflow to Re = 0.0083 the pattern relaxes to a new stationary deformed state whose final wave number distribution is shown by the solid squares in Fig. 4. Near outlet where the difference between $k(x)$ and k_c is maximum the distribution becomes slightly nonlinear. In this region the pattern is less stiff because the diffusion constant $D_\parallel(q)$ is locally smaller there. This effect is not properly described by the solution (3.6) of the phase equation (solid line in Fig. 4) which assumes a uniform value for $D_\parallel$. Further increase of the flow rate steepens the wave number profile (3.6) (dashed line in Fig. 4) and a local Eckhaus instability near the outlet of the channel occurs as soon as the Eckhaus threshold $k_E^+ = k_c + \sqrt{\epsilon/3\xi_0^2}$ is crossed. If the computer run had been started with an undisturbed roll pattern of $\bar{k} < k_c$ (e.g. 24 rolls instead of 27) an analogous scenario would have happened near the entrance of the channel at the lower Eckhaus boundary $k_E^- = k_c - \sqrt{\epsilon/3\xi_0^2}$. Consequently, for stationary deformed patterns to appear maximum *and* minimum local wave numbers must not cross the Eckhaus boundaries, i.e.

$$\begin{Bmatrix} k_{max} \\ k_{min} \end{Bmatrix} \equiv k(x = \begin{Bmatrix} \Gamma \\ 0 \end{Bmatrix}) = k_c + q \pm \frac{\Gamma}{2} \frac{\omega_c + \Omega}{D_{\|}} \lessgtr k_c \pm \sqrt{\epsilon/3\xi_0^2} \equiv \begin{Bmatrix} k_E^+ \\ k_E^- \end{Bmatrix}, \qquad (3.8a)$$

or equivalently

$$(\omega_c + \Omega) < \frac{2}{\Gamma} D_{\|} \left\{ \sqrt{\epsilon/3\xi_0^2} - |q| \right\}. \qquad (3.8b)$$

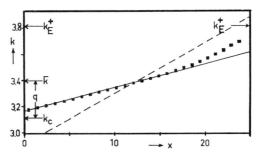

Fig. 4. Wave number distribution (solid squares) resulting at Re = 0.0083 from an initial pattern with 27 rolls at Re = 0 under phase pinning b.c. The solid line represents the corresponding solution (3.6) of the phase equation (3.5). Near outlet k almost reaches the upper Eckhaus boundary k_E^+. After increasing of the flow rate to Re = 0.0167 the Eckhaus threshold is crossed at the outlet, the system becomes locally unstable there, and a pair of rolls is annihilated. Dashed line denotes the stationary Eckhaus–unstable solution (3.6) for Re = 0.0167.

Into this stability condition enters the flow rate Re via $\omega_c + \Omega$, the Rayleigh number Ra via ϵ, the wave number of the basic undisturbed structure via q, and the channel length Γ. For q = 0.277, Γ = 25 and ϵ = 0.215 eq. (3.8) predicts a critical Reynolds number of Re_1 = 0.014. Beyond this threshold a local Eckhaus instability occurs near the outlet and annihilates two convective rolls there. The new average wave number $\bar{k}$ decreases to $\bar{k}' = 25\pi/\Gamma = \pi$ and the resulting new k–distribution being shifted downwards by $\bar{k} - \bar{k}' = 2\pi/\Gamma$ lies completely within the stable Eckhaus band. Thus, by annihilation of two rolls the stationary pattern has restabilized. According to eqn. (3.8) the critical flow rate for this new 25–roll state (with q' = $\bar{k}' - k_c$ = 0.025) is Re_2 = 0.028.

Since the new wave number profile $\bar{k}'(x)$ is practically concentric with the Eckhaus band a further increase of Re beyond Re_2 causes k_E^+ and k_E^- to be crossed *simultaneously*. That causes *permanent* roll creation near the inlet and roll annihilation

near the outlet and a traveling wave pattern in the bulk as shown in Fig. 5.
By imposing q=0 in (3.8) we obtain an estimate for the critical flow rate for the onset of roll propagation

$$\text{Re}_{TW} \simeq \frac{2.126}{(\sigma+0.3432)} \sqrt{\frac{\epsilon}{\tau_0}}. \tag{3.9}$$

Here we used $\xi_0^2 = 0.148$ and $\tau_0 = \frac{\sigma+0.5117}{19.65\sigma}$. This is only an upper limit for the transition because nonlinearities in the wave number distribution k(x) are not taken into account.

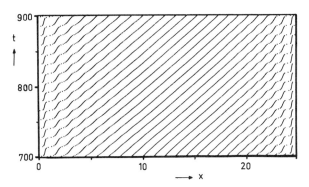

Fig. 5. Time evolution of the nodes of the vertical velocity field w(x,z = 0.5,t) under phase pinning b.c. For these parameters ($\epsilon = 0.215$, Re $= 0.0833$) a stationary pattern is Eckhaus unstable both at inlet and outlet. Creation and annihilation of rolls at inlet and respectively outlet results, and downstream pattern propagation in the bulk occurs.

IV. SUMMARY

We have investigated the Rayleigh–Bénard problem subjected to a plane Poiseuille shear flow perpendicular to the roll chain. This convective structure can be realized experimentally in a long narrow conduit if a weak throughflow is imposed. While channel sidewalls are essential for transversal rolls to appear we have used a 2 d description in the vertical crossection perpendicular to the roll axes which neglects sidewall effects. Results obtained by an amplitude equation are presented and compared with computer simulations of the full 2d hydrodynamic field equations. The main issue of our analysis was the influence of different inlet and outlet b.c. on the nonlinear convective structure.

If convective motion is suppressed at the cell apertures a more or less extended region near the inlet of the channel remains free of convection whereas the remaining part of the duct is filled with traveling rolls. The bulk wavelength of this localized traveling wave structure is uniquely selected by the control parameter combination independent of channel length and history of the system. Increasing the flow rate broadens the conductive region to the debit of the convective one. If a critical Reynolds number is exceeded any convection is "blown" out of the duct even though the basic conductive state is unstable. This is because the system has become convectively unstable and initially localized perturbations are carried out of the cell. At the same time the system becomes very sensitive against perturbations. With a continous source of noise, e.g., inlet turbulence the disturbances are amplified downstream and generate "noise sustained" convective structures. These states disappear as soon as the noise is removed. This amplification mechanism seems to be an interesting tool to measure experimental noise. If the system is absolutely unstable noise does not play an important role because the supercritical convection structure is not appreciably affected by small perturbations.

We also have considered b.c. that pin the phase of the convective rolls at the cell apertures. The standing deformed patterns with dilated rolls near the inlet and compressed rolls near the outlet appear for small flow rates; they are explained by an inhomogeneous phase equation obtained by a Reynolds number expansion of the amplitude equation. This phase equation is equivalent to that one discussed earlier by Pocheau et al[27]. For increasing throughflow the stationary deformed convection pattern becomes unstable due to a local Eckhaus instability. Roll generation at inlet and annihilation at outlet occurs finally resulting in a transition to traveling rolls.

REFERENCES

1. For a review until 1976 see: R.E. Kelly, in *Physicochemical Hydrodynamics*, ed. by D. Spaulding, Advanced Publication (1977), p. 65- 79.
2. Y. Kamotani and S. Ostrach, J. Heat Transfer **98**, 62 (1976).
3. G. J. Hwang and C. L. Liu, Can. J. Chem. Engng. **54**, 521 (1976).
4. J. M. Luijkx, J. K. Platten, and J. C. Legros, Int. J. Heat Mass Transfer **24**, 1287 (1981).
5. K. C. Chiu and F. Rosenberger, Int. J. Heat Mass Transfer **30**, 1645 (1987).
6. M. T. Ouazzani, J. K. Platten, and A. Mojtabi, Int. J. Heat Mass Transfer **33**, 1417 (1990).
7. S. Trainoff and G. Ahlers, private communication.
8. F. B. Lipps, J. Atmos. Sci. **28**, 3 (1971).
9. K. Fukui and M. Nakajima, Int. J. Heat Mass Transfer **26**, 109 (1983).
10. M. T. Ouazzani, J. P. Caltagirone, G. Meyer, and A. Mojtabi, Int. J. Heat Mass Transfer **32**, 261 (1989).
11. G. Evans and R. Greif, Int. J. Heat Mass Transfer **32**, 895 (1989).
12. H. W. Müller, M. Lücke, and M. Kamps, Europhys. Lett. **10**, 451 (1989).
13. H. W. Müller, M. Lücke, and M. Kamps, in preparation.
14. J. M. Luijkx, PhD–Thesis, University of Mons (Belgium) (1983).
15. J. K. Platten and J. C. Legros, *Convection in Liquids*, Springer, Berlin (1984).
16. A. Tsameret and V. Steinberg, Europhys. Lett. **14**, 331 (1991).
17. K. L. Babcock, G. Ahlers, and D. S. Cannell, preprint.
18. P. Büchel, D. Roth, M. Lücke, and R. Schmitz, unpublished.
19. K. S. Gage and W. H. Reid, J. Fluid Mech. **33**, 21 (1968).
20. S. H. Davis, J. Fluid Mech. **30**, 465 (1967).
21. K. Stork and U. Müller, J. Fluid Mech. **54**, 599 (1972).
22. P. Huerre, in *Propagation in Systems Far from Equilibrium*, ed. by J.E. Wesfreid, H. R. Brand, P. Manneville, G. Albinet, and N. Boccara, Springer, Berlin (1988), p. 340.
23. J. E. Welch, F. H. Harlow, J. P. Shannon, and B. J. Daly, Los Alamos Scientific Laboratory Report No. LA- 3425, (1966).
24. M. Lücke, M. Mihelcic, B. Kowalski, and K. Wingerath, in *The Pysics of Structure Formation: Theory and Simulation*, ed. by W. Güttinger and G. Dangelmayr, Springer, Berlin (1987), p. 97.
25. C. W. Hirt, B. D. Nichols, and N. C. Romero, Los Alamos Scientific Laboratory Report No. LA- 5652, (1975).
26. R. J. Deissler, in *Patterns, Defects and Materials Instabilities*, ed. by D. Walgraef and N. M. Ghoniem, Kluwer Academic Publ., Holland (1990), p. 83.
27. A. Pocheau, V. Croquette, P. Le Gal, and C. Poitou, Europhys. Lett. **3**, 915 (1987).

TAYLOR VORTEX FLOW WITH SUPERIMPOSED RADIAL MASS FLUX

Karl Bühler

Institut für Strömungslehre und Strömungsmaschinen
Universität Karlsruhe
D-7500 Karlsruhe 1, Kaiserstr. 12, Germany

The Taylor vortex flow is considered between two concentric porous cylinders. This offers the possibility to superimpose a radial mass flux on the Taylor vortices. We investigate the influence of the radial mass flux analytically within the linear stability theory, based on the small gap approximation.
The results obtained concern the stability and the spatial and temporal behavior of the vortex motion. A radial sink flow stabilizes the basic circumferential flow while a radial source flow has a destabilizing effect on the onset of Taylor vortices. The presence of a superimposed radial flow violates the "principal of exchange of stability", so that a time-dependent motion occurs. The sign of the flow direction in the vortices changes periodically with time. These results lead to further investigations of the nonlinear problem by numerical simulation.

1 INTRODUCTION

The stability behavior of the flow between two concentric rotating cylinders was first investigated theoretically and experimentally by Taylor [1]. Since this problem has received considerable interest in nature and technology, further questions have been considered. Many investigations were done on the influence of an axial throughflow. A report of some aspects of this problem is given by Bühler and Polifke [2]. The influence of a finite gap geometry and variations of the gap width are also of interest. Many aspects of these different problems were reported in the papers [3, 4, 5]. A recently new problem is the flow between two rotating porous cylinders with a superimposed radial flow, as shown in principal in Fig.1. The stability and the time behavior of this flow is of interest in science and technology . The present work is concerned with an analytical investigation to understand the physical aspects of the onset of Taylor vortices and their time behavior.

2 BASIC FLOW

The basic flow in the cylindrical gap of Fig.1 is given by the circular Couette-flow in the case of a rotating inner cylinder and a source-sink flow in radial direction. If rotation and a radial mass flux is present, the basic flow results as an exact solution of the Navier-Stokes equations. The distribution of the circumferential velocity $u(y, Re_o)$ is shown in Fig.2. The amplitude of the circular Couette-flow increases with a radial source flow and decreases in case of radial sink flow. The strength of the radial mass flux is given by the Reynolds number Re_o while their sign characterizes the direction of the flow. v_o is the radial velocity at the inner cylinder, $s = R_2 - R_1$ is the gap width and ν is the kinematic viscosity. The circumferential velocity at the inner cylinder is $U = R_1\omega_1$.

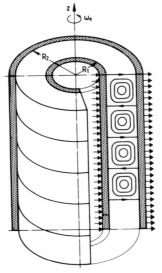

Figure 1. Principal sketch of the toroidal Taylor vortices in the cylindrical gap with a superimposed radial mass flux.

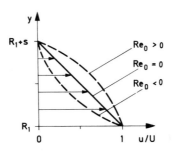

Figure 2. Circumferential velocity distribution of the basic flow, Reynolds number $Re = v_o s/\nu$. Source flow for $Re_o > 0$, sink flow for $Re_o < 0$.

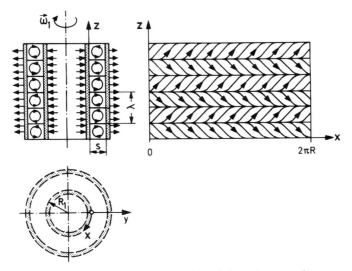

Figure 3. Toroidal Taylor vortices and local cartesian coordinate system.

3 LINEAR STABILITY THEORY

We apply the linear theory in the small gap limit to explain the stability and time-behavior of the Taylor vortices. The analysis is based on the idea of Oswatitsch [6] and further developments by Bühler [2, 7, 8]. The problem is described with local cartesian coordinates as shown in Fig.3 and the centrifugal force is incorporated. The Coriolis force can be neglected in the small gap limit. The nondimensional perturbation equations lead to the following parameters:

$$Re = \frac{R_1 \omega_1 s}{\nu}, \quad Re_o = \frac{v_o s}{\nu}, \quad Ta^2 = Re^2 \delta \qquad (1)$$

with $\delta = s/R_1$ as nondimensional gap width. The perturbation equations allow a solution in normal mode form:

$$(u'', v'', w'', p'') = (\hat{u}, \hat{v}, \hat{w}, \hat{p}) \cdot e^{(i\pi y' + i\alpha z' + \sigma t')} \qquad (2)$$

with $\alpha = 2\pi \frac{s}{\lambda}$ as wave number for the periodicity in z' direction.
After eliminating $\hat{w}$ and $\hat{p}$, the algebraic equations follow for the perturbation amplitudes of $\hat{u}$ and $\hat{v}$:

$$(\sigma + \pi^2 + \alpha^2 + i\pi Re_o)\hat{u} + \frac{d}{dy'}(\frac{u}{U})\hat{v} = 0 \qquad (3)$$

$$-Ta^2 \cdot 2\frac{u}{U} \cdot \hat{u} + (\sigma + \sigma\frac{\pi^2}{\alpha^2} + 2\pi^2 + \frac{\pi^4}{\alpha^2} + \alpha^2 + i\pi Re_o + iRe_o\frac{\pi^3}{\alpha^2})\hat{v} = 0 \qquad (4)$$

We approximate the basic flow in the same way as Oswatitsch [6] in the middle of the gap. The gradient of the circumferential velocity is given by

$$[\frac{d}{dy'}\lim_{v_o \to 0}\frac{u}{U}]_{y'=\frac{1}{2}} \cong -1 + \frac{1}{8}Re_o^2 \qquad (5)$$

and the circumferential velocity is

$$2\frac{u}{U}\Big|_{y'=\frac{1}{2}} \cong 1 + \frac{1}{2}Re_o \tag{6}$$

The gradient of the circumferential velocity is independent of the direction of the radial flow. The amplitude of the circumferential velocity varies linear with the Reynolds number of the radial flow. A radial source flow increases the amplitude while a radial sink flow decreases the amplitude of the circumferential velocity in the middle of the gap.

Within this approximation, the amplitude equations are the following:

$$(\sigma + \pi^2 + \alpha^2 + i\pi Re_o)\hat{u} - \hat{v} = 0 \tag{7}$$

$$-Ta^2(1 + \frac{1}{2}Re_o)\hat{u} + [\sigma(1 + \frac{\pi^2}{\alpha^2}) + 2\pi^2 + \frac{\pi^4}{\alpha^2} + \alpha^2 + i\pi(1 + \frac{\pi^2}{\alpha^2})Re_o]\hat{v} = 0 \tag{8}$$

This homogenous system of equations for $\hat{u}(\frac{1}{2})$ and $\hat{v}(\frac{1}{2})$ has a solution, if the coefficient determinant is zero.

For the case $Re_o = 0$ the solution of (7) and (8) has a real eigenvalue σ. The corresponding solution of the neutral-stable state with $\sigma = 0$ represents the steady Taylor vortex flow. This result is given in [7].

We consider now the general case with a superimposed radial flow with $Re_o \neq 0$. The solution of the equations (7) and (8) is then time-dependent. The "principle of exchange of stability" is violated. The time-behavior is described by the imaginary part of the eigenvalue $\sigma = \beta + i\gamma$. As solution condition follows:

$$Ta^2 = \frac{1}{1 + \frac{1}{2}Re_o}\{(\pi^2 + \alpha^2)(\frac{\pi^4}{\alpha^2} + 2\pi^2 + \alpha^2) - (\frac{\pi^2}{\alpha^2} + 1)(\gamma + \pi Re_o)^2 +$$

$$+ i[(\gamma + \pi Re_o)(\frac{\pi^4}{\alpha^2} + 2\pi^2 + \alpha^2) + (\pi^2 + \alpha^2)(\frac{\pi^2}{\alpha^2} + 1)(\gamma + \pi Re_o)]\} \tag{9}$$

The existence of a real solution requires the vanishing of the imaginary part in (9). This leads to the following eigenvalue γ:

$$\gamma = -\pi Re_o \tag{10}$$

Within this linear perturbation theory only the linear parts of the eigenvalues are important. This yields the result:

$$Ta^2 = (1 - \frac{1}{2}Re_o)[(\pi^2 + \alpha^2)(\frac{\pi^4}{\alpha^2} + 2\pi^2 + \alpha^2) - (\frac{\pi^2}{\alpha^2} + 1)2\pi\gamma Re_o] \tag{11}$$

For the critical wave number $\alpha = \frac{\pi}{\sqrt{2}}$ follows:

$$Ta_c^2 = Re^2\frac{s}{R_1} = \frac{27}{4}\pi^4 - \frac{27}{8}\pi^4 Re_o \tag{12}$$

The quadratic term in Re_o drops out from symmetry conditions. This solution clearly demonstrates the influence of the direction of the radial flow.

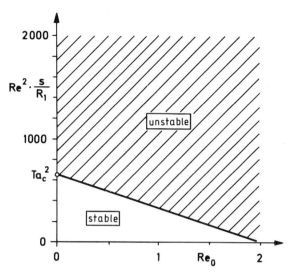

Figure 4. Stability behavior of the Taylor vortex flow with a superimposed radial source flow. The wavelength is $\alpha = \pi$.

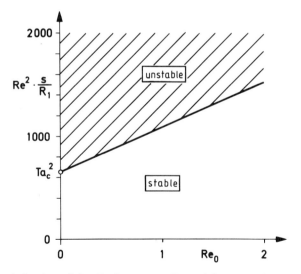

Figure 5. Stability behavior of the Taylor vortex flow with a superimposed radial sink flow. The wavelength is $\alpha = \pi$.

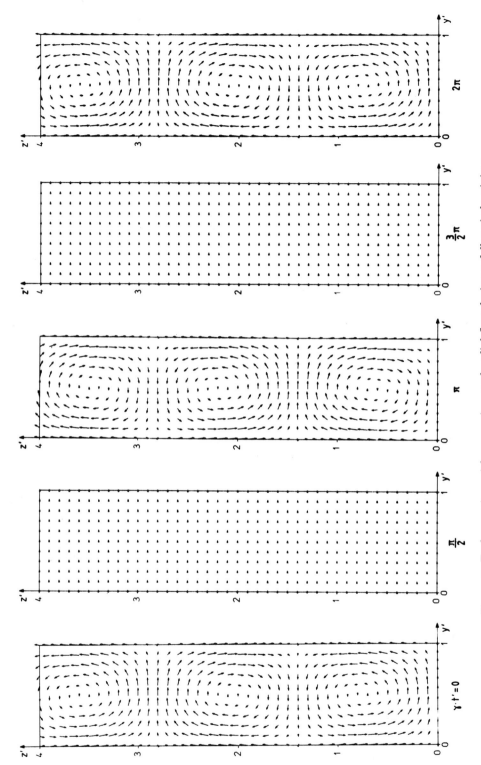

Figure 6. Taylor vortices with a superimposed radial flow during a full periode of time. Dimensionless gap width $\delta = \frac{s}{R_1} = 0.1$, wavelength $\alpha = \frac{\pi}{\sqrt{2}}$.

4 RESULTS

In this chapter we discribe the results on the stability and time-dependence of the Taylor vortex flow with superimposed radial flow. Because of the approximations in these linear theory considerations, the results are of qualitative nature, but nevertheless the physical aspects of the solutions are very interesting. A radial source flow has a destabilizing effect on the onset of Taylor vortices. Fig.4 shows the decreasing critical Taylor number with increasing Reynolds number of the radial flow. A radial sink flow therefore stabilizes the onset of Taylor vortices. Fig.5 shows the increasing critical Taylor number with increasing Reynolds number of the radial sink flow. The influence of the radial flow on the stability behavior results from the change of the amplitude of the basic flow. The convective terms of the radial flow in the perturbation equations lead to a time-dependent oscillatory solution. This time behavior is demonstrated in Fig.6 by the flow structure in the meridional plane at different times of one periode.

5 CONCLUSIONS

With approximations in the linear theory it is shown, that a radial flow has an important influence on the stability and time-behavior of Taylor vortices. A radial source flow destabilizes the onset of the Taylor vortex flow while a radial sink flow has a stabilizing effect on the onset of the Taylor vortex flow.

References

[1] Taylor G.I., Stability of a viscous liquid contained between two rotating cylinders. Proc.Roy.Soc.(A) 223, 289-343 (1923)

[2] Bühler K., Polifke N., Dynamical behavior of Taylor vortices with superimposed axial flow. Nonlinear Evolution of Spatio-Temporal Structures in Dissipative Continuous Systems Ed. by F.H. Busse and L.Kramer, Plenum Press, New York, 21-29 (1990)

[3] Zierep J., Oertel jr. H., Convective Transport and Instability Phenomena, Karlsruhe: Braun 1982

[4] Bühler K., Kirchartz K.R., Wimmer M., Strömungsmechanische Instabilitäten, Strömungsmechanik und Strömungsmaschinen 40, 99-126 (1989)

[5] Bühler K., Coney J.E.R., Wimmer M., Zierep J., Advances in Taylor vortex flow: A Report on the Fourth Taylor Vortex Flow Party Meeting. Acta Mechanica 62, 47-61 (1986)

[6] Oswatitsch K., Physikalische Grundlagen der Strömungslehre. Handbuch der Physik, Bd. VIII/1, Hrsg. S. Flügge, Berlin: Springer, 1959

[7] Bühler K., Ein Beitrag zum Stabilitätsverhalten der Zylinderspaltströmung mit Rotation und Durchfluß. Strömungsmechanik und Strömungsmaschinen 32, 35-44 (1982)

[8] Bühler K., Der Einfluß einer Grundströmung auf das Einsetzen thermischer Instabilitäten in horizontalen Fluidschichten und die Analogie zum Taylor-Problem. Strömungsmechanik und Strömungsmaschinen 34, 67-76 (1984)

VORTEX PATTERNS BETWEEN CONES AND CYLINDERS

Manfred Wimmer

Institut für Strömungslehre, Universität Karlsruhe
Kaiserstr. 12, D - 7500 Karlsruhe 1, Germany

Pattern forming systems have been intensively studied during the last years. A rich variety of flow patterns and structures can for instance be observed for unstable flows near rotating bodies (cf. Wimmer, M. 1988). As an example we will have a look at the vortices occurring in the gap between rotating cylinders and cones. Fig. 1 shows the possible combinations of these bodies. The flow between cylinders is relatively well known since G. I. Taylor's (1923) famous publication. Within the gap we obtain a one-dimensional, circular Couette flow which becomes unstable for higher Reynolds numbers, showing counter-rotating Taylor vortices simultaneously on the whole length of the cylinder. Between coaxial cones we have already a fully developed three-dimensional basic flow because of the linear alteration of the centrifugal forces. Taylor vortices will appear first at the larger radius extending to the smaller radius while increasing the angular velocity. For different initial- and boundary conditions a great number of flow patterns occurs, already reported by the author (Wimmer, M. 1983, 1988).

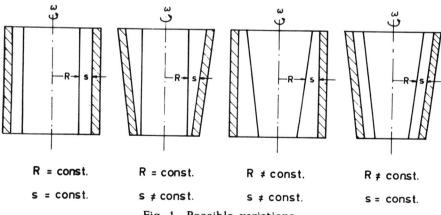

Fig. 1 Possible variations

The investigation of flow properties between cylinder-cone combinations is therefore the consequent next step. There was still the open question: What happens if a cylinder rotates in a conical container or a cone in a cylindrical shell. The Taylor number $T = R\, s^3\, \omega^2/\nu^2$ can give an answer, since it contains the influence of dynamics and geometry. If we keep the angular velocity ω and the kinematic viscosity ν constant for all arrangements shown in Fig. 1 (conditions which can easily be accomplished during an experiment), then the dynamical part ω^2/ν^2 is a constant, too, and the Taylor number only depends upon the geometric parameters radius R and gap width s. These parameters have different alterations in each case, as it is clearly displayed in Fig. 1. There are combinations, where both parameters are constant or only one of them; and for a cone in a cylinder neither R nor s are constant. Since the Taylor number depends linearly on R and to the third power on s, it is apparent, that the different behaviour of R and s must have an influence on the onset and development of the vortices.

The flow between cones and cylinders is not only of academic interest, but has also a direct technical application. Furthermore, the different behaviour of the vortices plays a fundamental part in the process of wave-number selection. Since the selection of wave numbers is accomplished by a spatial variation of the Reynolds number, conical ramps are used to combine the subcritical and the supercritical part of the gap, as it was reported by Dominguez-Lerma et al. (1986). To produce such spatial gap variations, the conical part can be machined in the outer stationary shell (ramps) or in the inner rotating body (tapered cylinder); i. e. we have exactly the same geometric combinations like in the present paper and as it was described previously by the author (Wimmer, M. 1985). Even if the vertex angles of the conical parts are much smaller during the arrangement for the wave-number selection than during the present experiments, the main effects discussed below are still valid for the procedure of the wave-number selection and should be taken into account.

Let us start with the less complicated case of a rotating cylinder in a conical container. Despite of a constant radius and a constant angular velocity, one obtains an imbalance of the centrifugal forces caused by the varying gap width, resulting in a three-dimensional basic flow. For higher angular velocities the first Taylor vortices appear at the larger gap width in accordance with the Taylor number, while at the smaller part the undisturbed basic flow still exists. By a further increase of the angular velocity, vortices occur also in this part of the gap as soon as the critical values are fulfilled. Finally, the whole gap is filled with vortices of different sizes. According to the linear alteration of the gap width, the wavelength of the vortices is linearly altered, too. Such a linear increase of the vortex extension with growing gap width was also confirmed by Abboud's (1988) calculations.

Somewhat more complicated is the case of a rotating cone in a cylinder, where neither the radius nor the width of the gap remains constant. Fig. 2 shows the principle of the test arrangement. It demonstrates that the experiments have not been carried out with a perfect cone. We know from experience that disturbances are preferably generated at the cone's apex and travelling through the gap, disturbing the flow which should be observed. That is why the apex has been cut, resuling in a truncated cone, with the advantage of a suspension on both sides. The ring at the top is interchangeable, and hence we have the possibility to apply different end plates; e. g. stationary end plates, completely bridging the gap, or bridging it only partly — as it is displayed - and rotating end plates, again completely or partly bridging the gap. Because of these altered boundary conditions — together with different initial conditions — a rich variety of flow patterns can be generated.

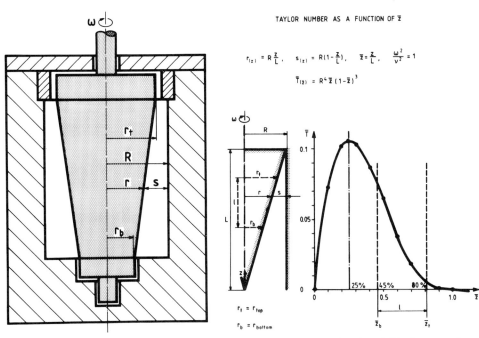

Fig. 2 Test arrangement   Fig. 3 Curve of the local Taylor number

In the gap the undisturbed basic flow is again fully three-dimensional. At the largest radius the medium is deflected outwards, goes down in spirals near the outer cylinder and up again in the vicinity of the rotating cone still in spirals; thus forming a closed flow. The strength of the meridional swirl is not uniform because of the different cross sections in the gap. This will have some consequences on the vortices. For higher angular velocities, vortices will occur in certain parts of the gap, whereas in other parts the undisturbed basic flow is still preserved. Thus, different kinds of flow — stable and unstable ones — exist side by side in the gap between cones and cylinders. For the present experimental set-up, first vortices are visible at middle axial latitudes — i. e. at middle sized gap widths. The vortices do not appear at the bottom, where the gap width is largest - as it could be anticipated because of $T \sim s^3$, and they do not appear first at the top, where the radius is largest producing the greatest centrifugal forces, but they will be generated at that place where the product $r\,s^3$ has it's maximum according to the Taylor number.

207

A very simple calculation confirms this statement. As it is displayed in Fig. 3, we now consider a perfect cone of the length L and the basis radius R in a cylinder of the same radius. In the local Taylor number the geometric parameters r and s are now functions of the axial position z. If - as was mentioned above - the dynamical part ω^2/ν^2 remains constant - say 1 - and z is made dimensionless to $\bar{z}$, then the Taylor number takes the form $\overline{T}(\bar{z}) = R^4 \bar{z} (1-\bar{z})^3$ We obtain the maximum of this function - the spot of the generation of the first vortices - at a distance of 25% from the cone's apex. The diagram of Fig. 3 displays the curve of the local Taylor number with it's maximum at $\bar{z} = 0.25$. Because of the reason already discussed above, we do not have a perfect cone. That is why the truncated cone of the length l is also dislpayed in Fig. 3 by broken lines, as well as that regime which is valid for the truncated cone as part of the complete curve of the local Taylor number. We see that for the present case the maximum is outside of the gap. Furthermore, we can deduce that for increasing angular velocities the first vortices have to appear at $\bar{z}_b$ - i. e. at the bottom where the width of the gap is largest.

In the present case, the first vortices are, however, observed for middle sized gap widths. An explanation for this discrepancy gives the following consideration. The vortices are indeed generated at the largest gap width; that means, at the smallest radius with the smallest centrifugal forces. They are, therefore, very weak and are taken upwards along with the meridional flow. Hence, they come in regions, where locally the critical values are not yet reached and decay. Only after a further increase of the angular velocity, vortices can be produced here, too, which are stronger now. During they are slowly travelling upwards, they finished their development, and that is why first vortices are observed at middle axial latitudes.

While propagating through the gap, the vortices keep their form as closed rings perpendicular to the axis of rotation. A new pair of vortices is generated in the lower part of the gap, whilst it decays again at higher latitudes. Hence, we have the situation that patterns in form of closed vortex rings propagate through the flow field in a closed flow system without any influence from outside, like for coaxial cones. How strong is the movement that makes the vortices travel through the gap? Experiments show that it depends on the vertex angle of the cone and the Reynolds number. Results of measurements are displayed by Fig. 4. We see that the velocity of the travelling vortices decreases linearly with growing Reynolds number. At about Re = 350, the vortices are vigorous enough to compensate the influence of the meridional flow, and the movement stops. Thus, a steady state of flow is established by increasing the Reynolds number.

For the combination of a rotating cone in a cylinder not even the steady state is unique, but depends upon different parameters as for instance different end plates, different vertex angles of the cone and different rates of acceleration of the inner body to angular velocities to enable a supercritical flow. The different modes can be obtained either directly out of the undisturbed basic flow by different initial conditions - i. e. rates of accelerations - or by a change from one mode into another. Hence, the first appearance and disappearance occur at different Reynolds numbers so that different regimes of existence are occupied.

For a quasi-steady acceleration one obtaines a flow configuration with six vortices. This mode occurs first at the lowest critical Reynolds number and exists until the beginning of turbulence. For a higher rate of acceleration a mode with eight vortices is observed, and for again another acceleration rate a mode with four vortices appears. This one occupies only a small regime of existence and changes to the more stable mode with six votices by increasing the Reynolds number. For two stationary end plates, symmetric boundary conditions are given. Since Taylor vortices preferably appear in pairs, we obtain *normally* an even number of vortices, as described above. However, there also exists a mode with only three vortices. Such an *anomalous* mode can only be produced by very high acceleration rates, so that the flow has no time to develop. Now, at the moment of the appearance of the first vortex the entire flow becomes frost-bound, so to say. Since we have now an odd number of vortices, one of them is forced to rotate in the "wrong" way. This leads to a radial outwards jet at the stationary top end plate.

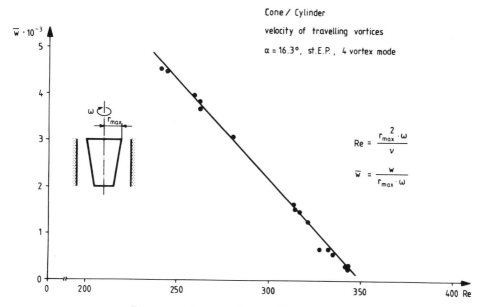

Fig. 4   Velocity of travelling vortices

For other end plates, one stationary and one rotating, the boundary conditions are asymmetric, resulting in *now* an odd number of vortices. For such an arrangement, three out of four possible modes have an odd number of vortices namely three, five and seven vortices, and one as an anomalous mode has six vortices. In this case we have the rare situation for Taylor vortices that two vortex cells are forced to rotate in the same direction. Considering an arrangement with a stationary end plate at the bottom and a rotating one at the top, we obtain for slightly supercritical Taylor numbers two pairs of Taylor vortices at the bottom, where the gap is larger, while another vortex is induced at the top by the rotating end plate. The sense of rotation of these vortices is fixed, resulting in an outwards jet at the top and an inwards jet at the bottom; the adjacent Taylor vortices are counter-rotating ones. The fluid between these systems can not remain in a state of rest, but executes a rotation induced by the lower neighbouring Taylor vortex. Thus, the first and the second vortex from top (both no real Taylor vortices, but induced vortices) rotate in the same direction divided by a shear layer.

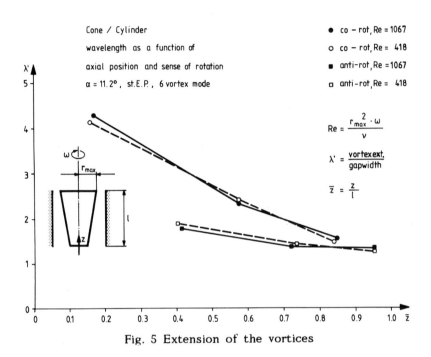

Fig. 5 Extension of the vortices

For all vortex modes, described so far, we notice a different extension of the vortices. During the experiments it turned out, that the size of the vortices depends upon the local axial position, the sense of rotation of a vortex; i. e. whether it rotates in the direction of the meridional flow or opposite to it, and naturally upon the width of the gap and the rate of acceleration. Thus, we have a dependence of the wavelength on *four* different parameters. Measurements of the extension of the vortices are given by Fig. 5 for stationary end plates, a six vortex mode and various Reynolds numbers. Vortices rotating in direction of the meridional flow are treated separately from those rotating against it. We see very clearly, that the extension of the co-rotating vortices decreases much stronger than the counter-rotating ones. The influence of the different Reynolds numbers is negligible.

Finally, there exists another unusual mode in the gap between cones and cylinders. It is an unsteady mode caused by a disturbance asymmetric to the axis of rotation. The vortices take now the form of a helix, coiled around the rotating cone, while the whole system moves downwards. All the described modes of flow are solution branches for different initial- and boundary conditions. The study of the flow between cone-cylinder combinations is therefore an excellent method to analyse the behaviour of Taylor vortices for different parameters.

References

Wimmer, M. (1988) Prog. aerospace Sci. 25/1, 43
Taylor, G. I. (1923) Phil. Trans. A 223, 289
Wimmer, M. (1983) Z. Andew. Math. Mech. 63, 299
Dominguez-Lerma, M. A; Cannell, D. S; Ahlers, G. (1986) Phys. Rev. A 34/6, 4956
Wimmer, M. (1985) Z. Angew. Math. Mech. 65, 255
Abboud, M. (1988) Z. Angew. Math. Mech. 68, 275

INSTABILITY OF TAYLOR-COUETTE FLOW OF HELIUM II

Chris J. Swanson and Russell J. Donnelly

Department of Physics
University of Oregon
Eugene, Oregon 97403

The study of Taylor-Couette flow has been useful in understanding classical fluid flows and non-linear phenomena because of its simplicity and its versatility. The Taylor-Couette system promises to be equally useful in understanding the behavior of liquid helium II, a non-classical fluid.

The Taylor transition in Taylor-Couette flow of liquid helium was first studied with stability theory by Chandrasekhar and Donnelly[1] who recognized the importance of the problem as a rigorous test of the equations of motion in helium. Subsequently experiments and other theories have examined various aspects of the flow from the first appearance of quantized vortices to turbulent flow. Unfortunately no clear picture has emerged to help us understand these various aspects.

Recently Donnelly and LaMar reviewed both experimental and theoretical work on helium Taylor-Couette flow.[2] First, they noted that the equations of flow of helium II have evolved somewhat since their inception and consequently the earliest stability analysis[1] was based on equations in which an important term was unknown. On the experimental side they found that instability in the flow of helium II occurs when the inner *or* outer cylinder is rotating. This contrasts with the classical case where rotation of the outer cylinder is stable against infinitesmal perturbations and the flow is laminar until a direct transition to turbulent flow occurs. With regard to rotations of the inner cylinder the data from experiment to experiment was not found to agree well because of the different temperatures and geometries used.

During preparation of the review, Carlo Barenghi and Chris Jones of the University of Newcastle upon Tyne decided to undertake a fresh look at the stability theory using the modern Hall-Vinen-Bekarevich-Khalatnikov equations of motion.[3] Barenghi and Jones found the instability of helium to be markedly different from the classical case. First for the rotation of the outer cylinder the superfluid is unstable to long wavelength disturbances at all non-zero Reynolds numbers, as was suggested by earlier experiments. For inner cylinder rotation an instability similar to the classical Taylor transition occurs at finite Reynolds numbers but is strongly temperature and geometry dependent. As one increases temperature to 2.172 (beyond which helium is a classical fluid) the critical Reynolds number approaches the classical critical Reynolds number for the transition to Taylor vortex flow. As the temperature is decreased from 2.172 the critical Reynolds number increases and the critical wave number decreases. The effect of lowering the temperature is to stablize the flow. Eventually, as the temperature continues to decrease the critical wave number becomes zero as in the case of outer cylinder rotation, and end effects become important.

Barenghi and Jones also found that the critical Reynolds number is geometry dependent. For decreasing radius ratio, $\eta = R_1/R_2$, both the critical Reynolds number and the wave number of the instability decrease. Thus for experiments done with smaller radius ratios the effects of the ends become very important. For our experiment we have chosen a radius ratio of .979 and have studied the higher temperature range to best allow for comparison of our results with the stability analysis. Furthermore we have concentrated exclusively on inner cylinder rotation.

To completely specify the experimental problem, the pertinent parameters are Re, η, Γ, d, and the temperature T. The gap is $d = .0472$ cm, the height $h = 9.436$ cm., and the aspect ratio is $\Gamma = h/d = 199$. The Reynolds number is $Re = \Omega_1 R_1 d/\nu_n$ where ν_n is the kinematic viscosity and Ω_1 is the inner cylinder angular velocity. The critical Reynolds number for the Taylor transition is denoted Re_c. The temperature plays an important role in many of the force terms in the equations of motion such as the quantized vortex tension term.

Experimental measurement techniques of the critical Reynolds number are limited by the cryogenic requirements. Our technique to probe the fluid was to examine the density of quantized vortex lines. In rotating superfluid helium atomic sized vortex lines with quantized circulation are threaded throughout the fluid except at very low rotational velocities. At low rotational velocities below some critical value, Re_v, no vortex lines exist. Above this critical, (but still below the Taylor transition) an array of equally spaced vortex lines enter the container and align themselves parallel to the rotation vector. Our experimental technique is to measure the density of vortex lines as a function of angular velocity of the inner cylinder. A break in the line density indicates a transition to another flow regime.

The density of vortex lines is described by "Feynman's rule" $n = |\text{curl } \vec{v_s}|/\kappa$ where $\kappa = h/m$ is the quantum of circulation about a single vortex line. Here h is Plank's constant and m is the mass of the helium atom. At rotation rates above Re_v but still below the Taylor transition, Re_c, enough vortex lines exist to treat them as a continuum and the velocity profile is very nearly the same as the classical laminar velocity,

$$\vec{v} = (Ar + B/r)\hat{\phi} . \qquad (1)$$

Consequently we expect

$$n = \frac{2A}{\kappa} . \qquad (2)$$

We have indeed found that the line density obeys this rule when Re is within a certain range. Above this range the vortex array becomes unstable and the average line density deviates from (2).

To detect the quantized vortices we use second sound resonances within the annular cavity. Second sound is a longitudinal temperature wave peculiar to helium. It behaves much like ordinary (first) sound and has a temperature dependent wave speed ranging from 0 to 2000 cm/s. It is useful as a vorticity probe since it is attenuated by vortex lines aligned perpendicular to the direction of propagation. The attenuation can be related to the line density by a experimentally determined constant called the mutual friction coeficient.

The second sound is excited by AC Joule heating of a 20 micron, high resistance, Evenohm wire attached lengthwise along the inner cylinder. The resulting temperature waves propagate axially and azimuthally at twice the input frequency. The second sound is detected by applying a DC bias current to a thin painted strip of Aquadag carbon compound which is strongly temperature dependent. The temperature wave causes the resistance of the carbon to fluctuate producing a weak AC signal. The signal is amplified and transferred through slip rings to a phase sensitive lockin amplifier. The attenuation of the second sound can be known very accurately by measuring the change in resonance amplitude. The best resolution which we were able to achieve was about 20 cm. of line per cubic centimeter which translates into a detection of vortex core present in a concentration of a part in 10^{12}.

To detect the instability we excited an axial resonance mode. Below the Taylor transition the temperature waves were travelling mostly parallel to the vortex array. Thus no attenuation from vortex lines was expected. Above the transition the vortex lines bend and are easily detected by the second sound. There is a region of high shear at the top and bottom of the annulus which creates end vorticity. End vorticity was present in significant amounts but did not obscure the presence of the transition nor was it as abundant as the vorticity in the bulk flow.

The transition was marked by a break in the slope of line density as a function of Reynolds number sometimes accompanied by a dicontinuous jump in the line density. The slope changed by a factor of 1.5 to 3 depending on temperature and was easily recognizable. A plot of line density vs. Re at 2.1 K is shown in figure 1. The critical Reynolds number in this case was determined to be 344. Each point was taken by averaging the resonant amplitude over a specified number of rotations of the inner cylinder. The Reynolds number was then increased and another average was taken. The squares represent line density as Re was being stepwise increased, and the triangles represent line density as Re was being stepwise decreased.

In this and most of the other runs no hysteresis was found. In fact, various protocols for waiting and accelerating between points yeilded no change in the hysteresis. By contrast, in classical Taylor-Couette flow, there is a criteria set forth by Park et. al. for passing through the Taylor transition without hysteresis.[4] Namely, only when

$$a^* = \frac{R_1 d^2 L}{v^2} \frac{d\Omega_1}{dt} < 6 \qquad (3)$$

will the hysteresis be negligible. Due to the very small kinematic viscosity of helium our dimensionless accelerations were in all cases at least two orders of magnitude greater than the above criterion. The lack of hysteresis may provide an important clue in understanding the dynamics of the superfluid transition.

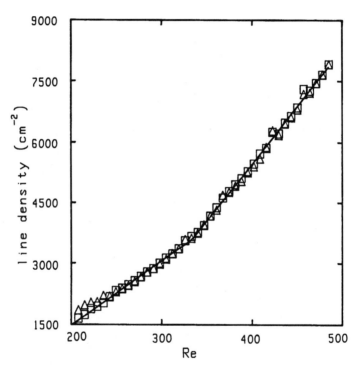

FIG. 1. Vortex line density as a function of inner cylinder Reynolds number at T = 2.10 K. The squares and trianges correspond to data taken as the Reynolds number was being stepped up and stepped down respectively. The line is a least squares fit to the data.

The line through the data in figure 1 is a non-linear, least squares, five parameter fit. The fitting function consisted of two lines joined smoothly by the Fermi function,

$$n = (aRe+b)f(Re) + (cRe+d)(1-f(Re)) \qquad (4)$$

where

$$f(Re) = \frac{1}{(1+e^{(Re-Re_c)/\Delta})}.$$

The critical Reynolds number Re_c and the parameters a, b, c, and d were free parameters while the sharpness Δ of the exponential was fixed at 5. We found the fit to be relatively insensitive to the sharpness within a range of 3 to 8 beyond which the fitting routine would usually not converge.

Only data within a certain range of Re was suitable to fit. At very low and very high Reynolds numbers the slope differs from either of the two slopes shown in figure 1. Below Re = 200, the line density begins to level off at 0 indicating the absence of vortex lines. At Reynolds numbers two to three times above the critical, further instabilities set in. Selecting too many or too few points could have an effect on the fitting parameter Re_c. In those cases where the choice was not clear, a larger uncertainty was given to the critical value.

The temperature dependence of the critical Reynolds number is plotted in figure 2. The open squares are our experimental data points and the solid triangles are calculations from the stability analysis of Barenghi and Jones.[5] The error bars on the graph represent 90% confidence intervals of the fit except for cases mentioned above. The theoretical values are slightly higher than those of the experiment. This error could be due to a variety of sources. First, many of the temperature dependent parameters used in the stability analysis are known only to within 5%. Furthermore, due to shrinkage of the cylinders the physical dimensions of the apparatus are only accurate to .2%. Since the critical Reynolds number depends strongly on radius ratio, an error in η of .2% would correspond to a shift in Re_c of 20. Considering these errors the agreement is surprisingly good for temperatures at and above 2.1 K verifying the correctness of the equations of motion and boundary conditions.

Of particular interest in figure 2 is the downward turn of the data as $T \to 2.172$. The rightmost triangle is the calculated critical Reynolds number at 2.172 for ordinary, classical helium I. The approach of the experimental critical value to the classical one shows that the transition is the well known Taylor transition. We presume that much of the character of that transition is retained as long as the temperature is not too far from 2.172.

As the temperature is lowered below 2.1 K the transition is substantially altered. According to the stability analysis, the critical wave number $k = 2\pi d/\lambda$ goes to zero and the critical Reynolds number sharply drops as well. In a finite geometry it is impossible to reach $k = 0$, and the lowest acheivable wave number is determined by the ascpect ratio. It is possible that these long wavelength disturbances are Ekhaus unstable bringing the critical Reynolds number up to a non-zero value.

Experimentally we find that between 2.1 K and 2.0 K, Re_c slowly turns down until at 2.0 K there is a sharp decrease. Accompanying this decrease is an increased difficulty in determining Re_c. The line density becomes slightly curved as opposed to

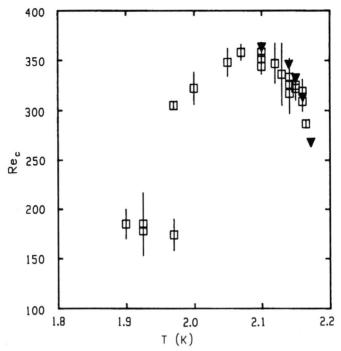

FIG. 2. Temperature dependence of the critical Reynolds number is shown. The squares are experimental data points with errors from the fitting routine. The solid triangles are calculated values from the stability analysis of Barenghi and Jones. The lowest triangle is the critical Reynolds number for He I at 2.172 K.

linear and discontinuous, and the other breaks in slope infringe on the range of Re that can be used to fit. Below 2.0 K the break in the slope of line density may or may not correspond to the "Taylor transition". If Re_c goes to zero it is likely that we are seeing some other instability in the flow. However, as mentioned above, if the small wave numbers are Ekhaus unstable, it is possible that the break does correspond to the Taylor transition. Further experiments and non-linear theory are needed to clarify the details.

As with its classical counter part, Taylor-Couette flow of liquid helium II promises to be a testing ground for superfluid dynamics. Using this system we have shown the success of the modern equations of motion and the boundary conditions by verifying the linear stability theory of Barenghi and Jones.[3] The theoretical guidance has proven invaluable in directing our experimental study. Further theory both linear and non-linear is clearly called for. Experiments to study the structure of the flow by varying the diagnostics and geometry immediately suggest themselves. Also variations on the basic experiment such as the addition of axial flow using a heater would be a good next step. Finally liquid helium Taylor-Couette flow is particularly well suited to study high Reynolds number turbulence. The kinematic viscosity is almost two orders of magnitude lower than that of water for instance. These considerations point to the opening of a new field in helium II fluid dynamics.

REFERENCES

[1] S. Chandrasekhar and R. J. Donnelly, Proc. R. Soc. London, Ser. A **241**, 9 (1957).
[2] R. J. Donnelly and M. M. LaMar, J. Fluid Mech. **186**, 163 (1988).
[3] C. F. Barenghi and C. A. Jones, J. Fluid Mech. **197**, 551 (1988).
[4] K. Park, G. L. Crawford, and R. J. Donnelly, Phys. Rev. Lett. **47**, 1448 (1981).
[5] C. F. Barenghi and C. A. Jones (Private communication).

EFFECTS OF RADIAL TEMPERATURE GRADIENT ON MHD STABILITY OF COUETTE FLOW BETWEEN CONDUCTING CYLINDERS - A WIDE GAP PROBLEM

H.S. Takhar

Dept. of Engineering, Manchester University, Manchester
M13 9PL (UK)

M.A. Ali

Dept. of Mathematics, Bahrain University, PO Box 32038,
Bahrain (Middle East)

V.M. Soundalgekar

31A-12, Brindavan Society, Thane (400601) India

INTRODUCTION

Stability of Couette flow under different conditions and assuming a narrow gap annulus has received great attention in the past. Notable among these are studies by Taylor (1923), Chandrasekhar (1953, 1954, 1961), Edwards (1958), Becker and Kaye (1962), Lai (1962), Kurzweg (1963), Harris and Reid (1964), Krueger, Gross and DiPrima (1966), Hassard, Cheng and Ludford (1972), Bahl (1972), Soundalgekar, Takhar, Smith (1981), Takhar, Smith and Soundalgekar (1985). Takhar, Ali and Soundalgekar (to be published) presented the study of MHD stability of Couette flow on taking into account the presence of radial temperature gradient and the axial magnetic field, in a narrow-gap annulus. The corresponding stability of Couette flow in a wide-gap annulus has been studied by very few researchers because of its complex nature. Notable among these are studies by Chandrasekhar (1958). Chandrasekhar and Elbert (1962), Walowit, Tsao and DiPrima (1964), Sparrow, Munro and Jonsson (1964), Astill and Chung (1976), Takhar, Ali and Soundalgekar (1988), Chandrasehkar (1958) derived results for $\eta = \frac{1}{2}$ (i.e. $R_1 = \frac{1}{2} R_2$) whereas in other papers, Chandrasekhar and Elbert (1962) simplified the numerical procedure by considering the corresponding adjoint eigenvalue problem. Walowit et al. (1964) simplified the method of solution of an eigenvalue problem by giving an algebraic series solution instead of Chandrasekhar's trigonometric series solution. Sparrow et al., Takhar et al. solved the eigenvalue problem numerically using the Runge-Kutta method, whereas Astill and Chung solved it by a finite-difference method. The only paper which deals with MHD stability of wide-gap problem is the one by Chang and Sartory (1967). It deals with the stability of the flow of an electrically conducting fluid in a wide-gap of permeable, perfectly conducting cylinders. In the narrow gap case, there are many papers on the MHD stability of Taylor flows with both conducting and non-conducting walls of the two concentric cylinders. So Ali, Soundalgekar and Takhar (to be published) solved this MHD stability of Taylor flow for both conducting and non-conducting impermeable cylinders separated by a wide-gap. This eigenvalue problem was solved numerically following the method of Harris and Reid (1964) and Sparrow et al. (1964). In Chang and Sartory's (1967) paper, the basic velocity was assumed to be A/r, where A is a constant. We have assumed the basic velocity of the form Ar + B/r to solve this eigenvalue problem. Hence a comparison is not possible between our results and those of Chang and Sartory (1967).

How does the radial temperature gradient existing between two cylinders, separated by a wide-gap affect the stability of the flow? This is also an important topic from application point of new. Hence it is now proposed to consider the stability of Couette flow by assuming the existence of following conditions: i) perfectly conducting walls, ii) presence of an axial magnetic field, iii) the inner cylinder rotating or both the cylinders co-rotating or counter-rotating, iv) the presence of a radial temperature gradient and v) the radius ratio ranging between 0.1 to 0.95.

In Section 2, the mathematical problem is formulated, the method of solution is described and the numerical values of the critical wave-number and Taylor number are derived. This is followed by a discussion. In Section 3, the conclusions are set out.

MATHEMATICAL ANALYSIS

Consider the flow of an electrically conducting viscous liquid due to the rotation of either the inner cylinder alone or due to both the cylinders co-rotating or counter-rotating. If (u, v, w) are the velocity components along the (r, θ, z) directions with (f, g, h) as the components of the applied magnetic field, then the steady state solutions are given for the case of an applied magnetic field as follows:

$$u = w = 0, \quad v = V(r)$$
$$f = g = 0, \quad h = H = \text{constant} \tag{1}$$
$$V = Ar + B/r \tag{2}$$

where,

$$A = \frac{\Omega_2 R_2^2 - \Omega_1 R_1^2}{R_2^2 - R_1^2}, \quad B = \frac{R_1^2 R_2^2 (\Omega_1 - \Omega_2)}{R_2^2 - R_1^2}.$$

The steady temperature T satisfies the following equation

$$\frac{d^2T}{dr^2} + \frac{1}{r}\frac{dT}{dr} = 0 \tag{3}$$

TABLE I. CRITICAL VALUES OF TAYLOR NUMBERS AND WAVE
NUMBERS FOR VARIOUS VALUES OF η, μ, Q AND N

η	μ	Q	N	a_c	T_c	η	μ	Q	N	a_c	T_c
0.95	0.0	0	0.1	3.128	1.475 (7)*	0.95	0.1	0	0.1	3.125	1.494 (7)
			0.05	3.128	1.495 (7)				0.05	3.124	1.519 (7)
			0.0	3.128	1.517 (7)				0.0	3.124	1.545 (7)
			-0.05	3.127	1.539 (7)				-0.05	3.124	1.572 (7)
			-0.1	3.127	1.561 (7)				-0.1	3.122	1.600 (7)
		100	0.1	3.135	1.494 (7)			100	0.1	3.132	1.513 (7)
			0.05	3.135	1.515 (7)				0.05	3.132	1.539 (7)
			0.0	3.135	1.537 (7)				0.0	3.132	1.566 (7)
			-0.05	3.135	1.560 (7)				-0.05	3.131	1.593 (7)
			-0.1	3.135	1.583 (7)				-0.1	3.131	1.622 (7)
		300	0.1	3.149	1.533 (7)			300	0.1	3.146	1.553 (7)
			0.05	3.149	1.555 (7)				0.05	3.146	1.580 (7)
			0.0	3.150	1.578 (7)				0.0	3.146	1.607 (7)
			-0.05	3.150	1.602 (7)				-0.05	3.146	1.636 (7)
			-0.1	3.150	1.626 (7)				-0.1	3.146	1.666 (7)
		500	0.1	3.163	1.572 (7)			500	0.1	3.158	1.592 (7)
			0.05	3.163	1.595 (7)				0.05	3.159	1.620 (7)
			0.0	3.163	1.619 (7)				0.0	3.159	1.649 (7)
			-0.05	3.164	1.644 (7)				-0.05	3.160	1.680 (7)
			-0.1	3.164	1.669 (7)				-0.1	3.161	1.711 (7)
		700	0.1	3.175	1.611 (7)			700	0.1	3.171	1.632 (7)
			0.05	3.176	1.635 (7)				0.05	3.171	1.661 (7)
			0.0	3.177	1.660 (7)				0.0	3.173	1.692 (7)
			-0.05	3.178	1.686 (7)				-0.05	3.174	1.725 (7)
			-0.1	3.178	1.713 (7)				-0.1	3.175	1.756 (7)
		1000	0.1	3.193	1.670 (7)			1000	0.1	3.188	1.692 (7)
			0.05	3.194	1.696 (7)				0.05	3.190	1.723 (7)
			0.0	3.195	1.723 (7)				0.0	3.191	1.755 (7)
			-0.05	3.197	1.750 (7)				-0.05	3.193	1.789 (7)
			-0.1	3.198	1.779 (7)				-0.1	3.195	1.824 (7)
0.95	0.5	0	0.1	3.118	2.049 (7)*	0.95	-0.1	0	0.1	3.134	1.487 (7)
			0.05	3.117	2.136 (7)				0.05	3.133	1.505 (7)
			0.0	3.117	2.231 (7)				0.0	3.133	1.523 (7)
			-0.05	3.117	2.334 (7)				-0.05	3.133	1.542 (7)
			-0.1	3.117	2.447 (7)				-0.1	3.132	1.561 (7)
		100	0.1	3.125	2.075 (7)			100	0.1	3.141	1.506 (7)
			0.05	3.125	2.164 (7)				0.05	3.141	1.525 (7)
			0.0	3.125	2.261 (7)				0.0	3.141	1.543 (7)
			-0.05	3.126	2.376 (7)				-0.05	3.141	1.562 (7)
			-0.1	3.126	2.483 (7)				-0.1	3.140	1.582 (7)
		300	0.1	3.137	2.127 (7)			300	0.1	3.155	1.546 (7)
			0.05	3.138	2.220 (7)				0.05	3.155	1.565 (7)
			0.0	3.140	2.321 (7)				0.0	3.155	1.584 (7)
			-0.05	3.141	2.433 (7)				-0.05	3.155	1.604 (7)
			-0.1	3.142	2.555 (7)				-0.1	3.155	1.625 (7)
		500	0.1	3.149	2.179 (7)			500	0.1	3.169	1.585 (7)
			0.05	3.151	2.276 (7)				0.05	3.169	1.605 (7)
			0.0	3.153	2.382 (7)				0.0	3.169	1.625 (7)
			-0.05	3.155	2.499 (7)				-0.05	3.170	1.646 (7)
			-0.1	3.158	2.627 (7)				-0.1	3.170	1.668 (7)
		700	0.1	3.160	2.231 (7)			700	0.1	3.182	1.624 (7)
			0.05	3.163	2.332 (7)				0.05	3.182	1.645 (7)
			0.0	3.166	2.444 (7)				0.0	3.183	1.689 (7)
			-0.05	3.169	2.566 (7)				-0.05	3.183	1.689 (7)
			-0.1	3.173	2.700 (7)				-0.1	3.184	1.711 (7)
		1000	0.1	3.176	2.310 (7)			1000	0.1	3.200	1.684 (7)
			0.05	3.180	2.418 (7)				0.05	3.201	1.706 (7)
			0.0	3.184	2.536 (7)				0.0	3.202	1.729 (7)
			-0.05	3.189	2.667 (7)				-0.05	3.203	1.753 (7)
			-0.1	3.194	2.811 (7)				-0.1	3.203	1.777 (7)

TABLE I (Cont)

η	μ	Q	N	a_c	T_c	η	μ	Q	N	a_c	T_c
0.95	0.5	0	0.1	3.216	1.931 (7)*	0.95	-0.75	0	0.1	3.498	2.850 (7)
			0.05	3.216	1.964 (7)				0.05	3.501	2.874 (7)
			0.0	3.216	1.962 (7)				0.0	3.501	2.898 (7)
			-0.05	3.216	1.978 (7)				-0.05	3.504	2.923 (7)
			-0.1	3.216	1.995 (7)				-0.1	3.509	2.949 (7)
		100	0.1	3.225	1.954 (7)			100	0.1	3.507	2.879 (7)
			0.05	3.224	1.970 (7)				0.05	3.509	2.903 (7)
			0.0	3.224	1.987 (7)				0.0	3.512	2.928 (7)
			-0.05	3.224	2.003 (7)				-0.05	3.515	2.953 (7)
			-0.1	3.224	2.020 (7)				-0.1	3.517	2.979 (7)
		300	0.1	3.240	2.003 (7)			300	0.1	3.523	2.937 (7)
			0.05	3.240	2.019 (7)				0.05	3.526	2.962 (7)
			0.0	3.241	2.036 (7)				0.0	3.530	2.988 (7)
			-0.05	3.241	2.053 (7)				-0.05	3.533	3.014 (7)
			-0.1	3.241	2.071 (7)				-0.1	3.536	3.040 (7)
		500	0.1	3.256	2.051 (7)			500	0.1	3.539	2.995 (7)
			0.05	3.256	2.068 (7)				0.05	3.558	3.021 (7)
			0.0	3.256	2.086 (7)				0.0	3.562	3.047 (7)
			-0.05	3.257	2.104 (7)				-0.05	3.566	3.074 (7)
			-0.1	3.257	2.122 (7)				-0.1	3.569	3.101 (7)
		700	0.1	3.270	2.099 (7)			700	0.1	3.555	3.053 (7)
			0.05	3.271	2.117 (7)				0.05	3.558	3.080 (7)
			0.0	3.271	2.135 (7)				0.0	3.562	3.107 (7)
			-0.05	3.272	2.154 (7)				-0.05	3.566	3.135 (7)
			-0.1	3.272	2.173 (7)				-0.1	3.569	3.163 (7)
		1000	0.1	3.291	2.172 (7)			1000	0.1	3.577	3.140 (7)
			0.05	3.292	2.191 (7)				0.05	3.581	3.168 (7)
			0.0	3.292	2.221 (7)				0.0	3.585	3.196 (7)
			-0.05	3.293	2.230 (7)				-0.05	3.589	3.225 (7)
			-0.1	3.294	2.251 (7)				-0.1	3.593	3.255 (7)

numbers in the parenthesis denote powers of 10

TABLE II. CRITICAL VALUES OF TAYLOR NUMBERS AND WAVE NUMBERS FOR VARIOUS VALUES OF η, μ, Q AND N

η	μ	Q	N	a_c	T_c	η	μ	Q	N	a_c	T_c
0.85	0.0	0	0.1	3.132	7.007 (5)*	0.85	0.1	0	0.1	3.127	7.205 (5)
			0.05	3.131	7.111 (5)				0.05	3.126	7.337 (5)
			0.0	3.131	7.217 (5)				0.0	3.126	7.474 (5)
			-0.05	3.130	7.326 (5)				-0.05	3.126	7.616 (5)
			-0.1	3.129	7.439 (5)				-0.1	3.125	7.764 (5)
		100	0.1	3.190	7.841 (5)			100	0.1	3.185	8.060 (5)
			0.05	3.192	7.967 (5)				0.05	3.186	8.221 (5)
			0.0	3.193	8.097 (5)				0.0	3.188	8.389 (5)
			-0.05	3.194	8.232 (5)				-0.05	3.189	8.564 (5)
			-0.1	3.195	8.371 (5)				-0.1	3.191	8.746 (5)
		300	0.1	3.277	9.565 (5)			300	0.1	3.268	9.828 (5)
			0.05	3.280	9.744 (5)				0.05	3.273	1.006 (6)
			0.0	3.284	9.930 (5)				0.0	3.278	1.029 (6)
			-0.05	3.288	1.012 (6)				-0.05	3.283	1.054 (6)
			-0.1	3.292	1.032 (6)				-0.1	3.288	1.081 (6)
		500	0.1	3.334	1.137 (6)			500	0.1	3.324	1.168 (6)
			0.05	3.342	1.162 (6)				0.05	3.332	1.199 (6)
			0.0	3.348	1.187 (6)				0.0	3.341	1.231 (6)
			-0.05	3.356	1.213 (6)				-0.05	3.350	1.265 (6)
			-0.1	3.363	1.240 (6)				-0.1	3.360	1.301 (6)
		700	0.1	3.376	1.327 (6)			700	0.1	3.362	1.362 (6)
			0.05	3.385	1.358 (6)				0.05	3.374	1.402 (6)
			0.0	3.395	1.391 (6)				0.0	3.386	1.444 (6)
			-0.05	3.405	1.425 (6)				-0.05	3.400	1.489 (6)
			-0.1	3.417	1.461 (6)				-0.1	3.414	1.536 (6)
		1000	0.1	3.415	1.628 (6)			1000	0.1	3.397	1.670 (6)
			0.05	3.429	1.672 (6)				0.05	3.414	1.726 (6)
			0.0	3.444	1.719 (6)				0.0	3.433	1.786 (6)
			-0.05	3.460	1.768 (6)				-0.05	3.454	1.850 (6)
			-0.1	3.477	1.820 (6)				-0.1	3.476	1.919 (6)

TABLE II (Cont)

η	μ	Q	N	a_c	T_c	η	μ	Q	N	a_c	T_c
0.85	0.5	0	0.1	3.119	1.278 (6)*	0.85	-0.1	0	0.1	3.138	7.029 (5)
			0.05	3.119	1.361 (6)				0.05	3.138	7.115 (5)
			0.0	3.119	1.457 (6)				0.0	3.138	7.202 (5)
			-0.05	3.119	1.567 (6)				-0.05	3.137	7.292 (5)
			-0.1	3.119	1.695 (6)				-0.1	3.137	7.384 (5)
		100	0.1	3.168	1.418 (6)			100	0.1	3.199	7.864 (5)
			0.05	3.173	1.519 (6)				0.05	3.200	7.969 (5)
			0.0	3.179	1.636 (6)				0.0	3.201	8.076 (5)
			-0.05	3.187	1.772 (6)				-0.05	3.201	8.186 (5)
			-0.1	3.195	1.933 (6)				-0.1	3.202	8.299 (5)
		300	0.1	3.233	1.705 (6)			300	0.1	3.287	9.592 (5)
			0.05	3.249	1.845 (6)				0.05	3.291	9.740 (5)
			0.0	3.268	2.010 (6)				0.0	3.293	9.892 (5)
			-0.05	3.290	2.206 (6)				-0.05	3.297	1.005 (6)
			-0.1	3.318	2.445 (6)				-0.1	3.300	1.021 (6)
		500	0.1	3.270	2.000 (6)			500	0.1	3.349	1.140 (6)
			0.05	3.297	2.184 (6)				0.05	3.354	1.160 (6)
			0.0	3.329	2.406 (6)				0.0	3.360	1.181 (6)
			-0.05	3.369	2.676 (6)				-0.05	3.365	1.202 (6)
			-0.1	3.418	3.013 (6)				-0.1	3.371	1.224 (6)
		700	0.1	3.289	2.303 (6)			700	0.1	3.392	1.330 (6)
			0.05	3.326	2.538 (6)				0.05	3.400	1.356 (6)
			0.0	3.372	2.825 (6)				0.0	3.408	1.383 (6)
			-0.05	3.430	3.183 (6)				-0.05	3.417	1.411 (6)
			-0.1	3.505	3.640 (6)				-0.1	3.425	1.440 (6)
		1000	0.1	3.295	2.773 (6)			1000	0.1	3.436	1.632 (6)
			0.05	3.348	3.096 (6)				0.05	3.448	1.668 (6)
			0.0	3.416	3.499 (6)				0.0	3.460	1.706 (6)
			-0.05	3.504	4.018 (6)				-0.05	3.473	1.746 (6)
			-0.1	3.623	4.703 (6)				-0.1	3.487	1.787 (6)
0.85	0.5	0	0.1	3.275	9.877 (5)*	0.85	-0.75	0	0.1	3.819	1.633 (6)
			0.05	3.276	9.958 (5)				0.05	3.824	1.647 (6)
			0.0	3.276	1.004 (6)				0.0	3.829	1.661 (6)
			-0.05	3.276	1.013 (6)				-0.05	3.834	1.676 (6)
			-0.1	3.276	1.021 (6)				-0.1	3.840	1.690 (6)
		100	0.1	3.347	1.094 (6)			100	0.1	3.877	1.755 (6)
			0.05	3.348	1.104 (6)				0.05	3.883	1.771 (6)
			0.0	3.350	1.114 (6)				0.0	3.889	1.787 (6)
			-0.05	3.351	1.124 (6)				-0.05	3.895	1.803 (6)
			-0.1	3.352	1.134 (6)				-0.1	3.901	1.819 (6)
		300	0.1	3.458	1.311 (6)			300	0.1	3.971	2.002 (6)
			0.05	3.461	1.324 (6)				0.05	3.979	2.021 (6)
			0.0	3.464	1.337 (6)				0.01	3.986	2.041 (6)
			-0.05	3.467	1.365 (6)				-0.05	3.994	2.060 (6)
			-0.1	3.470	1.365 (5)				-0.1	4.002	2.080 (6)
		500	0.1	3.540	1.533 (6)			500	0.1	4.044	2.253 (6)
			0.05	3.546	1.550 (6)				0.05	4.054	2.276 (6)
			0.0	3.550	1.568 (6)				0.0	4.063	2.299 (6)
			-0.05	3.555	1.586 (6)				-0.05	4.072	2.323 (6)
			-0.1	3.560	1.604 (6)				-0.1	4.081	2.347 (6)
		700	0.1	3.606	1.762 (6)			700	0.1	4.104	2.510 (6)
			0.05	3.613	1.784 (6)				0.05	4.114	2.537 (6)
			0.0	3.619	1.806 (6)				0.0	4.126	2.564 (6)
			-0.05	3.626	1.828 (6)				-0.05	4.137	2.592 (6)
			-0.1	3.633	1.851 (6)				-0.1	4.148	2.620 (6)
		1000	0.1	3.682	2.119 (6)			1000	0.1	4.177	2.906 (6)
			0.05	3.691	2.148 (6)				0.05	4.190	2.939 (6)
			0.0	3.701	2.163 (6)				0.0	4.203	2.973 (6)
			-0.05	3.711	2.208 (6)				-0.05	4.217	3.007 (6)
			-0.1	3.722	2.240 (6)				-0.1	4.230	3.043 (6)

numbers in the parenthesis denote powers of 10

and the boundary conditions are given by

$$T = T_1 \text{ at } r = R_1, \qquad T = T_2 \text{ at } R_2 \tag{4}$$

where R_1, R_2 are the radii of the inner and the outer cylinders respectively. The solution of the system of equations (3) and (4) is given by

$$T = T_2 - \frac{(T_2 - T_1) \ln (r/R_2)}{\ln (R_1/R_2)}. \tag{5}$$

Assuming the perturbed flow as described by the variables

$$u', v' + V(r), w', p' + P(r), \rho' + \rho, f', g', H + h', T + T',$$

with axisymmetric disturbances about the z-axis, we can derive the equations governing the marginal state, in terms of normal modes as

$$(DD^* - \lambda^2)^2 u = 2 \frac{\lambda^2}{\nu} \frac{V}{r} v - \alpha \frac{\lambda^2}{\nu} \frac{V^2}{r} T - \frac{\lambda \mu_e}{4\pi\rho\nu} H (DD^* - \lambda^2) f \tag{6}$$

$$(DD^* - \lambda^2) v = \frac{2A}{\nu} u - \frac{\lambda \mu_e}{4\pi\rho\nu} H g \tag{7}$$

$$(DD^* - \lambda^2) f = \frac{\lambda}{e} H u \tag{8}$$

$$(DD^* - \lambda^2) g = \frac{\lambda}{e} H v - \frac{2B}{er^2} f \tag{9}$$

$$(DD^* - \lambda^2) T = - \frac{\rho c_p (T_2 - T_1)}{r k \ln (R_1/R_2)} u \tag{10}$$

where $\quad D = \dfrac{d}{dr}, \quad D^* = \dfrac{d}{dr} + \dfrac{1}{r}$.

$$DD^* = \frac{d^2}{dr^2} + \frac{1}{r}\frac{d}{dr} - \frac{1}{r^2}, \qquad D^*D = \frac{d^2}{dr^2} + \frac{1}{r}\frac{d}{dr}$$

On introducing the following non-dimensional variables

$$x = \frac{r - R_1}{d}, \quad a = \lambda d, \quad \eta = \frac{R_1}{R_2}, \quad \mu = \frac{\Omega_2}{\Omega_1}, \quad \xi = \frac{r}{R_2}, \quad G(x) = \frac{V}{r\Omega_1} = A_1 + \frac{B_1}{\xi^2}$$

$$\bar{u} = \frac{2Ad^2}{\nu} u, \quad \theta = -\rho c_p \frac{2kR_2 A \ln \eta}{\nu(T_2 - T_1)} T, \tag{11}$$

$$\bar{f} = \frac{2 e A d}{\nu a H} f, \quad \bar{g} = \frac{e}{aHd} g, \quad A_1 = \frac{\mu - \eta^2}{1 - \eta^2}, \quad B_1 = \eta^2 \frac{1 - \mu}{1 - \eta^2}$$

into the equations (6)-(10), we get

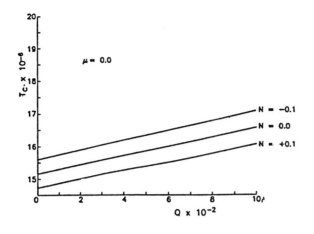

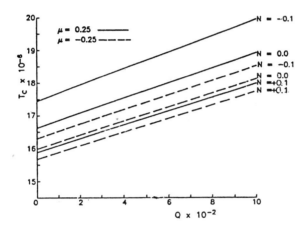

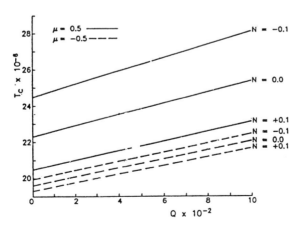

Fig.1. Variation of T_c with Q; $\eta = 0.95$

$$\left[(DD^* - a^2)^2 + \frac{\mu_e H^2 d^2}{4\pi\rho\nu e} a^2\right] \bar{u} = \frac{4A\Omega_1 d^4}{\nu^2} a^2 G(x) v +$$

$$\frac{\alpha \rho c_p(T_2-T_1) d^4\Omega_1^2 \xi a^2 [G(x)]^2}{\nu k \ln \eta} \theta \qquad (12)$$

$$(DD^* - a^2) v = \bar{u} - \frac{\mu_e H^2 d^2}{4\pi\rho\nu e} a^2 \bar{g} \qquad (13)$$

$$(DD^* - a^2) \bar{f} = \bar{u} \qquad (14)$$

$$(DD^* - a^2) \bar{g} = \frac{B_1}{A_1} \frac{\nu}{e} \frac{1}{\xi^2} \bar{f} + v \qquad (15)$$

$$(D^*D - a^2) \theta = \frac{1}{\xi} \bar{u} \qquad (16)$$

where $D = \frac{d}{dx}$ and $D^* = (\frac{d}{dx} + \frac{1-\eta}{\xi})$.

These equations (12)-(16) further reduce to

$$[DD^* - a^2)^2 + Q a^2] u = -a^2 \text{Ta} [G(x) v + N \xi (G(x))^2 \cdot \theta] \qquad (17)$$

$$(DD^* - a^2) v = u - Qa^2 g \qquad (18)$$

$$(DD^* - a^2) g = \frac{B_1}{A_1 \xi^2} \text{Pm} f + v \qquad (19)$$

$$(D^*D - a^2)\theta = \frac{1}{\xi} u \qquad (20)$$

$$\left.\begin{array}{l} \text{Ta} = \frac{-4A\Omega_1 d^2}{\nu^2}, \text{Ra} = \frac{-\alpha\rho c_p (T_2 - T_1)d^4\Omega_1^2}{\nu k \ln\eta} \\[6pt] \text{Pr} = \frac{\nu\rho C_p}{k}, \text{Pm} = \frac{\nu}{e}, Q = \frac{\mu_e H^2 d^2}{4\pi\rho\nu e} \\[6pt] N = \frac{\text{Ra}}{\text{Ta}} = \frac{\text{Pr } \alpha (T_2 - T_1)}{4A_1 \ln\eta} \end{array}\right\} \qquad (21)$$

Here Ta is the Taylor number, Ra the Rayleigh number, Pr the Prandtl number, Pm the magnetic Prandtl number, Q the magnetic field parameter and N is the ratio of the Rayleigh and Taylor numbers respectively and the overbar is deleted.

Now, for many electrically conducting fluids, the magnetic Prandtl number is very small e.g. for mercury, Pm = 1.45 x 10^{-7}. Hence we can neglect the terms containing Pm in above equations (17)-(21). Thus eliminating v between (17), (18) and (19), we get the following equations:

$$[(DD^* - a^2)^2 + Qa^2] u = -a^2 \text{Ta} [G(x) (DD^* - a^2) g + N \xi (G(x))^2 \theta] \qquad (22)$$
$$[(DD^* - a^2)^2 + Qa^2] g = u \qquad (23)$$
$$(DD^* - a^2) \theta = \frac{1}{\xi} u \qquad (24)$$

TABLE III. CRITICAL VALUES OF TAYLOR NUMBERS AND WAVE NUMBERS
FOR VARIOUS VALUES OF η, μ, Q AND N

η	μ	Q	N	a_c	T_c	η	μ	Q	N	a_c	T_c
0.75	0.0	0	0.1	3.136	2.029 (5)*	0.75	0.1	0	0.1	3.131	2.139 (5)
			0.05	3.136	2.060 (5)				0.05	3.130	2.182 (5)
			0.0	3.136	2.093 (5)				0.0	3.130	2.228 (5)
			-0.05	3.136	2.126 (5)				-0.05	3.129	2.275 (5)
			-0.1	3.136	2.161 (5)				-0.1	3.129	2.325 (5)
		100	0.1	3.274	2.709 (5)			100	0.1	3.263	2.852 (5)
			0.05	3.277	2.762 (5)				0.05	3.268	2.925 (5)
			0.0	3.281	2.816 (5)				0.0	3.273	3.001 (5)
			-0.05	3.284	2.873 (5)				-0.05	3.278	3.082 (5)
			-0.1	3.288	2.932 (5)				-0.1	3.284	3.167 (5)
		300	0.1	3.402	4.210 (5)			300	0.1	3.381	4.422 (5)
			0.05	3.414	4.321 (5)				0.05	3.397	4.576 (5)
			0.0	3.427	4.438 (5)				0.0	3.414	4.740 (5)
			-0.05	3.440	4.562 (5)				-0.05	3.433	4.916 (5)
			-0.1	3.454	4.692 (5)				-0.1	3.453	5.105 (5)
		500	0.1	3.448	5.902 (5)			500	0.1	3.417	6.188 (5)
			0.05	3.470	6.097 (5)				0.05	3.445	6.457 (5)
			0.0	3.494	6.306 (5)				0.0	3.476	6.749 (5)
			-0.05	3.519	6.528 (5)				-0.05	3.510	7.067 (5)
			-0.1	3.546	6.766 (5)				-0.1	3.548	7.414 (5)
		700	0.1	3.456	7.775 (5)			700	0.1	3.413	8.137 (5)
			0.05	3.489	8.083 (5)				0.05	3.453	8.559 (5)
			0.0	3.525	8.416 (5)				0.0	3.501	9.025 (5)
			-0.05	3.563	8.774 (5)				-0.05	3.554	9.539 (5)
			-0.1	3.606	9.162 (5)				-0.1	3.613	1.011 (6)
		1000	0.1	3.428	1.089 (6)			1000	0.1	3.366	1.136 (6)
			0.05	3.478	1.143 (6)				0.05	3.429	1.210 (6)
			0.0	3.535	1.202 (6)				0.0	3.501	1.292 (6)
			-0.05	3.597	1.266 (6)				-0.05	3.586	1.385 (6)
			-0.1	3.668	1.337 (6)				-0.1	3.685	1.491 (6)
0.75	0.25	0	0.1	3.126	2.592 (5) *	0.75	-0.1	0	0.1	3.146	2.019 (5)
			0.05	3.125	2.680 (5)				0.05	3.146	2.044 (5)
			0.0	3.125	2.774 (5)				0.0	3.145	2.069 (5)
			-0.05	3.125	2.875 (5)				-0.05	3.144	2.095 (5)
			-0.1	3.124	2.984 (5)				-0.1	3.144	2.122 (5)
		100	0.1	3.250	3.438 (5)			100	0.1	3.288	2.695 (5)
			0.05	3.258	3.583 (5)				0.05	3.291	2.736 (5)
			0.0	3.266	3.741 (5)				0.0	3.294	2.779 (5)
			-0.05	3.276	3.913 (5)				-0.05	3.296	2.823 (5)
			-0.1	3.286	4.102 (5)				-0.1	3.299	2.868 (5)
		300	0.1	3.349	5.279 (5)			300	0.1	3.428	4.184 (5)
			0.05	3.375	5.581 (5)				0.05	3.437	4.272 (5)
			0.0	3.404	5.918 (5)				0.0	3.447	4.363 (5)
			-0.05	3.436	6.298 (5)				-0.05	3.457	4.458 (5)
			-0.1	3.474	6.727 (5)				-0.1	3.468	4.558 (5)
		500	0.1	3.365	7.320 (5)			500	0.1	3.486	5.861 (5)
			0.05	3.409	7.842 (5)				0.05	3.503	6.016 (5)
			0.0	3.461	8.440 (5)				0.0	3.552	6.177 (5)
			-0.05	3.522	9.130 (5)				-0.05	3.541	6.348 (5)
			-0.1	3.593	9.933 (5)				-0.1	3.562	6.527 (5)
		700	0.1	3.340	9.542 (5)			700	0.1	3.506	7.718 (5)
			0.05	3.404	1.035 (6)				0.05	3.533	7.961 (5)
			0.0	3.480	1.130 (6)				0.0	3.561	8.218 (5)
			-0.05	3.573	1.243 (6)				-0.05	3.592	8.491 (5)
			-0.1	3.687	1.377 (6)				-0.1	3.624	8.781 (5)
		1000	0.1	3.263	1.316 (6)			1000	0.1	3.499	1.081 (6)
			0.05	3.355	1.455 (6)				0.05	3.540	1.123 (6)
			0.0	3.472	1.622 (6)				0.0	3.586	1.168 (6)
			-0.05	3.622	1.828 (6)				-0.05	3.635	1.217 (6)
			-0.1	3.817	2.082 (6)				-0.1	3.689	1.269 (6)

TABLE III (Cont)

η	μ	Q	N	a_c	T_c	η	μ	Q	N	a_c	T_c
0.75	-0.25	0	0.1	3.177	2.183 (5)*	0.75	-0.5	0	0.1	3.410	3.233 (5)
			0.05	3.176	2.204 (5)				0.05	3.411	3.262 (5)
			0.0	3.176	2.225 (5)				0.0	3.413	3.290 (5)
			-0.05	3.175	2.247 (5)				-0.05	3.414	3.319 (5)
			-0.1	3.175	2.269 (5)				-0.1	3.416	3.349 (5)
		100	0.1	3.330	2.900 (5)			100	0.1	3.593	4.120 (5)
			0.05	3.332	2.934 (5)				0.05	3.597	4.162 (5)
			0.0	3.334	2.969 (5)				0.0	3.601	4.204 (5)
			-0.05	3.336	3.005 (5)				-0.05	3.606	4.248 (5)
			-0.1	3.338	3.042 (5)				-0.1	3.610	4.292 (5)
		300	0.1	3.491	4.463 (5)			300	0.1	3.808	5.962 (5)
			0.05	3.499	4.534 (5)				0.05	3.817	6.038 (5)
			0.0	3.506	4.608 (5)				0.0	3.828	6.114 (5)
			-0.05	3.514	4.684 (5)				-0.05	3.838	6.193 (5)
			-0.1	3.522	4.762 (5)				-0.1	3.849	6.274 (5)
		500	0.1	3.573	6.205 (5)			500	0.1	3.940	7.932 (5)
			0.05	3.587	6.328 (5)				0.05	3.955	8.048 (5)
			0.0	3.602	6.456 (5)				0.0	3.971	8.168 (5)
			-0.05	3.617	6.589 (5)				-0.05	3.987	8.292 (5)
			-0.1	3.633	6.728 (5)				-0.1	4.003	8.418 (5)
		700	0.1	3.169	8.118 (5)			700	0.1	4.030	1.004 (6)
			0.05	3.641	8.309 (5)				0.05	4.052	1.020 (6)
			0.0	3.662	8.510 (5)				0.0	4.074	1.038 (6)
			-0.05	3.686	8.719 (5)				-0.05	4.096	1.055 (6)
			-0.1	3.711	8.938 (5)				-0.1	4.119	1.074 (6)
		1000	0.1	3.649	1.128 (6)			1000	0.1	4.126	1.344 (6)
			0.05	3.683	1.161 (6)				0.05	4.157	1.370 (6)
			0.0	3.720	1.195 (6)				0.0	4.190	1.398 (6)
			-0.05	3.758	1.232 (6)				-0.05	4.223	1.426 (6)
			-0.1	3.800	1.270 (6)				-0.1	4.257	1.455 (6)

numbers in the parenthesis denote powers of 10

TABLE IV. CRITICAL VALUES OF TAYLOR NUMBERS AND WAVE NUMBERS FOR VARIOUS VALUES OF η, μ, Q AND N

η	μ	Q	N	a_c	T_c	η	μ	Q	N	a_c	T_c
0.5	0.0	0	0.1	3.164	7.168 (4)*	0.5	0.1	0	0.1	3.152	9.283 (4)
			0.05	3.163	7.301 (4)				0.05	3.151	9.441 (4)
			0.0	3.163	7.439 (4)				0.0	3.151	9.949 (4)
			-0.05	3.162	7.582 (4)				-0.05	3.150	1.013 (5)
			-0.1	3.161	7.731 (4)				-0.1	3.150	1.072 (5)
		100	0.1	3.463	1.740 (5)			100	0.1	3.397	2.208 (5)
			0.05	3.482	1.802 (5)				0.05	3.431	2.352 (5)
			0.0	3.502	1.867 (5)				0.0	3.469	2.515 (5)
			-0.05	3.524	1.938 (5)				-0.05	3.514	2.701 (5)
			-0.1	3.548	2.014 (5)				-0.1	3.566	2.915 (5)
		300	0.1	3.463	4.456 (5)			300	0.1	3.272	5.464 (5)
			0.03	3.535	4.748 (5)				0.05	3.382	6.116 (5)
			0.0	3.620	5.076 (5)				0.0	3.529	6.926 (5)
			-0.05	3.719	5.448 (5)				-0.05	3.726	7.945 (5)
			-0.1	3.834	5.869 (5)				-0.1	3.990	9.242 (5)
		500	0.1	3.304	7.813 (5)			500	0.1	3.016	9.269 (5)
			0.05	3.434	8.552 (5)				0.05	3.178	1.085 (6)
			0.0	3.605	9.423 (5)				0.0	3.441	1.298 (6)
			-0.05	3.833	1.046 (6)				-0.05	3.886	1.594 (6)
			-0.1	4.128	1.168 (6)				-0.1	4.586	1.996 (6)
		700	0.1	3.107	1.157 (6)			700	0.1	2.774	1.331 (6)
			0.05	3.284	1.299 (6)				0.05	2.951	1.619 (6)
			0.0	3.556	1.475 (6)				0.0	3.308	2.046 (6)
			-0.05	3.987	1.692 (6)				-0.05	4.161	2.693 (6)
			-0.1	4.605	1.956 (6)				-0.1	5.575	3.581 (6)
		1000	0.1	2.825	1.763 (6)			1000	0.1	2.492	1.950 (6)
			0.05	3.036	2.049 (6)				0.05	2.651	2.485 (6)
			0.0	3.473	2.431 (6)				0.0	3.091	3.393 (6)
			-0.05	5.002	3.212 (6)				-0.05	4.605	4.965 (6)
			-0.1	5.570	3.514 (6)				-0.1	6.205	1.536 (7)

TABLE IV (Cont)

η	μ	Q	N	a_c	T_c	η	μ	Q	N	a_c	T_c
0.5	-0.1	0	0.1	3.191	6.839 (4)*	0.5	-0.5	0	0.1	4.791	2.055 (5)
			0.05	3.191	6.922 (4)				0.05	4.796	2.072 (5)
			0.0	3.189	7.008 (4)				0.0	4.801	2.089 (5)
			-0.05	3.188	7.095 (4)				-0.05	4.806	2.106 (5)
			-0.1	3.188	7.185 (4)				-0.1	4.811	2.123 (5)
		100	0.1	3.546	1.650 (5)			100	0.1	5.053	3.189 (5)
			0.05	3.558	1.688 (5)				0.05	5.064	3.222 (5)
			0.0	3.571	1.728 (5)				0.0	5.077	3.256 (5)
			-0.05	3.585	1.770 (5)				-0.05	5.089	3.291 (5)
			-0.1	3.599	1.813 (5)				-0.1	5.101	3.326 (5)
		300	0.1	3.677	4.194 (5)			300	0.1	5.442	5.838 (5)
			0.5	3.731	4.731 (5)				0.05	5.468	5.921 (5)
			0.0	3.791	4.562 (5)				0.0	5.495	6.007 (5)
			-0.05	3.586	4.769 (5)				-0.05	5.522	6.094 (5)
			-0.1	3.929	4.994 (5)				-0.1	5.548	6.184 (5)
		500	0.1	3.648	7.360 (5)			500	0.1	5.718	8.906 (5)
			0.05	3.759	7.803 (5)				0.05	5.759	9.064 (5)
			0.0	3.891	8.294 (5)				0.0	5.804	9.227 (5)
			-0.05	4.044	8.839 (5)				-0.05	5.846	9.395 (5)
			-0.1	4.223	9.444 (5)				-0.1	5.889	9.567 (5)
		700	0.1	3.555	1.097 (6)			700	0.1	5.914	1.233 (6)
			0.05	3.732	1.182 (6)				0.05	5.975	1.259 (6)
			0.0	3.958	1.279 (6)				0.0	6.037	1.286 (6)
			-0.05	4.241	1.388 (6)				-0.05	6.100	1.314 (6)
			-0.1	4.577	1.511 (6)				-0.1	6.163	1.343 (6)
		1000	0.1	3.356	1.698 (6)			1000	0.1	6.112	1.806 (6)
			0.05	3.630	1.873 (6)				0.05	6.204	1.853 (6)
			0.0	4.034	2.077 (6)				0.0	6.299	1.901 (6)
			-0.05	4.576	2.313 (6)				-0.05	6.396	1.952 (6)
			-0.1	5.192	2.579 (6)				-0.1	6.497	2.005 (6)

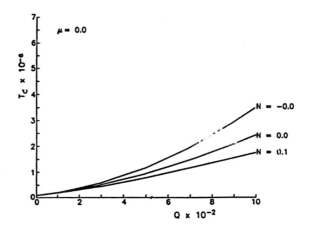

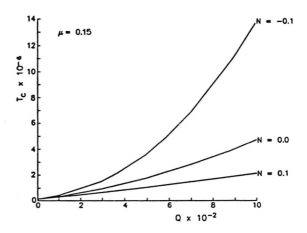

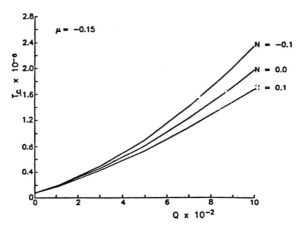

Fig.2. Variation of T_c with Q; $\eta = 0.5$

For convenience, we define Taylor number and the magnetic field parameter, following sparrow et al. (1964) and Astill and Chung (1976), as follows:

$$\overline{Ta} = \frac{\Omega_1^2 R_2^4}{\nu^2}, \quad \overline{Q} = \frac{\mu_e H^2 R_2^2}{4\pi\rho\nu e} \tag{25}$$

and $\bar{N} = \Pr \alpha (T_2 - T_1)$.

Then in view of (25), equations (22) - (24) reduce to the following:

$$[(DD^* - a^2)^2 + \bar{Q}(1-\eta)^2 a^2] u = a^2 \overline{Ta} \, 4 A_1 (1-\eta)^4 . [G(x).(DD^* - a^2) g$$

$$+ \frac{\bar{N} \xi}{4 A_1 \ln \eta} (G(x))^2 \theta] \tag{26}$$

$$[(DD^* - a^2)^2 + \bar{Q}(1-\eta)^2 a^2] g = u \tag{27}$$

$$(DD^* - a^2)\theta = \frac{1}{\xi} u . \tag{28}$$

The boundary conditions for the case of conducting walls are given by

$$u = Du = D^*g = (DD^* - a^2) g = \theta = 0 \text{ at } x = 0, 1 . \tag{29}$$

Equations (26)-(28) with boundary conditions (29) define an eigenvalue problem of the form

$$S[\overline{Ta}, a, \mu, \eta, \bar{N}, \bar{Q}] = 0 . \tag{30}$$

For fixed values of μ, η, $\bar{N}$, and $\bar{Q}$, we determine the minimum positive $\overline{Ta}$ over all real positive 'a' for which there exists a non-trivial solution of the eigenvalue problem. The critical value of $\overline{Ta}$ determined the velocity Ω_1, at the onset of instability, and the corresponding critical value of 'a' determined the spacing of the vortices in the z-direction.

METHOD OF SOLUTION

To solve the eigenvalue problem (30), we adopt the numerical procedure described by Harris and Reid (1964) for an isothermal non-magnetic Couette flow. The procedure starts by rewriting the governing equations as a system of first order equations. For that purpose, we put

$$\left. \begin{array}{l} A = u, \, B = Du, \, C = (DD^* - a^2)u, \\ E = D^*(DD^* - a^2) u, \, F = g, \, G = D^*g \\ H = (DD^* - a^2)g, \, I = D^*(DD^* - a^2)g, \, J = \theta, \, K = D\theta. \end{array} \right\} \tag{31}$$

In view of (31), we can easily prove the following relations:

TABLE V. CRITICAL VALUES OF TAYLOR NUMBERS AND WAVE NUMBERS FOR VARIOUS VALUES OF η, μ, Q AND N

η	μ	Q	N	a_c	T_c	η	μ	Q	N	a_c	T_c
0.2	0.0	0	0.1	3.266	2.871 (5)*	0.2	0.01	0	0.1	3.254	3.441 (5)
			0.05	3.265	2.950 (5)				0.05	3.255	3.570 (5)
			0.0	3.263	2.035 (5)				0.0	3.253	3.709 (5)
			-0.05	3.262	3.123 (5)				-0.05	3.252	3.859 (5)
			-0.1	3.261	3.217 (5)				-0.1	3.251	4.021 (5)
		100	0.1	3.673	1.316 (6)			100	0.1	3.560	1.540 (6)
			0.05	3.747	1.425 (6)				0.05	3.653	1.712 (6)
			0.0	3.838	1.553 (6)				0.0	3.772	1.923 (6)
			-0.05	3.947	1.704 (6)				-0.05	3.925	2.185 (6)
			-0.1	4.077	1.881 (6)				-0.1	4.119	2.517 (6)
		300	0.1	3.595	4.036 (6)			300	0.1	3.288	4.548 (6)
			0.05	3.941	4.740 (6)				0.05	3.620	5.601 (6)
			0.0	4.514	5.682 (6)				0.0	4.326	7.178 (6)
			-0.05	5.330	6.938 (6)				-0.05	5.535	9.519 (6)
			-0.1	6.282	8.589 (6)				-0.1	6.915	1.288 (7)
		500	0.1	3.320	6.984 (6)			500	0.1	2.966	7.641 (6)
			0.05	3.871	8.711 (6)				0.05	3.393	1.010 (7)
			0.0	5.081	1.126 (7)				0.0	4.877	1.438 (7)
			-0.05	6.761	1.485 (7)				-0.05	7.296	2.116 (7)
			-0.1	8.517	1.974 (7)				-0.1	9.684	3.134 (7)
		700	0.1	3.048	9.925 (6)			700	0.1	2.725	1.063 (7)
			0.05	3.642	1.299 (7)				0.05	3.121	1.478 (7)
			0.0	5.451	1.797 (7)				0.0	5.236	2.306 (7)
			-0.05	8.037	2.522 (7)				-0.05	8.871	3.684 (7)
			-0.1	10.564	3.530 (7)				-0.1	12.168	5.809 (7)
		1000	0.1	2.745	1.422 (7)			1000	0.1	2.483	1.491 (7)
			0.05	3.261	1.962 (7)				0.05	2.784	2.182 (7)
			0.0	5.848	2.987 (7)				0.0	5.630	3.851 (7)
			-0.05	9.793	4.525 (7)				-0.05	11.028	6.784 (7)
			-0.1	11.626	6.704 (7)				-0.1	15.526	1.143 (8)
0.2	-0.01	0	0.1	3.279	2.553 (5)*	0.2	-0.1	0	0.1	4.237	3.353 (5)
			0.05	3.278	2.609 (5)				0.05	4.243	3.394 (5)
			0.0	3.277	2.668 (5)				0.0	4.250	3.436 (5)
			-0.05	3.276	2.729 (5)				-0.05	4.257	3.479 (5)
			-0.1	3.274	2.793 (5)				-0.1	4.263	3.523 (5)
		100	0.1	3.780	1.177 (6)			100	0.1	5.197	9.695 (5)
			0.05	3.844	1.255 (6)				0.05	5.228	9.907 (5)
			0.0	3.918	1.343 (6)				0.0	5.260	1.013 (6)
			-0.05	4.003	1.442 (6)				-0.05	5.292	1.036 (6)
			-0.1	4.100	1.556 (6)				-0.1	5.326	1.059 (6)
		300	0.1	3.908	3.688 (6)			300	0.1	6.077	2.544 (6)
			0.05	4.239	4.169 (6)				0.05	6.181	2.639 (6)
			0.0	4.701	4.796 (6)				0.0	6.292	2.741 (6)
			-0.05	5.294	5.580 (6)				-0.05	6.409	2.850 (6)
			-0.1	5.983	6.556 (6)				-0.1	6.532	2.966 (6)
		500	0.1	3.725	6.463 (6)			500	0.1	6.500	4.415 (6)
			0.05	4.313	7.720 (6)				0.05	6.712	4.640 (6)
			0.0	5.282	9.406 (6)				0.0	6.931	4.885 (6)
			-0.05	6.520	1.161 (7)				-0.05	7.168	5.152 (6)
			-0.1	7.860	1.445 (7)				-0.1	7.421	5.442 (6)
		700	0.1	3.447	9.334 (6)			700	0.1	6.718	6.510 (6)
			0.05	4.172	1.163 (7)				0.05	7.036	6.924 (6)
			0.0	5.664	1.493 (7)				0.0	7.386	7.381 (6)
			-0.05	7.603	1.937 (7)				-0.05	7.768	7.885 (6)
			-0.1	9.584	2.517 (7)				-0.1	8.182	8.440 (6)
		1000	0.1	3.083	1.361 (7)			1000	0.1	6.792	9.980 (6)
			0.05	3.819	1.788 (7)				0.05	7.299	1.079 (7)
			0.0	6.067	2.470 (7)				0.0	7.876	1.170 (7)
			-0.05	9.090	3.410 (7)				-0.05	8.512	1.273 (7)
			-0.1	11.966	4.654 (7)				-0.1	9.206	1.387 (7)

numbers in the parenthesis denote powers of 10

i) $C = DB + mB - (m^2 + a^2)A$,   ii) $E = DC + mC$

ii) $(DD^* - a^2)^2 u = DE - a^2 C$,   iv) $G = DF + mF$

v) $H = DG - a^2 F$ vi) $I = DH + mH$

vii) $(DD^* - a^2)^2 g = DI - a^2 H$

viii) $(D^*D - a^2)\theta = DK + mK - a^2 J$ (32)

where $m = \dfrac{1-\eta}{\xi}$.

Equations (26)-(28) can now be transformed into a system of first order equations as follows:

$$DA = B, \quad DB = C - mB + (m^2 + a^2) A ,$$

$$DC = E - mC, \quad DE = a^2 C - \bar{Q}(1-\eta)^2 a^2 A + a^2 \overline{Ta} \cdot 4A_1$$

$$(1-\eta)^4 [G(x)H + \dfrac{\bar{N}\xi (G(x))^2 J}{4A_1 \ln\eta}], \quad (33)$$

$$DF = G - mF, \quad DG = H + a^2 F, \quad DH = I - mH ,$$
$$DI = a^2 H - \bar{Q}(1-\eta)^2 a^2 F + A ,$$
$$DJ = K, \quad DK = -mK + a^2 J + \dfrac{1}{\xi} A .$$

The boundary conditions for the case of contucting walls require that

$$[A = B = G = H = J] = 0 \text{ at } x = 0, 1 \quad . \quad (34)$$

We now define five linearly independent solutions A_i, F_i and J_i ($i = 1, 2, 3, 4, 5$) of equations (26)-(28) by imposing the intial conditions

$$[A_i = B_i = G_i = H_i = J_i] = 0 \ (i = 1, 2, 3, 4, 5) \quad (35)$$

and

$$[C_i, E_i, F_i, G_i, H_i] = \begin{cases} (1, 0, 0, 0, 0), & i = 1 \\ (0, 1, 0, 0, 0), & i = 2 \\ (0, 0, 1, 0, 0), & i = 3 \\ (0, 0, 0, 1, 0), & i = 4 \\ (0, 0, 0, 0, 1), & i = 5 \end{cases} \quad (36)$$

at $x = 0$.

A solution of equations (26)-(28), which automatically satisfies the boundary conditions at $x = 0$ can now be written in the form

$$u = \sum_{i=1}^{5} M_i A_i, \quad g = \sum_{i=1}^{5} N_i F_i, \quad \theta = \sum_{i=1}^{5} P_i J_i \quad (37)$$

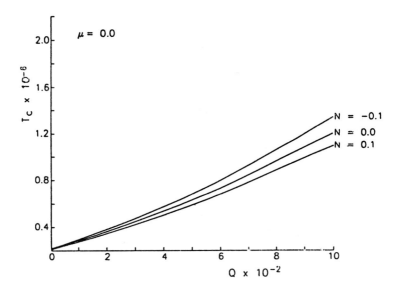
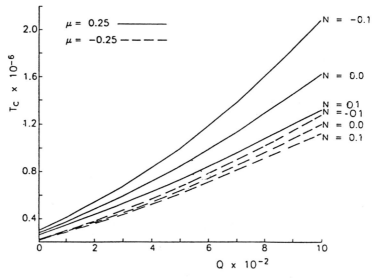

Fig.3. Variation of T_c with Q; $\eta = 0.75$

where M_i, N_i and P_i are constants. The boundary conditions at $x = 1$ require that

$$\left.\begin{array}{l}\sum_{i=1}^{5} M_i A_i = 0, \quad \sum_{i=1}^{5} M_i B_i = 0, \quad \sum_{i=1}^{5} N_i G_i = 0 \\ \\ \sum_{i=1}^{5} N_i H_i = 0, \quad \sum_{i=1}^{5} P_i J_i = 0 \end{array}\right\} \quad (38)$$

For a non-trivial solution, it is necessary that the determinant of the above system vanishes i.e.

$$S(\overline{T}a, a, \mu, \eta, \overline{N}, \overline{Q}) = \begin{vmatrix} A_i \\ B_i \\ G_i \\ H_i \\ J_i \end{vmatrix} = 0, \text{ at } x = 0, 1. \quad (39)$$

Equation (30) is the characteristic equation from which the critical Taylor number $\overline{T}_c$ and the corrseponding wave number a_c can be obtained for any assigned values of μ, $\overline{Q}$ and $\overline{N}$. The solution procedures followed to integrate the fifty equations defined by (33) and to evaluate the determinant (39) are the same as described in Harris and Reid (1964).

DISCUSSION

Since for a given fluid and fixed R_2, the values of $\overline{T}_c$, $\overline{Q}$ and $\overline{N}$ determine respectively the critical angular velocity Ω_{1c} the strength of the magnetic field and the quantity of heating, we find that the stability effect of various values of $\overline{Q}$, $\overline{N}$, μ and η are best measured through a comparison with the stability condition of a Couette flow with $\overline{Q} = 0$ (no axial magnetic field is present), $\overline{N} = 0$ (isothermal flow), $\mu = 0$ (stationary outer cylinder) and $\eta \sim 1$ (narrow gap between the cylinders). For such flow, it has been reported previously that the critical Taylor number $T_c = 3390$. (If we take $\eta = 0.95$, then the corresponding value of $\overline{T}_c$ equals 1.517×10^7 and we will denote this number by $\widetilde{T}_c$). The calculated changes on $\widetilde{T}_c$ due to the action of various factors are presented in Tables I-VI and in Figs. 1-4 for the case of conducting walls (in these Tables and Figures, we have dropped the bars from $\overline{T}_c$, $\overline{Q}$ and $\overline{N}$). From a general inspection of the results presented, we may state that for various values of η, the type of reaction of $\overline{T}_c$ to the presence of heating and axial magnetic field is the same as that described for the narrow gap case (i.e., $\eta \sim 1$). However, the quantitative influences of $\overline{Q}$, $\overline{N}$ and μ are strongly dependent on the value of η.

A close examination of the Tables and Figures reveals the following.

1. In general, a fixed negative temperature gradient between the cylinders (i.e. temperature of the outer cylinder is less than that of the inner one)

TABLE VI: CRITICAL VALUES OF TAYLOR NUMBERS AND WAVE NUMBERS FOR VARIOUS VALUES OF η, μ, Q AND N

η	μ	Q	N	a_c	T_c	η	μ	Q	N	a_c	T_c
0.1	0.0	0	0.1	3.345	2.265 (6)*	0.1	0.001	0	0.1	3.340	2.431 (6)
			0.05	3.341	2.348 (6)				0.05	3.337	2.529 (6)
			0.0	3.340	2.437 (6)				0.0	3.335	2.636 (6)
			-0.05	3.338	2.532 (6)				-0.05	3.332	2.751 (6)
			-0.1	3.335	2.635 (6)				-0.1	3.332	2.875 (6)
		100	0.1	3.877	1.121 (7)			100	0.1	3.821	1.191 (7)
			0.05	3.991	1.257 (7)				0.05	3.940	1.351 (7)
			0.0	4.137	1.427 (7)				0.0	4.099	1.555 (7)
			-0.05	4.325	1.642 (7)				-0.05	4.312	1.819 (7)
			-0.1	4.565	1.917 (7)				-0.1	4.588	2.168 (7)
		300	0.1	4.170	3.179 (7)			300	0.1	4.017	3.331 (7)
			0.05	4.711	3.943 (7)				0.05	4.565	4.211 (7)
			0.0	5.613	5.094 (7)				0.0	5.559	5.596 (7)
			-0.05	6.889	6.874 (7)				-0.05	7.018	7.845 (7)
			-0.1	8.350	9.608 (7)				-0.1	8.643	1.144 (8)
		500	0.1	4.040	5.182 (7)			500	0.1	3.859	5.375 (7)
			0.05	4.858	6.865 (7)				0.05	4.647	7.281 (7)
			0.0	6.603	9.752 (7)				0.0	6.550	1.074 (8)
			-0.05	9.173	1.148 (8)				-0.05	9.463	1.717 (8)
			-0.1	11.763	2.323 (8)				-0.1	12.194	2.829 (8)
		700	0.1	3.820	7.082 (7)			700	0.1	3.656	7.293 (7)
			0.05	4.733	9.862 (7)				0.05	4.480	1.039 (8)
			0.0	7.329	1.519 (8)				0.0	7.276	1.676 (8)
			-0.05	11.281	2.534 (8)				-0.05	11.725	2.978 (8)
			-0.1	14.765	4.310 (8)				-0.1	15.131	5.294 (8)
		1000	0.1	3.537	9.756 (7)			1000	0.1	3.391	9.973 (7)
			0.05	4.383	1.435 (8)				0.05	4.120	1.499 (8)
			0.0	8.148	2.461 (8)				0.0	8.095	2.717 (8)
			-0.05	14.224	4.617 (8)				-0.05	14.874	5.498 (8)
			-0.1	18.443	8.454 (8)				-0.1	19.325	1.039 (9)
0.1	-0.001	0	0.1	3.349	2.130 (6)*	0.1	-0.1	0	0.1	8.071	4.100 (6)
			0.05	3.347	2.201 (6)				0.05	8.080	4.140 (6)
			0.0	3.345	2.277 (6)				0.0	8.088	4.181 (6)
			-0.05	3.342	2.358 (6)				-0.05	8.096	4.223 (6)
			-0.1	3.340	2.444 (6)				-0.1	8.105	4.265 (6)
		100	0.1	3.932	1.062 (7)			100	0.1	8.605	6.657 (6)
			0.05	4.040	1.179 (7)				0.05	8.866	6.738 (6)
			0.0	4.175	1.324 (7)				0.0	8.887	6.825 (6)
			-0.05	4.346	1.502 (7)				-0.05	8.908	6.915 (6)
			-0.1	4.557	1.725 (7)				-0.1	8.929	7.007 (6)
		300	0.1	4.317	3.042 (7)			300	0.1	9.429	1.240 (7)
			0.05	4.843	3.711 (7)				0.05	9.994	1.259 (7)
			0.0	5.666	4.684 (7)				0.0	10.044	1.283 (7)
			-0.05	6.796	6.127 (7)				-0.05	10.095	1.307 (7)
			-0.1	8.106	8.270 (7)				-0.1	10.146	1.332 (7)
		500	0.1	4.221	5.002 (7)			500	0.1	10.646	1.874 (7)
			0.05	5.051	6.498 (7)				0.05	10.815	1.917 (7)
			0.0	6.656	8.942 (7)				0.0	10.901	1.962 (7)
			-0.05	8.945	1.300 (8)				-0.05	10.991	2.009 (7)
			-0.1	11.358	1.955 (8)				-0.1	11.083	2.058 (7)
		700	0.1	3.999	6.882 (7)			700	0.1	11.332	2.566 (7)
			0.05	4.971	9.385 (7)				0.05	11.458	2.637 (7)
			0.0	7.382	1.391 (8)				0.0	11.589	2.711 (7)
			-0.05	10.923	2.204 (8)				-0.05	11.725	2.789 (7)
			-0.1	14.290	3.587 (8)				-0.1	11.867	2.871 (7)
		1000	0.1	3.691	9.546 (7)			1000	0.1	12.013	3.694 (7)
			0.05	4.646	1.376 (8)				0.05	12.212	3.819 (7)
			0.0	8.201	2.250 (8)				0.0	12.422	3.951 (7)
			-0.05	13.685	3.970 (8)				-0.05	12.642	4.092 (7)
			-0.1	18.106	7.003 (8)				-0.1	12.873	4.241 (7)

numbers in the parenthesis denote powers of 10

has a greater effect on the stability of the flow than the equivalent positive temperature gradient. This difference between the effects of negative $\bar{N}$ and positive $\bar{N}$ is more pronounced in the presence of (i) a large gap between the cylinders (small η), (ii) co-rotating cylinders with high velocities ratio μ and (iii) strong magnetic field. However, for negative μ and large η, equal magnitudes of $+\bar{N}$ and $-\bar{N}$ destabilise and stabilise the flow respectively by nearly the same amount. These features are demonstrated in Figs. 1-4. These figures also indicate that the effect of heating on $\bar{T}_c$ is enhanced when it is accompanied by the above three factors, (i), (ii) and (iii). As an example in Table II, for $\eta = 0.85$, $\mu = 0.0$ and $\bar{Q} = 100$, the presence of a temperature gradient that corresponds to $\bar{N} = -0.1$ induces a variation in the $\bar{T}_c$ of $\bar{N} = 0$ by about 0.3% more than that induced by $\bar{N} = 0.1$ whereas, as shown in Table V, for $\eta = 0.2$, $\mu = 0.01$ and $\bar{Q} = 1000$, the $\bar{T}_c$ for $\bar{N} = 0$ varies by about 190% more when $\bar{N} = -0.1$ than when $\bar{N} = 0.1$. The dependence of the stability contribution of heating on the gap size between the cylinders can be realised if we note, for example, that as $\bar{N}$ is raised from 0 to 0.1 (when $\mu = 0$ and $\bar{Q} = 500$) the $\bar{T}_c$ suffers a reduction by about 4% when $\eta = 0.85$ (Table II) and by more than 35% when $\eta = 0.2$ (Table V). This effect of heating on $\bar{T}_c$ increases with co-rotating cylinders ($\mu > 0$) and decreases with counter-rotating cylinders ($\mu < 0$).

2. In the absence of heating, magnetic field and rotation of the outer cylinder (i.e. $\bar{N} = \bar{Q} = \mu = 0$), the critical angular velocity Ω_{1C} can be controlled by adjusting the ratio of the radii of the two cylinders. It is found that most stable state of the flow is achieved by having η close to one (i.e., narrow gap between the cylinders). However, as η then slightly decreases, the flow suffers a great reduction in its stability condition and this is reflected by the changes in $\bar{T}_c$. $\bar{T}_c$ loses about 20% of its original value as η drops by 0.1 from $\eta = 0.95$ to $\eta = 0.85$. This fast rate of destabilisation slows down as η gets smaller and when η equals about 0.5 (i.e. $R_1 = \frac{1}{2} R_2$) the flow reaches its minimum stable state.

The presence of the magnetic field affects the stability action of η in two ways.

(a) It increases the stability potential of the fluid for all value of η.
(b) It shifts the minimum value of $\bar{T}_c$ towards a value of η larger than 0.5, this shifting being greater, the greater the strength of the magnetic field.

Some other interesting and interconnected relations between the flow stability and μ can be obtained as follows:-

1. For any fixed $\bar{Q}$ and $\bar{N}$, the rate of stabilisation of the flow with larger magnitudes of positive μ is greater than that of negative μ. Consequently, rotating the two cylinders in the same direction with a given ratio μ of the angular velocities always contributes to a more

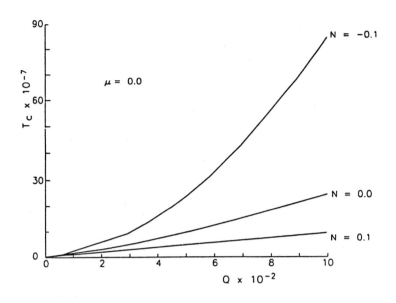
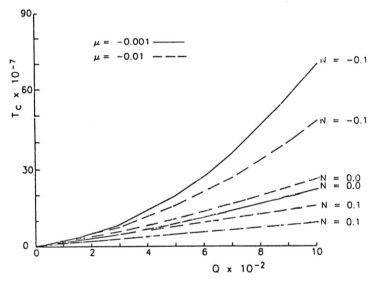

Fig.4. Variation of T_c with Q; $\eta = 0.1$

stable flow than rotating them with the same magnitude of μ in the opposite direction. As can be seen from Tables III, with $\bar{Q} = 0$, $\bar{N} = 0$ and $\eta = 0.75$, the $\bar{T}_c$ for $\mu = 0$ increases by about 33% when $\mu = 0.25$ and by less than 7% when $\mu = -0.25$.

2. As the negative value of μ is increased in magnitude, the tendencies of $\bar{Q}$ and $\bar{N}$ to bring about changes in the stability of the flow are reduced considerably. In Table II, for $\eta = 0.85$ and $\bar{N} = 0.1$, raising $\bar{Q}$ from 0 to 1000 causes $\bar{T}_c$ to increase by about 130% when $\mu = -0.1$ and by less than 80% when $\mu = -0.75$.

Finally, we discuss the effects of different values of η, μ, $\bar{Q}$ and $\bar{N}$ on the critical wave number a_c. From Tables I-VI, we may state the following related conclusions.

1. For fixed, μ, $\bar{Q}$, η, the a_c exhibits a gradual increase or decrease with larger magnitudes of negative or positive temperature gradients respectively. The rate of variation of a_c with $\bar{N}$ is greater in the presence of (i) large $\bar{Q}$, (ii) positive μ and (iii) small η. Also, with (i), (ii) and (iii) it becomes more evident that a given positive temperature gradient has a smaller effect on the value a_c of the isothermal flow than the equivalent negative temperature gradient. In Table III, the a_c value for $\eta = 0.75$, $\mu = 0.0$, $\bar{Q} = 1000$ and $\bar{N} = 0.0$ increases by about 3% when $\bar{N} = 0.1$ and by about 3.8% when $\bar{N} = -0.1$. Whereas, it is evident from Table V that the a_c value for $\eta = 0.2$, $\bar{Q} = 1000$ and $\bar{N} = 0.0$ increases by about (a) 53% when $\mu = 0.0$ and $\bar{N} = 0.1$, (b) 99% when $\mu = 0.0$ and $\bar{N} = -0.1$, (c) 56% when $\mu = 0.01$ and $\bar{N} = 0.1$ and (d) 176% when $\mu = 0.01$ and $\bar{N} = -0.1$.

2. For fixed η, μ and $\bar{N}$, the a_c slowly increases with higher values of $\bar{Q}$ provided that η is large. But as η gets smaller, the a_c value decreases with $\bar{Q}$ when (a) $\bar{N} \geq 0$, (b) $\mu \geq 0$ or slightly negative and (c) $\bar{Q} > \tilde{Q}$, where $\tilde{Q}$ depends on η. For $\eta \simeq 0.75$, $\tilde{Q} \simeq 700$. But as η decreases, $\tilde{Q}$ moves toward smaller values. For $\eta \simeq 0.5$, $\tilde{Q} = 300$ and for $\eta = 0.35$, $\tilde{Q} = 100$.

3. For fixed η, $\bar{Q}$ and $\bar{N}$, the increasing values of positive μ cause the a_c to decrease when $\bar{N}$ is positive and to increase when $\bar{N}$ is negative, provided that $\bar{Q} > \tilde{Q}$. This $\tilde{Q}$ is a function of η. For large η ($\eta \sim 0.85$) $\tilde{Q}$ is around 1000. But as η becomes smaller, $\tilde{Q}$ also becomes smaller. For $\eta \simeq 0.35$, $\tilde{Q} \simeq 100$. On the other hand, with increasing magnitudes of negative μ, the a_c value always increases except when η is small and $\bar{Q}$ is large. In that case, the a_c value increases when $\bar{N} \geq 0$ and decreases when $\bar{N} < 0$. In Table V, as μ goes from -0.01 to -0.1, the a_c value for $\eta = 0.2$, $\bar{Q} = 100$ increases by about 120% at $\bar{N} = 0.1$ and decreases by 23% at $\bar{N} = -0.1$.

3. Conclusions

If the contribution of each factor to the stability of the Couette flow is measured according to its effect on $\tilde{T}_c$ (i.e. $\bar{T}_c$ for $\mu = 0$, $Q = 0$, $N = 0$ and $\eta \sim 1$), then we find the following statements to be true.

1. Applying a magnetic field in the axial direction always leads to a more stable flow. The greater the strength of the magnetic field, the greater is the angular velocity Ω_1 needed to bring about instability in the flow.

2. The presence of a positive temperature gradient between the cylinders (temperature of the outer cylinder greater than that of the inner one) destabilizes the flow, while the negative temperature gradient stabilizes it.

3. The stability of flow is continuously enhanced when the two cylinders are rotated in the same or the opposite direction with an increasing magnitude of μ. However, in the case of counter-rotating cylinders, the rate of the stabilization is slower and it starts after the flow attains its maximum stable state. This minimum point is a function of η and $\bar{Q}$. For larger η ($\eta \sim 1$) it occurs at about $\mu = 0$. Nut as η becomes smaller and $\bar{Q}$ becomes larger, the minimum point moves gradually toward the negative values of μ.

4. In relation to the effect of various gap sizes between the cylinders on the critical Taylor number $\bar{T}_c$, we have established that the maximum and the minimum $\bar{T}_c$ are achieved when $\eta \sim 1$ (narrow gap size) and $\eta \sim 1/2$ (i.e. $R_1 \sim 1/2\, R_2$) respectively. $\bar{T}_c$ decreases at decelerating rate from $\eta \sim 1$ to $\eta \sim 1/2$. It then increases at an accelerating rate with smaller values of η.

REFERENCES

Astill, K.N., & Chung, K.C. 1976 A numerical study of the flow between rotating cylindrs. ASME. Pub. 76-FE-27, 1-8.

Bahl, S.K. 1972 The effect of radial temperature gradient on the stability of a viscous flow between two rotating coaxial cylinders. Trans. ASME. E: J. Appl. Mech. 39, 593-594.

Becker, K.M. & Kaye, J. 1962 The influence of a radial temperature gradient on the instability of fluid flow in an annulus with an inner rotating cylinder. Trans. ASME. C: J. Heat Transfer. 84, 106-111.

Chandrasekhar, S. 1953 The stability of viscous flow between rotating cylinders in the presence of a magnetic field. Proc. Roy. Soc. A, 216, 293-309.

Chandrasekhar, S. 1954 The stability of viscous flow between rotating cylinders. Mathematika. 1, 5-13.

Chandrasekhar, S. 1958 The stability of viscous flow between rotating cylinders in the presence of a magnetic field. Proc. Roy. Soc. A. 246, 301-311.

Chandrasekhar, S. 1961 Hydrodynamic and hydromagnetic stability. Oxford: Clarendon Press.

Chandrasekhar, S. & Elbert, D. 1962 The stability of viscous flow between rotating cylinders, II. Proc. Roy. Soc. A. 268, 145-152.

Chang, T.S. & Sartory, W.K. 1965 Hydromagnetic stability of dissipative vortex flow. Phys. Fluids. 8, 235-241.

Edmonds, F.N. 1958 Hydromagnetic stability of a conducting fluid in a circular magnetic field. Phys. Fluids. 1, 30-41.

Harris, D.L. & Reid, W.H. 1964 On the stability of viscous flow between rotating cylinders. Part 2. Numerical analysis. J. Fluid Mech. 20, 95-101.

Hassard, B.D., Chang, T.S. & Ludford, G.S. 1972 An exact solution on the stability of m.h.d. Couette flow. Proc. Roy. Soc. A. 327, 269-278.

Krueger, E.R., Gross, A. & DiPrima, R.C. 1968 On the relative importance of Taylor-Vortex and non-axisymmetric modes in the flow between rotating cylinder. J. Fluid Mech. 24, 521-538.

Kurzweg, U.H. 1963 The stability of Couette flow in the presence of an axial magnetic field. J. Fluid Mech. 17, 52-60.

Lai, W. 1962 Stability of a revolving fluid with variable density in the presence of a circular magnetic field. Phys. Fluids. 5, 560-566.

Soundalgekar, V.M., Takhar, H.S. & Smith, T.J. 1981 Effects of radial temperature gradient of viscous flow in an annulus with a rotating inner cylinder. Warme-Stoffubertragung. 15, 233-238.

Sparrow, E.M., Munro, W.D. & Jonsson, U.K. 1964 Instability of the flow between rotating cylinders : the wide-gap problem. J. Fluid Mech. 20, 35-46.

Takhar, H.S., Smith T.J. & Soundalgekar, V.M. 1985 Effects of radial temperature gradient on the stability of flow between two rotating concentric cylinders. J. Mathematical Analysis and Application. 111, 349-352.

Taylor, G.I. 1923 Stability of a viscous liquid contained between two rotating cylinders. Phil. Trans. Roy. Soc. A 223, 289-343.

Walowit, J., Tsao, S. & DiPrima, R.C. 1964 Stability of flow between arbitrarily spaced concentric cylindrical surfaces including the effect of a radial temperature gradient. Trans. ASME. E : J. Appl. Mech. 31, 585-593.

STRUCTURE AND PERTURBATION IN GÖRTLER VORTEX FLOW

Philippe Petitjeans and José Eduardo Wesfreid

Laboratoire PMMH - URA 857 CNRS - ESPCI
10, rue Vauquelin
75231 Paris cedex 05, France

INTRODUCTION:

The Görtler instability, like the Taylor-Couette and Dean instabilities, is a centrifugal instability that occurs in a boundary layer region of fluid flow on a concave wall. The unstable fluid motion near the wall becomes apparent by the formation of longitudinal counter-rotating rolls. In the case of flow on turbine blades, for instance, this type of flow pattern has great influence on the local momentum, mass, and heat transfer.

This investigation help to the discussion of the influence of Görtler rolls on the film cooling characteristic of turbine blades. First, we present experimental evidence of the interaction between injected jets and Görtler rolls in the concave section via fluorescein salt and argon laser flow visualization [1, 2]. Second, we compare the flow visualization pictures to the real velocity field obtained via laser Doppler anemometry. To explain results, we evoke the interpretation proposed by Liu and Sarry.

EXPERIMENTAL ARRANGEMENT AND METHODS

The experiments were carried out in a low velocity water tunnel (Figure 1). The tunnel was especially designed so that the laminar boundary layer on the test section would start to develop from the concave surface leading edge and would become unstable to a Görtler instability earlier than the Tollmien-Schlichting instability. The width and the height of the concave test section are 10 cm and 5 cm, respectively, and the radius of curvature R is 10 cm. Flow is generated by a constant level reservoir.

The streamwise velocity component is mesured by a laser Doppler anemometry system. The use of a fiber optic system allows to move the probe of the anemometer automatically and to scan the velocity field. The boundary layers developed upstream of the test section are eliminated at the entry to the test section through four deviation slots, one for each side. The deviation rate at each slot is controlled by independent valves to obtain the desirable flow in the concave test section. Details of the tunnel designed and its specifications have been reported by Peerhossaini [3].

For flow visualization, the experimental apparatus is equipped with dye injection facilities. The dye used is fluorescein sodium salt diluted in water to 0.1 grams per liter. Molecules of dye are excited by the 490 nm line of a argon ion laser, and the structures of the flow are visualized with a laser light sheet across the concave test section.

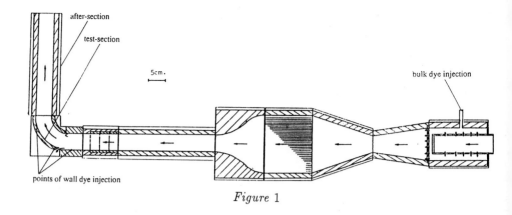

Figure 1

INFLUENCE OF JETS ON GÖRTLER ROLLS

In the experiment we inject jets of dye normal to the basic flow just before the concave test section, or 30 cm upstream. In the Figure 2, we see a dye line injected upstream which remains very thin in all the straight section, but which growths in remarkable fashion putting in evidence the centrifugal instability on the concave section: the dye jets spread around the rolls and turns with them. In the other pictures 3 and 4, we can observe more details of the flow, in particular, we see that the tracers particles does not make one turn along all the concave wall. Indeed, the transversal and radial perturbation velocity of the Görtler vortex are of order $O(Re^{-1/2})$, so very small compared to the streamwise flow velocity $O(1)$.

The interaction of normal jets-flow is well known experimentally, numerically, and theoretically [4, 5, 6]. A jet in a flat plate acts like a soft cylinder around which the free stream separates producing a weak horseshoes vortex at the foot of the jet by the lift of the boundary layer away from each other, and decay slowly downstream. The core of the jet produces a vortex pair (see figure 5). This vortex pair resembles a Görtler vortex pair but is smaller in size and lower in energy compared to the Görtler vortex [7]. This situation can be modelised as a sink on the downstream part of the jet and as a source at the upstream part. In our experiment, the jet is injected 30 cm before the beginning of the concave test section, perpendiculary to the flow. The jet to crossflow velocity ratio $R = U_j/U_0$ (where U_0 is the velocity of the flow, and U_j the normal velocity of the injection velocity jet) is estimed to about 4 from an extrapolation of the results of the reference [5].

In the injection region, the trajectory of the jet follows a law proportionnal to the square root of the distance from the injection point. The width and the height of the jet have the same order of magnitude. We observe that this agrees with numerical simulation [4], even in a turbulent boundary layer. In the entry of the concave test section, the height (and the width) are about 0.8 cm. At this point, a typical form of a cross section resembles a "kidney bean".

The Figure 6 is a cross section view of the flow in the concave part of the tunnel in which three "mushroom" formations can be observed. The one located at the center of the picture represents the jet perturbed by the Görtler rolls. The other two similar "mushrooms", located on each side, are naturally formed without any effect from the injected jet.

The latter being evidence that what we observe at the center is not a natural evolution of the jet itself. This confirms that the longitudinal vorticity induced by the injection of the jet is very small compared to that produced by the Görtler instability. Other results [7] have made this observation proving the usefulness of this method to create weak and controlled perturbations on Görtler rolls or any other longitudinal vortex.

Figure 2

Figure 3. one jet, and Figure 4 : two jets.

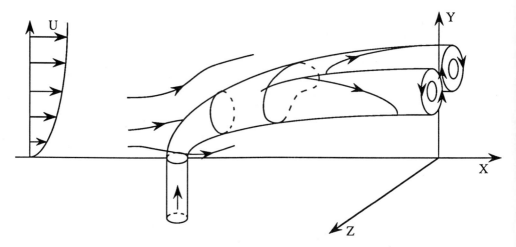

Figure 5. *Schematic view of the jet.*

Figure 6. *One jet at the center of the section*

VISUALIZATION AND VELOCITY FIELD

In this part, we compare the visualization (using fluorescein and argon laser sheet) and the streamwise velocity field (using laser Doppler anemometry).

In the Figure 7-a, we can see a "mushroom" (that can be seen in many examples of visualization in fluid mechanics) in a cross section of the flow in the concave test section. Here we try to understand the form of the "mushroom" and what are the relations with the velocity field.

To do so, we make a visualization in a cross section at 52° from the leeding edge (Figure 7-a). As the flow is stationnary, we scan in the same cross section and at the same conditions, the velocity field using the automatic anemometry system. We do streamwise velocity mesurements in about 2500 cross section points, and therefore, we transform these results as an image file by associated a gray level (between 0 and 255) to each velocity mesurements. So, we obtain a picture as representation of the real streamwise velocity field (Figure 7-b) where the isochrome lines corespond to lines of isovelocity (streamwise component), and we can superpose these two images (7-a and 7-b) in order to compare.

First, we can see the difference in respect with the height of the mass and velocity "mushrooms". The form is about similar and the position of the "mushrooms" corresponds.

This problem was analysed by Liu and Sarry [8] in a numerical study. They related the iso-concentration lines (the "mushrooms") with the iso-streamwise velocity lines as a function of the Schmidt number $Sc = \nu/D$ (where ν is the kinematic viscosity and D the coefficient of the mass diffusion). The iso-temperature lines were also related as a function of the Prandtl number $Pr = \nu/K$ (where K is the coefficient of the thermal diffusion).

They write similar convection diffusion equations for the momentum, and the concentration:

$$(\partial u/\partial t) + u.\nabla u = \nu \nabla^2 u + o.t.$$

where o.t. (other terms) includs the effect of curvature and centrifugal force. In the case of weak curvature, and if we don't consider the centrifugal instability or the process by which the rolls and the asociated "mushrooms" are formed, it can be supposed the existance of longitudinal rolls (whatever the mecanism which create them) and it can be neglected the o.t. terms. As this Liu and Sarry focalise in the effect of the nonlinear coupling in the $u.\nabla u$ terms.

$$(\partial Z_i/\partial t) + u.\nabla Z_i = (\nu/Sc)\nabla^2 Z_i$$

where Z_i is the mass fraction of the ith species

The similarity of these two equations leads to identify the velocity, and concentration profils when the Schmidt number is near unity. The numerical simulation of the reference [8] shows a similar behaviour as our experiments for the different size of the mass and the velocity of the "mushrooms" stem. Our experimental results and the numerical calculations by Liu and Sarry are the first to compare the simultanous pictures of the concentration and velocity field in the Görtler instability.

From the streamwise velocity field image we can obtain the streamwise velocity perturbative field by substraction of the mean boundary layer velocity field from the streamwise velocity field to obtain the image shown in Figure 7-c.

The contour lines show the positions where the amplitude of the perturbative velocity is maximum. Note that these zones are not at the same height. This means that the rolls are inclined as consequence of the nonlinear contribution to the velocity field of the rolls.

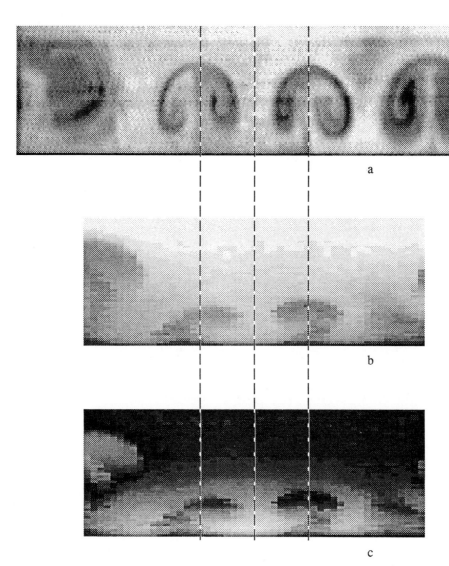

Figure 7 a, b and c.

7-a. Visualization of a cross section of the flow at 52° from the leeding edge, by the laser induced fluorescence technique (10 cm X 2.5 cm)

7-b. Picture representing the streamwise velocity: Clear zones high velocities while dark zones correspond low velocities (6.66 cm X 2.5 cm)

7-c. Picture representing the perturbative streamwise velocity (6.66 cm X 2.5 cm)

ACKNOWLEDGEMENT

The authors acknowledge A. Jimenez for suggestions and comments.

REFERENCES

[1] P. BRY and J. BERNARD, Some aspects of the turbine blade pressure side boundary layer: The point of view of the turbine designer, *Congrès Euromech 261, Nantes* (juin 1990)

[2] H. PEERHOSSAINI and J.E. WESFREID, Les tourbillons de Görtler et leur influence sur les turbines à gaz, *Bulletin de l'Association Technique Maritime et Aéronautique* **88** , 361-381 (1988)

[3] H. PEERHOSSAINI, L'instabilité d'une couche limite sur une paroi concave (Les tourbillons de Görtler), *Thèse d'Etat, Université Paris 6* (1986)

[4] J. ANDREOPOLOUS and W. RODI, Experimental investigation of jets in a crossflow, *J. Fluid. Mech.* **138** , 93-127 (1984)

[5] R.I. SYKES, W. S. LEWELLEN and S. F. PARKER, On the vorticity dynamics of a turbulent jet in a crossflow, *J. Fluid. Mech.* **168**, 393-413 (1986)

[6] G. P. HUANG, Modélisation et calcul de jets tridimentionnels en présence d'un écoulement transversal, *Thèse de l'Ecole Centrale de Lyon* (1989)

[7] J. M. CHOMAZ and M. PERRIER, A forced experiment on Görtler instability, *Preprint* (1991)

[8] J. T. C. LIU and A. S. SARRY, Concentration and heat transfer in nonlinear Görtler vortex flow and the analogy with longitudinal momentum transfer, *Proc. R. Soc. Lond. A.* **432**, 1 (1990)

TRANSITION TO TURBULENCE IN GÖRTLER FLOW

Wei Liu and J. Andrzej Domaradzki

Department of Aerospace Enginering
University of Southern California
Los Angeles,CA 90089-1191

INTRODUCTION

Counter-rotating streamwise vortices play an important role in transition to turbulence and in the dynamics of fully turbulent boundary layer flows [1]. However, the mechanisms by which they are generated are largely unknown and in turbulent flows they occur randomly in space and time, making their investigation difficult. The counter-rotating vortices generated in a boundary layer flow over a concave wall (Görtler vortices) share many similarities with vortical structures found in other types of boundary layers as documented by Blackwelder [1]. Since their generation mechanism is known and they form a deterministic, spatially regular pattern, they offer an attractive alternative in investigation of the dynamics of counter-rotating vortices in boundary layers.

The Görtler flow has been studied by many researchers since Görtler first discussed it in 1940 [2]. Most of the earlier research was concentrated on the linear stability theory and determining a neutral stability curve [2, 3, 4]. More recently Hall [5, 6] showed that the concept of a unique neutral curve is not tenable for the Görtler flow because its stability depends on how and where the boundary layer is perturbed. He also pointed out that the parallel flow assumption made in the earlier stability analyses is not correct except in the small-wavelength limit. Two-dimensional numerical simulations of Görtler flow were performed by Sabry and Liu [7, 8] and Park and Huerre[9]. The transition to turbulence in the Görtler flow was studied experimentally by, among others, Bippes [10] and Swearingen and Blackwelder [11].

The purpose of our work is to develop an efficient numerical code capable of modelling three-dimensional boundary layer flows over concave walls and investigate physics of such flows, in particular the mechanisms of transition to turbulence.

FORMULATION AND NUMERICAL METHOD

We employ Navier-Stokes equations for the incompressible flow in the vorticity form transformed into a natural curvilinear coordinate system for the flow over a curved boundary with the radius of curvature R.

In this curvilinear coordinate system the Navier-Stokes equations have the following form:

$$\frac{\partial}{\partial t}u = v\left(\frac{1}{h}\frac{\partial v}{\partial x} - \frac{\partial u}{\partial y}\right) - w\left(\frac{\partial u}{\partial z} - \frac{1}{h}\frac{\partial w}{\partial x} - \frac{\kappa}{h}u\right)$$
$$+ \nu\left[\frac{1}{h^2}\frac{\partial^2 u}{\partial x^2} + \frac{\partial^2 u}{\partial y^2} + \frac{\partial^2 u}{\partial z^2} - \frac{\kappa}{h}\frac{\partial u}{\partial z} - \frac{2\kappa}{h^2}\frac{\partial w}{\partial x} - \frac{\kappa^2}{h^2}u\right]$$
$$- \frac{1}{\rho}\frac{1}{h}\frac{\partial \Pi}{\partial x} \qquad (1)$$

$$\frac{\partial}{\partial t}v = w\left(\frac{\partial w}{\partial y} - \frac{\partial v}{\partial z}\right) - u\left(\frac{1}{h}\frac{\partial v}{\partial x} - \frac{\partial u}{\partial y}\right)$$
$$+ \nu\left[\frac{1}{h^2}\frac{\partial^2 v}{\partial x^2} + \frac{\partial^2 v}{\partial y^2} + \frac{\partial^2 v}{\partial z^2} - \frac{\kappa}{h}\frac{\partial v}{\partial z}\right]$$
$$- \frac{1}{\rho}\frac{\partial \Pi}{\partial y} \qquad (2)$$

$$\frac{\partial}{\partial t}w = u\left(\frac{\partial u}{\partial z} - \frac{1}{h}\frac{\partial w}{\partial x} - \frac{\kappa}{h}u\right) - v\left(\frac{\partial w}{\partial y} - \frac{\partial v}{\partial z}\right)$$
$$+ \nu\left[\frac{1}{h^2}\frac{\partial^2 w}{\partial x^2} + \frac{\partial^2 w}{\partial y^2} + \frac{\partial^2 w}{\partial z^2} - \frac{\kappa}{h}\frac{\partial w}{\partial z} + \frac{2\kappa}{h^2}\frac{\partial u}{\partial x} - \frac{\kappa^2}{h^2}w\right]$$
$$- \frac{1}{\rho}\frac{\partial \Pi}{\partial z} \qquad (3)$$

$$\frac{1}{h}\frac{\partial u}{\partial x} + \frac{\partial v}{\partial y} + \frac{\partial w}{\partial z} - \frac{\kappa}{h}w = 0 \qquad (4)$$

where x is the distance along the curved boundary, z is the ordinate in the direction normal to the boundary and y is in the spanwise direction, $\mathbf{u} = (u, v, w)$, $\kappa = 1/R$ is the curvature of the boundary, $h = 1 - \kappa z$, $\Pi = p + \frac{1}{2}\rho\mathbf{u}^2$ is the pressure head, and ν is the kinematic viscosity.

Equations (1)-(4) are rearranged to the form which is suitable for a fractional step method and solved using a modified version of the semi-implicit, pseudospectral numerical code FLOGUN developed originally by Orszag and Kells [12] for the simulation of channel flows.

We assume that the boundary layer is parallel which allows the application of periodic boundary conditions in the horizontal directions. In these directions Fourier modes are used in the numerical method and a uniform mesh in physical space. In the vertical, nonperiodic direction Chebyshev modes are used with mapping from the channel flow to the boundary layer flow geometry.

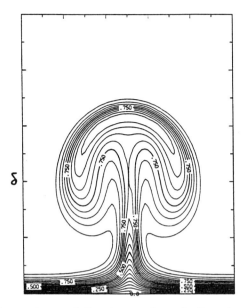

Figure 1. Iso-contours of the streamwise velocity u in $y - z$ plane.

RESULTS OF SIMULATIONS

Comparison with Experiments

For the comparison with the experimental results of Swearingen and Blackwelder [11] we use values of simulation parameters consistent with their experiment in which radius of curvature R was 3.2 m and the free streamwise velocity U_0 was $5m/s$. The initial condition in our simulations is a superposition of a Blasius mean flow and the most unstable mode of the linear stability theory for parallel Görtler flow. The amplitude of the initial Görtler vortices, $0.167U_0$, and the initial Görtler number, 4.985, in the simulations are equal to the experimental values [11] at the location $X_0 = 60$ cm from the leading edge. A random velocity field with the amplitude $0.5\%U_0$ was also added to the initial velocity field to simulate background noise which existed in the experiments. Our computational domain is a rectangular box. The box size is 2.0 cm in x direction, 1.8 cm in y direction, 24.85 cm in z direction. The spanwise wavelength used in the simulation coresponds to the distance between two low speed regions at the location where the detailed data were recorded. To compare the numerical results for the time evolving flow with the results obtained experimentally for the spatially evolving flow we introduce a convection velocity $U_c = 0.60U_0$ to convert temporal data into spatial ones.

In the laminar regime $X = 40-90$ cm the counter-rotating vortices pump fluid with a low u velocity away from the wall creating the characteristic "mushroom" structure of iso-contours of the streamwise velocity shown in fig. 1. The flow is two-dimensional and contains a low speed region of the streamwise velocity in the upwash region between the vortices (a peak region) and a high speed region in the downwash region (a valley region).

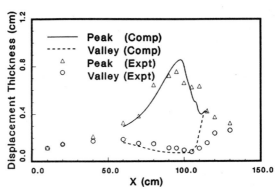

Figure 2. Development of the average boundary layer displacement thickness.

The development of the boundary layer thickness averaged over x direction is shown in fig. 2 where it is compared with the experimental results of Swearingen and Blackwelder [11]. The boundary layer thickness grows in the low speed region and decreases in the high speed region as the flow develops. At about 90 − 100 cm from the leading edge, the boundary layer thickness in the low speed region reaches its highest value and starts to decrease. At the end of the simulation the averaged quantities in the low speed region and in the high speed region have almost the same values and the flow is weakly turbulent.

Inflectional mean velocity profiles

Profiles of the streamwise velocity in the vertical direction z in the low speed region at different locations from the leading edge in the simulation are shown in fig. 3. The pumping action of the counter-rotating vortices gradually generates a highly unstable S-shaped profile with two inflectional points in the regions of intense, local horizontal shear layers. These regions move up and associated shear strengths increase with increasing distance from the leading edge. The inflectional profiles in the vertical direction exist not only in the low speed region between vortices, but also in the whole "mushroom" region. However the vertical shear $\frac{\partial u}{\partial z}$ has its maximum in the low speed region. At the later stages of the flow evolution the appearance of turbulence decreases the intensity of these shear layers.

A streamwise velocity profile in the spanwise direction y at different z locations is shown in fig. 4. In the side valley regions high speed fluid from the upper parts of the boundary layer is pushed down toward the wall and in the central peak region low speed fluid is pushed away from the wall through the action of the Görtler vortices. This process creates unstable profiles of the streamwise velocity in the spanwise direction y. At $z = 1.5\delta$, where δ is the Blasius boundary layer thickness, the mean velocity is essentially uniform. When the wall is approached the velocity profiles gradually change and become similar to two-dimensional wake profiles. There is a minimum on either side which is caused by low speed fluid being returned towards the wall by the action of the Görtler vortices at $z = 0.74 cm$. These velocity profiles have inflectional points

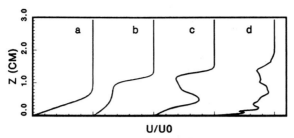

Figure 3. Profiles of the streamwise velocity in the vertical direction in the low speed region at different locations from the leading edge: (a) $X = 64cm$, (b) $X = 94cm$, (c) $X = 100cm$, (d) $X = 109cm$.

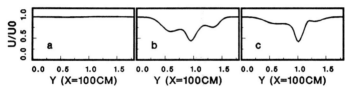

Figure 4. Profiles of the streamwise velocity in the spanwise direction at different vertical locations: (a) $z = 1.54cm$, (b) $z = 0.74cm$, (c) $z = 0.51cm$.

which are associated with the presence of the local vertically oriented shear layers. In the initial stages of the simulation, the intensity of these shear layers intensifies with time. However at the later stages of the evolution the intensity of these shear layers is decreased by the appearance of turbulence.

Linear Stability Analysis

In the last subsection we observed that inflectional velocity profiles of streamwise velocity both in the vertical and in the spanwise direction were formed during the transition process as seen in fig. 3,4. In general these inflectional profiles are inviscidly highly unstable and associated growth rates are larger than growth rates of other instabilities affecting the flow, e.g. Görtler instability in this case. Hence Görtler flow can be taken as a quasi-steady flow for inflectional instability analysis. We have performed standard stability analysis for these quasi-steady velocity profiles. Orr-Sommerfeld equation is used for this purpose.

The streamwise velocity profile in the vertical direction in the low speed region and the velocity profile in the spanwise direction at $z/\delta = 0.5$ at distance $X = 98$ cm from the leading edge, just before transition begins were used for the stability analysis. The frequencies of oscillations predicted by this 2-D stability analysis for the velocities w and v are 170 Hz and 200 Hz, respectively. Since the frequency deduced from the analysis of time series in the simulations is $200Hz$, which is the same as the frequency predicted by the stability analysis, this result supports the conclusion that the inviscid instability of inflectional velocity profiles of the streamwise velocity is responsible for the transition to turbulence in the Görtler flow.

The frequency of the oscillations in the outer region recorded by Swearingen and Blackwelder [11] is 130 Hz. Our numerical simulations do not predict this frequency. By checking stability analysis we note that the frequency 130 Hz corresponds to the wavelength 2.5 cm. Hence the reason that we can not pick up this frequency in the simulation is that the streamwise size of our computational box is 2.0 cm which is smaller than 2.5 cm suggested by the stability analysis.

Secondary Instabilities

In the following discussion we decompose the total velocity field as follows

$$\mathbf{U}(x,y,z) = \mathbf{U}_b(z) + \mathbf{U}_g(y,z) + \mathbf{u}(x,y,z) \tag{5}$$

where

$$\mathbf{U}_b(z) = \frac{1}{L_x L_y} \int_0^{L_x} \int_0^{L_y} \mathbf{U}(x,y,z) dx dy$$

$$\mathbf{U}_g(y,z) = \frac{1}{L_x} \int_0^{L_x} \mathbf{U}(x,y,z) dx - \mathbf{U}_b(z) \tag{6}$$

Here, $\mathbf{U}(x,y,z)$ is the total velocity, $\mathbf{U}_b(z)$ is a basic flow which is the average of total velocity over horizontal planes and is the function of the vertical coordinate z only, $\mathbf{U}_g(y,z)$ is the Görtler velocity obtained as the average of the total velocity field over lines in the x direction with the basic flow $\mathbf{U}_b$ subtracted, and $\mathbf{u}(x,y,z)$ is the perturbation to the Görtler flow. Note that for laminar Görtler vortices quantity $\mathbf{u}(x,y,z)$ vanishes and thus can serve as a useful measure of the departure of the flow from a purely two-dimensional laminar state. Nonzero velocity $\mathbf{U}_g(y,z)$ results from the primary Görtler instability of the base flow $\mathbf{U}_b(z)$, and $\mathbf{u}(x,y,z)$ is the velocity perturbation resulting from the secondary instability of the Görtler flow superimposed on the base flow $\mathbf{U}_b + \mathbf{U}_g$.

Figure 5 shows iso-contours of the rms values of the streamwise secondary perturbation component,

$$u_{rms}(y,z) = \sqrt{\frac{1}{L_x} \int_0^{L_x} u^2(x,y,z) dx} \tag{7}$$

instantaneous spanwise shear $\partial U/\partial y$, and vertical shear $\partial U/\partial z$ in $y - z$ plane at $X = 100 cm$.

Frequently regions of large vertical shears $\partial U/\partial z$ associated with the inflectional points on vertical profiles of streamwise velocity, are considered to be a primary cause of the transition process. However, after close examination of our results shown in fig. 5, we conclude that the large rms values of u correlates with the regions of the large spanwise shear $\partial U/\partial y$, but they have no obvious correlation with the vertical shear $\partial U/\partial z$. This result indicates that in the transition process the regions of large spanwise shear $\partial U/\partial y$ may play a greater role than the regions of large vertical shear $\partial U/\partial z$. This conclusion is consistent with the experimental results of Swearingen and Blackwelder [11] who showed large degree of correlation between the regions of large spanwise shear and the regions of large streamwise velocity fluctuations.

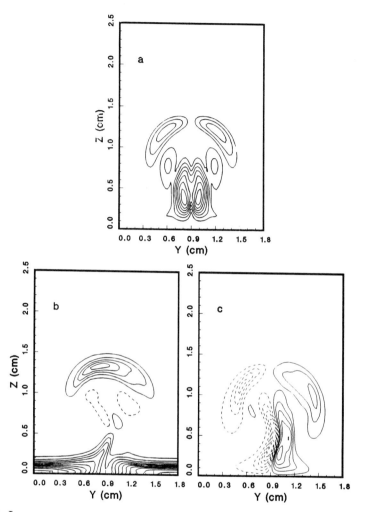

Figure 5. Iso-contours in $y - z$ plane at $X = 100cm$: (a) rms streamwise component of secondary perturbation u_{rms}/U_0, (b) Shear $\partial U/\partial z$, (c) Shear $\partial U/\partial y$.

Kinetic Energy Balance Analysis

The kinetic energy balance analysis is often used in studying turbulence phenomena. Usually in the kinetic energy balance analysis for turbulent boundary layer flows over concave walls assumption of homeogeneity in the horizontal plane is made and the mean flow is the function of z only as, for instance in Moser and Moin [13]. However, in our case the flow is not homogenous in the spanwise direction due to the presence of Görtler vortices. It is thus very important to account explicitly for this inhomogeneity in the energy balance analysis since otherwise some important features of the energetics of the flow may not be observed. For this reason we take as the mean flow in our case sum of the basic flow which is the function of z only and the Görtler flow which is the function of both y and z as defined in the last subsection. Turbulent velocity is identified as the perturbation $\mathbf{u}(x,y,z)$ to this mean field and averages are taken over lines in the x direction in which turbulent quantities are statistically homogeneous.

The kinetic energy production has the following form in the curvilinear coordinate system used :

$$[-\overline{uw}(\frac{\partial U_b}{\partial z} - \frac{\kappa}{h}U_b) - \overline{uw}(\frac{\partial U_g}{\partial z} - \frac{\kappa}{h}U_g) - \overline{uv}\frac{\partial U_g}{\partial y} - \overline{vw}\frac{\partial V_g}{\partial z} - \overline{vw}\frac{\partial W_g}{\partial y}] \qquad (8)$$

where the overbar denotes averaging over x direction.

We found that the maximum value of term $\overline{uv}\frac{\partial U_g}{\partial y}$, i.e. production due to the spanwise shear of the Görtler vortices, is at least one order of magnitude larger than the remaining production terms. Iso-contour plot of the production term $\overline{uv}\frac{\partial U_g}{\partial y}$ is shown

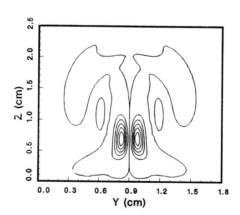

Figure 6. Iso-contours of the production term $-\overline{uv}\frac{\partial U_g}{\partial y}$ in the y-z plane (X=98 cm).

in fig. 6. In this figure the regions of large production of $\overline{uv}\frac{\partial U_g}{\partial y}$ correlate very well with the regions of large spanwise shear (fig. 5). This result provides additional evidence supporting the conclusion that spanwise shear plays a more important role in the initial phases of transition process than the vertical shear. It should also be noted that the maximum values of the production term near the wall are much larger than that in other locations. This is also the region where the process of transition begins.

CONCLUSIONS

Investigations of counter-rotating vortices in Görtler flow are very helpful in understanding dynamics of similar vortices in transitional and turbulent flows. A numerical code to simulate Görtler flow has been developed and it was demonstrated that the direct numerical simulation results for three-dimensional Görtler flow are in a good agreement with experimental results. Inflectional mean velocity profiles generated by counter-rotating vortices are found both in the spanwise and in the vertical directions. Standard stability analysis confirms that instabilities of these velocity profiles are the main mechanism of the transition to turbulence in such a flow. The numerical results show that the spanwise inflectional profile and associated shear $\partial U/\partial y$ play a more important role in the transition process than the vertical inflectional profile and shear $\partial U/\partial z$ which are conventionally considered the main cause of the transition . The numerical results indicate that after the transition process begins the velocity oscillations in the spanwise and vertical directions with distinct frequencies can be observed. The spanwise oscillations are associated with sinous motion of the unstable low speed streaks. The analysis of the kinetic energy balance equation which takes into account spanwise inhomogeneity of the mean flow shows that the production term arising from the spanwise shear induced by Görtler vortices is much larger than the production terms arising from the interaction of turbulence with the vertical shears.

References

[1] R. F. Blackwelder, Phys. Fluids **26**, 2807 (1983).
[2] H. Görtler, Nachr. Wiss. Ges. Göttingen Math. Phys. Kl. **2**, 1 (1940).
[3] G. Hämmerlin, J. Rat. Mech. Anal. **4**, 279 (1955).
[4] J. M. Floryan and W. S. Saric, AIAA J. **20**, 316 (1979).
[5] P. Hall, J. Fluid Mech. **124**, 475 (1982).
[6] P. Hall, J. Fluid Mech. **130**, 41 (1983).
[7] A. S. Sabry and J. T. C. Liu, Proc. of Symp. , (1988).
[8] A. S. Sabry and J. T. C. Liu, Report FM89-11, (1989).
[9] D. S. Park and P. Huerre, Submitted to J. Fluid Mech., (1991).
[10] H. Bippes, Heidel. Akad. Wiss., Naturwiss. Kl., Sitzungsber. **3**, 103 (1972).
[11] J. D. Swearingen and R. F. Blackwelder, J. Fluid Mech. **182**, 255 (1987).
[12] S. A. Orszag and L. C. Kells, J. Fluid Mech. **96**, 159 (1980).
[13] R. D. Moser and Moin P., J. Fluid Mech. **175**, 479 (1987).

EFFECT OF CURVATURE PLANE ORIENTATION ON VORTEX DISTORTION IN CURVED CHANNEL FLOW

H. Peerhossaini and Y. Le Guer

Thermofluids Research Group - Laboratoire de Thermocinétique - URA du CNRS 869

ISITEM - Université de Nantes - La Chantrerie, CP 3023, F-44087 Nantes -France

ABSTRACT

Flow in a succession of curved channels with varying curvature planes involves a torsion or twist. The effect of torsion on the secondary flow vortices is experimentally studied in the laminar regime. The experimental apparatus consists of a succession of curved channels in which the curvature plane rotates 90 degrees between each two adjacent curved elements. We have found only vortex tilting due to torsion effect.

I. INTRODUCTION

It is well established that counter-rotating streamwise vortices develop in the pressure driven flow in curved channels. Depending on the channel aspect ratio, a secondary flow pattern may invade the whole flow cross section (for small aspect ratios), or it may stay confined in the cross section extremities leaving the core flow intact (for large aspect ratios). At higher Dean numbers however, the core flow can be filled with smaller roll cells rotating in the opposite direction of the former secondary flow pattern. Eventhough both vortex structures are generally referred to as Dean vortices, their nature, wavelength, and the flow region where they appear are totally distinct. The former is a simple secondary flow due to an imbalance between centrifugal force and radial pressure gradient, while the latter is the result of an instability phenomenon with a well defined threshold.

In small aspect ratio channels, when the secondary flow is represented by streamlines in the plane of cross-section, it typically has the form of a pair of counter-rotating vortices. The vortices are symmetrically situated about the plane which contains the radius of curvature of the channel. Such a secondary flow pattern, arises due to the centrifugally-induced pressure gradient which drives the faster-moving fluids from the core outward to the concave outer wall. To satisfy the continuity, the low momentum fluid near the wall is then swept inward to the core. In a first approximation, at least three parameters may enter the analysis of the flow, namely : the Reynolds number, Re=Va/ν (where ν is the kinematic viscosity of the fluid and V is the mean flow velocity); the curvature ratio parameter $\gamma = a/R_c$ (where R_c is the mean radius of curvature and a is a typical length scale in the channel cross-section), and the aspect ratio $\beta = a/b$ (where a and b are two lengths which characterize the shape of the cross-section). A comprehensive analysis of this problem was given by Dean[1] for the first time. Dean's perturbation analysis of the problem is valid for small Dean number and large curvature ratio parameter. It is only in this case that the curvature ratio parameter γ does not appear independently and can be regrouped with the Reynolds number to define a single control parameter as: $Dn = Re \, (\gamma)^{1/2}$, the Dean number. In this definition, the Reynolds number is based on a and the flow mean velocity. For a large Dean number regime, the asymptotic boundary layer analysis is applicable[2].

The flow streamline pattern in the cross-section of a curved channel depends on the orientation of the curvature plane of the channel. Therefore, in a succession of curved channels, by changing the plane of curvature from one channel to the next, one can generate one type of trajectory in one channel, then destroy it and regenerate another type of trajectory in the next channel. Very complex trajectories can be produced this way and a fluid particle submitted to such a system will follow a chaotic motion (we term it as Lagrangian chaos) if the rotation of the curvature plane is properly chosen. Such a chaotic cross-sectional motion is expected to enhance the advection of passive scalars and therefore improve homogeneization in the volume and increase the efficiency of heat tranfer from the walls. A more detailed account of the main concepts of chaotic regime in a succession of Dean flows can be found in Jones *et al.*[3] and Peerhossaini and Le Guer[4]. Potential applications of such a system as a heat exchanger-blender for non-Newtonian fluids are addressed in Peerhossaini and Le Guer[5]. Jones *et al.*[3] have performed a numerical simulation of laminar flow through a twisted pipe of circular cross-section. Their results illustrate stretching of material lines, stirring of "blobs of material", chaotic three-dimensional trajectories and presence of islands of regular behavior in the pipe cross-section. In general, numerical experiments of Jones *et al.*[3] demonstrate that the chaotic trajectories we described above are attainable. This phenomenon has been termed "Chaotic Advection" by the latter authors. In an experiment Andereck *et al.*[6] have made a qualitative comparison of their results on tracer dispersion with those of reference 3.

Flow in a succesion of curved channels with varying curvature planes involves both a curvature and a torsion or twist. Eventhough curved channel flow has attracted and continues to attract much attention since the pioneering work of Dean, one area which has received relatively little attention is that of the distortion of the secondary flow pattern due to *torsion* effect . This subject is of more than academic interest, as many applications of curved channel flow have been encountered where rotation of the curvature plane is central to the basic physical phenomenon at the origin of the application. Examples can be cited from Lagrangian chaos or chaotic advection [3] to [5].

Torsion effects have been ignored for a long time perhaps because the main attention has been focussed on the practically important case of laminar flow in helically coiled tubes where helix pitch is small; therefore, the torsion effects enter the problem formulation at higher orders. It is interesting to know that, historically, Dean flow (curved pipe flow) was first considered in relation with helically coiled tubes: Grindly and Gibson[7] noticed the curvature effect on the flow in a coiled pipe in experiments on the air viscosity, and Eustice[8] demonstrated the existence of a secondary flow by dye injection into a coiled flexible tube. These experimental evidences prompted the theoretical works of Dean[1&9] who resolved the viscous incompressible flow equations through a pipe of circular cross section coiled in a circle. Therefore, small helix pitch justified the neglect of torsion effects on the secondary flow vortices.

Torsion effects on fully developed Dean flow in cicular cross section tubes have been recently addressed by Wang[9], Germano[11], Kao[12] and Tuttle[13]. Apart from controversies resulting from the orthogonality or non-orthogonality of the coordinates, Wang, Kao and Tuttle found a first order torsion effect, while Germano stated that only curvature can cause first order effect and the effect of torsion is of second order. Among the former three authors, Kao found the effect of torsion to give only an asymmetric configuration to the pair of Dean vortices, and that this conclusion is valid only for very small curvatures. Numerical simulations of Wang and Tuttle, however, showed that torsion effect can dominate the curavture effect up to a limit where the two recirculating cells become one cell (vortex coalescence).

As far as we know, there is no experimental verification of the above mentioned anlytical and/or numerical results. The only experiment reported on torsion effect is the one performed by Murakami et al.[14], a long time before these theoretical investigations. The experiment was carried out in a piping installation and attention was focussed mainly on the relationship between hydraulic loss due to two elbows located successively in a hydraulic power plant pipeline. Eventhough flow details were not reported, that investigation showed that for the successive elbows whose curvature planes make a 135° angle with each other, the two secondary flow vortices generated in the first elbow coalesce and become a large swirl downstream.

In the present work we are interested in the torsion effect on a pair of Dean vortices emerging from a curved channel and entering a similar one whose plane of curvature makes an angle with the previous channel, and this configuration is repeated for a succession of several curved channels. In section II we describe the experimental apparatus and procedures, and in section III preliminary results and discussion are presented.

II. EXPERIMENT

To carry out this investigation we have designed and constructed a specific water tunnel the main working parts of which are the curved channels, hereon we refer to as "elements". An element is a 90° bent channel of square cross-section of 40 ± 0.02 mm sides and inner and outer radii of $R_i = 200$ mm and $R_o = 240$ mm (Curvature ratio, $\beta = a/R_c = 0.18$) .Care has been taken in machining of the elements to guarantee a precision of up to 0.02 mm. This is necessary as the working part of the apparatus consists of several curved elements which have to be interchanged quite often to provide different curvature plane orientations required for generation of chaotic trajectories. The angle φ between the curvature plane of each pair of subsequent elements can vary in multiples of π/2 because of the square form of the element cross section . Based on a phenomenological argument, we have found that the most favorable angle φ for the generation of irregular trajectories is π/2. Though this geometry of the cross-section does not allow verification of the effect of a large number of possibilities for φ, the numerical results of Jones et al.[3] confirm that φ = π/2 is the optimum angle for obtaining an efficient stirring in the twisted pipes of circular cross-section. We presume that the same result stands for square cross-section too.

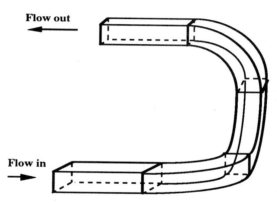

Fig. 1 Schematic diagram of the working section of the experimental apparatus

The flow generated either by a helical pump (for non-Newtonian working fluids) or from an elevated reservoir. Working fluid passes through a series of flow preparing elements before entering the succession of the curved elements. Between the flow preparing components and the working section, a transparent straight section is fitted where we measure velocity profiles by laser Doppler anemometry and visualize the flow by the laser induced fluorescence (LIF) technique. The flow rate through the circuit is measured by a magnetic flowmeter, with an accuracy of up to 1% below a flow speed of 0.3m/s. The succession of curved elements is followed by a transparent straight section where the integral effect of the curvature and torsion on mixing or heating can be examined. A schematic diagram of the working section of the experimental facility is shown in Fig. 1.

We have used the LIF technique to study the torsion effect on the Dean vortices due to the variation of the orientation of the curvature plane. The basic technique is explained elsewhere[15]. This technique consists of injecting a fluorescent dye in the fluid and illuminating sections of the flow by a light sheet of an appropriate wave length. The light energy, absorbed by the molecules of the dye travelling through the light sheet, excites the dye molecules which emit a light ray at a wavelength different from that of the incident light. In this experiment, a laser beam from a 5 Watt Argon ion source is focused on a cylindrical lens after passing through a 10 m monomode optic fiber which relates the laser source to the mobile LIF probe.The laser beam then expands to a light sheet when it passes through the cylindrical lens. We use the blue light (488 nm) ray which is appropriate for excitation of aqueous fluorescein solutions. Once excited, the fluorescein emits a green light at wavelength 514 nm. The probe assembly is mounted so that it may be pivoted and clamped in place to illuminate any angular position in each curved element or any position in the straight section. In all cases, the light sheet is perpendicular to the streamwise direction.

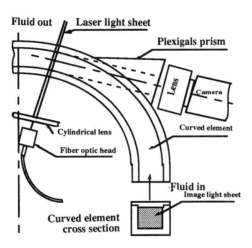

Fig. 2 This figure shows how the camera is oriented with respect to the convex wall

The flow cross sections are viewed by a CCD camera mounted on a specifically designed traversing mechanism to maintain the proper focus and a constant camera - light sheet orientation through all curved elements . Figure 2 shows schematically the traversing mechanism and how the camera is oriented from the convex wall of the curved element to view the illuminated planes at approximately 90° angle relative to the laser light sheet. Images viewed by the camera are visualized on a TV monitor screen and recorded by a video recorder. The same images are transferred directly to a digital image processing system where images can be recorded or analysed by the aid of a small computer. This provides a very flexible combination for qualitative and quantitative analysis of the transverse motion of a passive tracer and the effect of torsion in the test facilities.

An aqueous solution of Fluorescein (0.05 gr/l of water) from an elevated reservoir is introduced as a tracer to the flow through a special dye circuit. Fluorescein solutions can be added either at the entrance of the water tunnel (before the flowmeters) or at the transparent straight section which precedes the succession of the curved elements. Dye injected upstream of the flowmeter is mixed with water as it passes through the diffuser, the settling chamber and the nozzle; in doing so, the dye reveals the patterns in the whole flow cross-section, when it is visualized in the light sheet. On the contrary, injection of the dye at the straight section is performed through rakes of small injectors of 0.5 mm outer diameter. Each injector marks a streakline of dye which does not mix with the working fluid. As the flow is steady, each dye streak marks a trajectory and one can follow the stretching and deformation of each trajectory as it passes through different elements.

The velocity field in the flow is measured by a laser-Doppler anemometer (LDA). Two velocity components can be measured by the LDA system which consists of a separate air-cooled Argon ion laser source, and two separate measurement units, with front or back scattering arrangement. Translation of the measuring volume, data acquisition, and analysis are performed by an autonomous small computer.

III. RESULTS AND DISCUSSION

We begin the experiments with a fully-developed channel flow which emerges from a 2000 mm long straight section and enters the first curved element. Velocity profiles at the entrance to the first element are shown in Figure 3 for a large range of Dean numbers in the laminar regime. They show a parabolic shape (up to Dn = 247) indicating a fully-developed flow. Then we measure velocity profiles at the exit from each element (or at the entrance to the next one). We have fitted straight spacer channels 200 mm long between each two successive curved elements. The spacers are required for accomodating the anemometer, but they are short enough to not have any effect on the vortices.

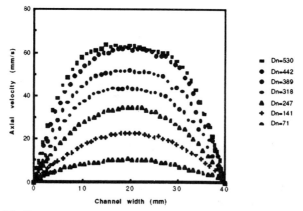

Fig.3 Axial velocity profile at the entrance of the first curved element

To get an image of the torsion effects, we first present the visualization results, then we discuss some preliminary velocity measurements.

Flow visualizations are performed at Dean number Dn = 62 corresponding to Re = 146. We have scanned the flow beginning from the first element.

In the first curved element we observe gradual development of the vortices which is completed by generation of a pair of fully formed vortices at an angular position of 75° from the entrance. These vortices are practically destroyed when they enter into the second curved element (with curvature plane turned 90°), and start to reconstruct themselves according to the rotation protocol imposed by the new environment. At the end of this element a pair of fully developed vortices with symmetry plane perpendicular to that of the previous pair is discernable. We then follow the new pair of vortices as they enter the third curved element (the curvature plane turns 90° again). Evolution of the vortices under torsion effect visualized at 2°, 15°, 30°, 45°, 60°, 75°, and 82° angular positions from the entrance to the third element is shown in Figur 4. We observe that up to 15° position, the torsion effect consists of attenuating the cell vorticity convected from the previous curved element. From 15° on the torsion effect on the vortices is manifested by tilting the symmetry plane and gradually orienting it to the diagonal direction. The symmetry plane is practically trapped at the cross section corner from 45° up to 82° position. At 82° the position, the vortices turn around and take an orientation reversed in comparison with their orientation in the entrance section. While the symmetry plane is trapped at the corner, we observe that it deforms under the turning effect of the vortices which try to take an orientation which conforms with the protocol of the new environment.

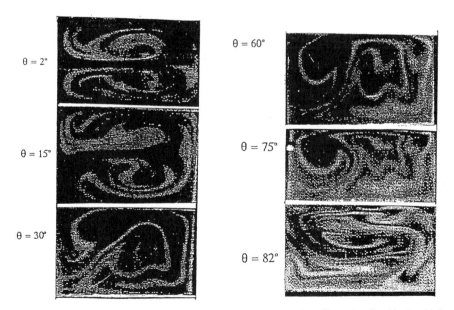

Fig. 4 Tilting of the symmetry plane of the Dean vortices due to the torsion effect, visualised in the third curved element

Diagonal orientation and subsequent deformation of the symmetry plane can be explained from Figure 5 on which we have schematically showed a flow cross section in the third curved element. On this figure we have superposed an *input* pair of vortices (coming from the second element) and a *virtual* pair of vortices which would have existed if the *input flow* to the element did not contain vortices. Streamline patterns show a competition between the two vortex configurations (due to *memory* and *protocol*) which results in *counter flow* and *coflow* regions. The result of this competition is a tilted pair of vortices similar to the one that we observe in Figure 4 between θ = 15° and 75°. In this regard the coalescence of the vortices and a single vortex flow described by Wang and Tuttle seems to be due to some stronger, perhaps nonlinear mechanism.

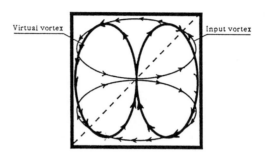

Fig. 5 Schematic diagram showing vortex tilting mechanism.

To compare our experimental results with those due to theory, we define a set of body centered coordinates as shown in Figure 6. If **G**(s) is the continuous curve which passes through the center of curved elements, and s is the arclength along this curve, the triad of unit vectors, **t, n, b**, the torsion τ' and the curvature κ can be given by the following equations (following Tuttle):

t = d**G**/ds;   **n** = (1/κ) (d**t**/ds);   **b** = **t** × **n**;   and d**b**/ds = -τ' **n**

We define a dimensionless torsion parameter τ = τ'a in the same way we introduced the curvature parameter γ = a/R = aκ . One therefore, can expand any dependent variable of the flow in terms of γ and τ as shown below:

$$F = F_{00} + \gamma F_{01} + \gamma^2 F_{02} + ... + \tau F_{10} + \tau\gamma F_{11} + \tau\gamma^2 F_{12} + ...$$

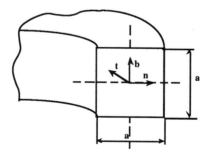

Fig. 6 Body centered coordinate system

The combined effect of curvature, torsion and Reynolds number can then be described by $\lambda = \tau/(\gamma \text{Re})$. The three parameters τ, γ and λ for the results of Wang[10], Tuttle[13], Mukarami et al.[14] (Reynolds number is not reported by Mukarami et al.), as well as the present investigation are tabulated below.

Author	τ	γ	Re	λ	Observation
Tuttle[12]	0.1	0.01	100	0.1	vortex coalescence
Wang[9]	0.02	0.1	20	0.01	two asymmetric vortices
Wang[9]	0.02	0.1	4.8	0.042	vortex coalescence
Mukarami et al[13]	0.675	0.5	?	?	vortex coalescence
Present work	0.18	0.18	146	0.007	two tilted vortices

From the table we see that compared to the results of Tuttle and Wang, our experiment is situated way below the vortex coalescence range of λ. Hence our two pair vortex flow agrees with their theoretical predictions. However, the cited theoretical results correspond to a smooth varaition of curvature plane, while in our experiments the curvature plane orientation varies abrubptly between each two elements.

CONCLUDING REMARKS

Basic design concepts and experimental procedures of a test facility for examining the torsion effects on the Dean vortices generated in a succession of curved channels with rotation of the plane of curvature were described in details. Preliminary results show that the effct of torsion is limited to tilting the symmetry plane of the vortices. This result , though new, is not in cotradiction with theory. On the other hand we did not observe the vortex coalescence reported before.

REFERENCES

1) W.R. Dean; Phil. Mag., 4, p. 208 (1927)

2) Y. Mori and Y. Uchida; Trans. JSME, 33, p. 1836 (1967)

3) S.W. Jones, O.M. Thomas and H. Aref;J. Fluid Mech., 209, p. 335 (1989)

4) H. Peerhossaini and Y. Le Guer; Thermofluids Report TFD-3, Laboratoire de Thermocinétique, ISITEM, Nantes, France (1989)

5) H. Peerhossaini and Y. Le Guer; TIFAN, ed. M. Lebouché and M. Lalland, 2, p. 153 (1989)

6) C.D. Andereck, S.S Courts and S. Tennakooon; in: Görtler Vortex Flows, ed. H. Peerhossaini and J.E. Wesfreid, p. 52, (1990)

7) J.H. Grindly and A.H. Gibson; Proc. Roy. Soc. London, A,80, p. 114 (1908)

8) J. Eustice; Proc. Roy. Soc. London, A, 85, p. 119, (1911)

9) W.R. Dean; Phil. Mag., 5, p. 673 (1928)

10) C.Y. Wang; J. Fluid Mech., 108, p. 185 (1981)

11) M. Germano; J. Fluid Mech., 125, p. 1 (1982)

12) H.C. Kao; J. Fluid Mech., 184, p. 335 (1987)

13) E.R. Tuttle, J. Fluid Mech., 219, p. 545 (1990)

14) M. Mukarami, Y. Shimiza and H. Shiragami; Bull. JSME, 12, p. 1369 (1969)

15) H. Peerhossaini and J.E. Wesfreid; Int. J. Heat & Fluid Flow, 9, p. 12 (1988)

SPLITTING, MERGING AND WAVELENGTH SELECTION OF VORTEX PAIRS IN CURVED AND/OR ROTATING CHANNELS

W.H. Finlay and Y. Guo

Dept. of Mechanical Engineering, University of Alberta

Edmonton, Alberta, Canada T6G 2G8

ABSTRACT

In channels that are rotating (about a spanwise axis) or curved, or both curved and rotating, steady two-dimensional vortices develop above a critical Reynolds number Re_c. The stability of these streamwise-oriented roll cells to spanwise-periodic perturbations of different wavelength than the vortices (i.e. Eckhaus stability) is examined numerically using linear stability theory and spectral methods. In curved and/or rotating channels, the Eckhaus stability boundary is found to be a small closed loop. Within the boundary, two-dimensional vortices are stable to spanwise perturbations. Outside the boundary, Eckhaus instability is found to cause the vortex pairs to split apart or merge together in a manner similar to that observed in recent experiments. For all channels examined, two-dimensional vortices are always unstable when $Re > 1.7 Re_c$. Usually the most unstable spanwise perturbations are subharmonic disturbances, which cause two pairs of vortices with small wavenumbers to be split apart by the formation of a new vortex pair, but cause two pairs of vortices with large wavenumber to merge into a single pair. In nonlinear flow simulations presented here and in experiments, most observed wavenumbers are close to those that are least unstable to spanwise perturbations.

INTRODUCTION

As the Reynolds number, Re, is increased above a critical value Re_c, the flow in curved and/or rotating channels undergoes a supercritical transition from one-dimensional ($1D$) Poiseuille type flow to two-dimensional ($2D$) streamwise-oriented vortices. Though these vortices bear some similarity to Taylor vortices, recent experimental studies in channels with large aspect ratio ($\Gamma \geq 40$) show there is considerable unsteadiness that is not present in Taylor vortex flow[1,2,3]. When viewed in a spanwise-streamwise plane, the vortices cause long streaks in experimental flow visualizations. These streaks are occasionally split apart by new streaks or merge together[2,3]. When viewed in cross-section, vortex pairs appear and disappear frequently[1]. We will refer to these phenomena as the splitting and merging of vortex pairs in this paper. These terms will always used to describe the behaviour of two or more vortex pairs, not individual vortex tubes. Splitting and merging of vortex pairs are not well understood. In this paper, these phenomena are examined using stability theory and nonlinear flow simulation.

The rotating channel geometry is defined by two parallel plates of infinite extent that are spaced a distance d apart. The flow is driven between the two plates by a streamwise pressure gradient. The curved channel geometry is the same as the Taylor-Couette geometry except that the flow is instead driven by an azimuthal pressure gradient and both inner and outer cylinders are stationary. The streamwise and spanwise directions of the flow are given by θ and z, while the direction normal to the walls is r. The curved-rotating channel geometry is obtained by spinning the entire curved channel about a spanwise axis. The Reynolds number is $Re = \overline{U}d/2\nu$, where $\overline{U}$ is the mean (bulk) streamwise velocity. The radius ratio of the two walls is $\eta = r_i/r_o$. The rotation number is defined as $Ro = \Omega d/2\nu$, where Ω is the rotation speed of the system about the z-axis. The spanwise wavenumber of the vortices is defined as $\alpha = \pi d/\lambda$, where λ is the spanwise vortex spacing.

THEORY AND NUMERICAL METHOD

We wish to examine the stability of steady, two-dimensional, streamwise-oriented vortices, with associated velocity field $\mathbf{u}^0$. We obtain $\mathbf{u}^0$ by numerically solving the incompressible, steady, Navier-Stokes equations using spectral methods. Periodic boundary conditions are imposed in the z direction, since the flow is assumed to have infinite span. We use a Fourier Galerkin method in the z direction and a Chebyshev tau method in the r direction. To eliminate aliasing error, the 3/2 rule is used to evaluate the nonlinear terms. Adequate resolution is insured by monitoring the energies in the highest modes. In our computation, the numbers of Fourier modes N and Chebyshev modes M vary from 16×16 to 20×26 ($N \times M$) depending on Re, η, Ro and the spanwise wavenumber α of the vortices. Normally we include only one pair of vortices in the computational box.

Once the $2D$ vortex flow $\mathbf{u}^0$ is found, its stability can be examined using linear stability analysis. In general the perturbation $\mathbf{u}'$ to the base flow $\mathbf{u}^0$ can be expressed as

$$\mathbf{u}'(r,\theta,z,t) = \tilde{\mathbf{u}}(r,z)\, exp\,[st + i(d\theta + bz)] \quad (1)$$

where d and b are the streamwise and spanwise wavenumbers of the perturbation. To avoid a singularity when $\eta = 1.0$ the nondimensional variables must be chosen carefully. In our formulation the perturbation wavenumber b is non-dimensionalized by the spanwise wavelength of the vortices $2\pi/\lambda$. Since we are only interested in the spanwise perturbation, we set $d = 0$ in equation (1).

The stability of $\mathbf{u}^0$ is determined by σ, the real part of the eigenvalue $s = \sigma + i\omega$. The flow pattern of the perturbation is given by

$$Real\{\tilde{\mathbf{u}}(r,z)\, exp(ibz)\} \, . \quad (2)$$

Fourier Galerkin and Chebyshev tau spectral methods are used to solve the above eigenvalue problem. Of the eigenvalues obtained, we are only interested in the eigenvalue with largest real part. We will use the term "Eckhaus[4] eigenvalue" to refer to this eigenvalue and "Eckhaus growth rate" to refer to the real part of this eigenvalue. More details of the method can be in reference[11].

The resulting codes were extensively verified by duplicating the results of previous authors, including the Eckhaus boundary for Taylor vortices[5], wavy instability ($b = 0, d \neq 0$) results for Taylor vortices[6], and wavy instability results obtained in channel flows with curvature or rotation[7,8].

ECKHAUS STABILITY BOUNDARY

The Eckhaus stability boundary is determined by a sign change in the Eckhaus growth rate σ as the parameters of the system are varied. Within the Eckhaus boundary, the Eckhaus growth rate is always negative. In the region neighboring the Eckhaus boundary, our numerical results show the eigenvalue with the maximum Eckhaus growth rate is always real. The stability boundary is thus determined by non-oscillatory perturbations. Fig. 1 shows the Eckhaus boundaries for several channels with either curvature or rotation. Similar results are found when both curvature and rotation are present. The Eckhaus stable region for the flow in curved and/or rotating channels is a small closed region tangent to the minimum of the neutral stability curve for the primary instability. In all cases calculated by us, the Eckhaus boundary is only a weak function of η and Ro.

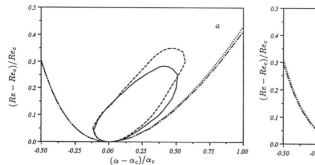
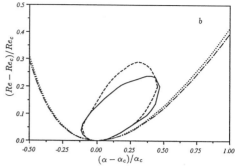

FIGURE 1. Eckhaus stability boundaries are shown for curved and/or rotating channel flow. The primary stability boundary is included for reference. (a) shown for curved channels are Eckhaus: ———, $\eta = 0.975$ ($Re_c = 114.26$, $\alpha_c = 1.98$); ----, $\eta = 0.7$ ($Re_c = 35.83$, $\alpha_c = 2.07$); and primary: —·—, $\eta = 0.975$; ·······, $\eta = 0.7$. (b) shown for rotating channels are Eckhaus: ———, $Ro = 0.005$ ($Re_c = 198.95$, $\alpha_c = 2.01$); ----, $Ro = 0.25$ ($Re_c = 44.30$, $\alpha_c = 2.46$); and primary: —·—, $Ro = 0.005$; ·······, $Ro = 0.25$.

When compared to other spanwise periodic flows known to the authors, curved and/or rotating channels exhibit significant differences. For example in Taylor-Couette flow, the Eckhaus boundary is an open region, i.e., for any Re, there is always a band of stable wavenumber α. On the Eckhaus boundary, the spanwise wavenumber of perturbations, b, approaches zero when Re is not very high[5]. When Re is high, the boundary is given by $b = 0.5$ (Paap & Riecke[9]). Paap & Riecke[9] refer to this as a short-wavelength instability, in order to distinguish it from the long-wavelength nature of the classical Eckhaus instability. Numerical calculations done by us for Taylor-Couette flow show that as α moves away from the Eckhaus boundary, b approaches 0.5 when Re is not very high. As Re increases, this happens very quickly. This is consistent with the results given by Paap & Riecke[9]. Our results also show that for any Re, α and b, the eigenvalue with maximum real part is always real in the Taylor-Couette problem. All parameters of the system are thus real and the Eckhaus stability criterion given by Eckhaus[4] is valid in the region near Re_c.

In channels with either curvature or rotation or both, the situation is more complicated. We have studied the case of the curved channel with $\eta = 0.975$ in most detail. In this case, the value of $b(Re)$ on the right side of the Eckhaus boundary increases monotonically from 0.0 at $Re = Re_c$ and reaches a constant value of 0.5 when $Re \geq 1.2Re_c$. On the left side, $b = 0$ determines the boundary for $Re < 1.2Re_c$, but for higher Re, $b = 0.5$ determines the left boundary. Clearly here the boundary defined by the instability of $2D$ vortices to spanwise perturbations is not the classical Eckhaus stability boundary ($b \to 0$) found in Taylor-Couette flow. When $Re \leq 1.125Re_c$, only the left side of the boundary is of the classical Eckhaus type. For simplicity, we still call the entire boundary an Eckhaus boundary.

The Eckhaus stability criterion[4] does not apply to the right side of the Eckhaus boundary even in the region close to Re_c. On the left side, it is valid with reasonable accuracy up to $Re < 1.04Re_c$. In contrast, in Taylor-Couette flow, the Eckhaus stability criterion is valid with reasonable accuracy up to $1.1Re_c$ for both sides of the boundary[5]. For the channel geometries, in the neighborhood of the right side of the Eckhaus boundary, the eigenvalue with Eckhaus growth rate is not real for all b when $Re > 1.1Re_c$. For some b, a complex conjugate pair has the maximum real part. Near the left side of the Eckhaus boundary, the eigenvalue with Eckhaus growth rate is real for all b. We believe that $b \neq 0$ and non-real eigenvalues near the right side of the boundary are the reason why the Eckhaus criterion is not valid here.

Beyond the top of the Eckhaus boundary, if Re/Re_c is not too high ($Re \leq 3.5Re_c$) and α is not too small or too large ($1.8 \leq \alpha \leq 5.0$), the eigenvalue with the maximum Eckhaus growth rate is given by $b = 0.5$ and is real.

Results similar to those discussed above are found in all channels we examined. In all cases, $2D$ vortices are unstable to spanwise perturbations when $Re > 1.7 Re_c$ (they are often unstable at even lower Re). Since the Eckhaus stable region is small and most experiments have been done outside this region, the instability associated with the most unstable mode, which usually has $b = 0.5$, is an important instability in curved and/or rotating channel flows.

SPLITTING AND MERGING OF VORTICES

In channel flow experiments, vortex pairs are sometimes observed to merge together (reducing the number of vortices across the channel) or to be split apart by the formation of new vortex pairs [1,2,3]. We believe that the splitting and merging of vortex pairs are associated with the instability of $2D$ vortices to spanwise perturbations. Our linear stability results indicate that no $2D$ vortex flow is stable to spanwise perturbations when $Re > 1.7 Re_c$. In order to understand how these $2D$ vortex pairs lose their stability to spanwise perturbations and split apart or merge together, we examine the flow pattern of the most unstable mode of linear stability theory. The flow pattern of the most unstable mode is given by the eigenfunction (2), which has the largest growth rate σ. Outside the Eckhaus stability boundary the most unstable mode occurs at $b = 0.5$ and grows at the rate $\exp(\sigma t)$. Fig. 2 and 3 show two vortex flows ($\alpha = 2.0$ in fig. 2 and $\alpha = 4.0$ in fig. 3), and the linear superposition of each base flow with its most unstable mode at $Re = 2.0 Re_c$ in the curved channel with $\eta = 0.975$. At this Re, all spanwise wavenumbers of the base flow are unstable to Eckhaus instability. In these figures, $z/\lambda = 0, 1.0$ and 2.0 are the centers of the base vortex pairs, where the fluid flows from the concave wall to convex wall (the inflow region).

Fig. 2(b) shows the base flow (fig. 2a) superimposed with the most unstable mode, whose kinetic energy has grown to 20% of the base flow's kinetic energy. It can be seen that one new pair of secondary vortices begins to form at $z/\lambda = 1.5$ near the concave wall. The original vortex pairs centered at $z/\lambda = 1.0$ and 2.0 spread apart from $z/\lambda = 1.5$ and are squeezed toward $z/\lambda = 0.5$ and 2.5. As the energy level of the perturbation increases, the new vortex pair grows bigger and moves away from the concave wall. This is a splitting event, since a new vortex pair appears between the base vortex pairs centered at $z/\lambda = 1.0$ and 2.0. The two base vortex pairs centered at $z/\lambda = 1.0$ and 2.0 will be split apart by the new pair, yielding three pairs.

Fig. 3(b) shows the base flow (fig. 3a) perturbed by 50% (kinetic energy) of the most unstable mode. Two neighboring vortices (centered at $z/\lambda = 1.5$) of the two base vortex pairs become smaller and weaker while the other two become bigger and stronger. As the amplitude of the perturbation is increased further, the two vortices on either side of $z/\lambda = 1.5$ eventually disappear while the other two vortices form into one pair. Thus two pairs merge into one pair. The large perturbation amplitudes used to produce figs. 2(b) and 3(b) are not linear amplitudes, but are used for visual clarity. The qualitative effects of the perturbations are independent of their amplitude.

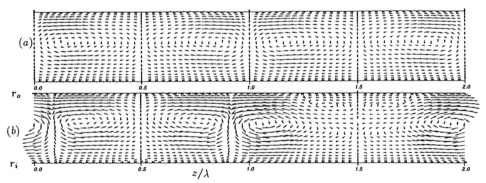

FIGURE 2. $2D$ vortices (a), and $2D$ vortices + the most unstable mode (b) in a curved channel ($\eta = 0.975$) at $Re = 2.0 Re_c$, $\alpha = 2.0$ projected onto the (r, z) plane. In (b), the kinetic energy of the most unstable mode is 20% of the base flow's kinetic energy. The most unstable mode has $b = 0.5$.

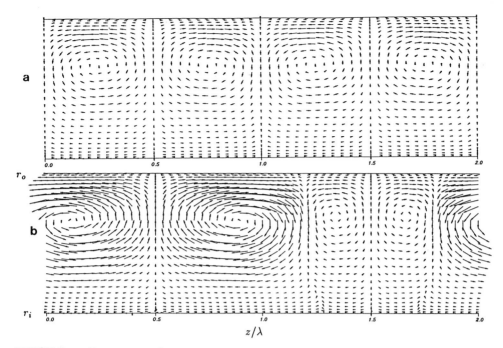

FIGURE 3. Same as in figure 2 but at $\alpha = 4.0$. In (b), the percentage of the most unstable mode is 50%.

Over a wide parameter range, the most unstable modes are found to have two different effects on the base vortex flow, associated either with splitting or merging. There is not a clear cut wavenumber α that divides the effects of the most unstable modes into vortex splitting or merging regions. For example at $\alpha = 3.0$ in the curved channel ($\eta = 0.975$, $Re = 2Re_c$), a small vortex pair at $z/\lambda = 0.5$ in the most unstable mode tends to cause the appearance of a new pair in the base flow at $z/\lambda = 0.5$, while single vortices at $z/\lambda = 0.0$, 1.0 and 2.0 in the most unstable mode tend to cause merging of the two vortex pairs in the base flow on either side of $z/\lambda = 1.5$ and this merging effect in return allows room for the appearance of the new pair. Thus when this eigenfunction is imposed on the base flow, splitting and merging of vortex pairs are seen to occur at the same time, but at spanwise locations separated by one wavelength. For general α and b, we find no strict relation between b and the number of vortices in the most unstable modes. It depends on both b and α. But it is always true that base vortex pairs with large wavenumbers merge whereas small wavenumbers split. Since base vortices with $\alpha < 1.5$ are not likely to occur experimentally because of the low growth rate of the primary instability at this α, subharmonic splitting and merging mechanisms will be the dominant feature of splitting and merging of vortex pairs in channels with either curvature or rotation or both. Since all $2D$ vortices are unstable to spanwise perturbations at high enough Re, the splitting and merging of vortex pairs, as suggested by figs. 2 and 3 will continually occur.

Support for our above results can be found from existing flow visualizations in curved and/or rotating channels given by various authors[1,2,3] as well as in nonlinear simulations discussed later in this paper. Splitting and merging of vortex pairs is prevalent in these experiments. When visualized with reflective flakes, in plan view the splitting of vortex pairs is indicated by two new bright streaks which represents a vortex pair first appearing between two existing streaks. The existing streaks are then split (spread) apart by the two new streaks. This is seen for example in fig. 6(d), (f) of reference 2 in channels with rotation or in fig. 14(d) of ref. 3 in a channel with both curvature and rotation. The disappearance of vortex pairs is indicated by adjacent bright streaks which occasionally merge together[2,3]. We suggest that subharmonic splitting and merging are the mechanisms behind these phenomena.

The actual splitting and merging processes observed in experiments are complicated and nonlinear. Ligrani & Niver[2] provide observations of these processes in a curved channel ($\eta = 0.979$). During the splitting of vortex pairs, according to their observation, new vortex pairs are formed first near the concave wall between other pairs, then followed by a readjustment of spanwise wavenumber. When viewed in the radial-spanwise plane, the new vortex pairs appear to "pop" out of the concave wall. Ligrani & Niver called them secondary vortices. This observation is consistent with our result in fig. 2(b). The formation of the secondary vortices observed by Ligrani & Niver[1] may thus be due to Eckhaus instability.

WAVENUMBER SELECTION

Our results suggest that the splitting and merging processes are sensitive to the rate at which local perturbations develop, i.e. to the local Eckhaus growth rate σ. Vortices which yield smaller σ produce a region where the perturbations take a longer time to develop, so that splitting or merging processes occurring elsewhere will suppress the splitting or merging of vortex pairs with low Eckhaus growth rate. Thus vortices with relatively lower growth rate are more likely to be observed in experiments. Plots of the Eckhaus growth rate σ(at $b = 0.5$) vs. α at different Re show that $\sigma(\alpha)$ has a minimum when Re is not too small (e.g. $Re > 2.2Re_c$ in the curved channel with $\eta = 0.975$). When contours of σ are plotted in an (α, Re) plane this minimum appears as a valley and we call it the Eckhaus valley. As Re increases, both sides of the valley become steeper. Based on our above discussion, observed wavenumbers should be close to this valley. For example, in the curved channel ($\eta = 0.975$) with $Re < 2.2Re_c$, σ varies little with α and here Eckhaus instability does not play a major role in the wavenumber selection process. Thus when $Re < 2.2Re_c$, without any other nonlinear selection mechanism the observed wavenumbers should be close to the ones with maximum primary growth rate, since these vortices develop most rapidly from the $1D$ Poiseuille type flow. Fig. 4 shows the Eckhaus valley for $\eta = 0.975$ in comparison with the wavenumbers observed by Kelleher et al. [10] in a curved channel with $\eta = 0.979$. Also shown are the wavenumbers with maximum primary growth rate and maximum pressure gradient[7]. For $Re < 2.2Re_c$, the observed wavenumbers are close to the ones with maximum primary growth rate. For $Re \geq 2.2Re_c$, the observed wavenumbers are close to the Eckhaus valley. This is consistent with our discussion.

A similar comparison can be found in fig. 5 at $Re = 472.5$ in channels with rotation. The wavenumbers were measured[2] at three different downstream locations

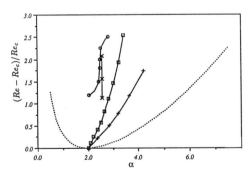

FIGURE 4. Eckhaus valley (○), the curves of maximum primary growth rate (□) and maximum pressure gradient[7] in a curved channel with $\eta = 0.975$, and the wavenumbers of $2D$ vortices observed by Kelleher et al. [10] (×) in $\eta = 0.979$. The primary stability boundary[7] (········) is included for reference.

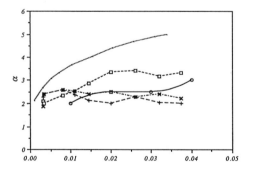

FIGURE 5. Eckhaus valley (○), the curve of maximum primary growth rate (········) and the observed wavenumbers of $2D$ vortices of Alfredsson & Persson[2] in a rotating channel at different downstream locations y/d for $Re = 472.5$: □ , $y/d = 40$; ×, $y/d = 80$; + , $y/d = 120$.

$y/d = 40, 80$ and 120. The observed wavenumbers at $y/d = 40$ are closer to those with maximum primary growth rate than the Eckhaus valley. As the downstream distance increases to 80 and 120, the observed wavenumbers become much closer to the Eckhaus valley. This suggests that near the entrance of the channel where the flow is still developing from $1D$ Poiseuille type flow to $2D$ vortices, the primary instability plays an important role in wavenumber selection process. Once the $2D$ vortices are more fully developed, the Eckhaus instability sets in and the wavenumbers of the vortices are selected by the Eckhaus valley.

Results similar to those given above are found in channels with both curvature and rotation, although the Eckhaus valley is not presented here (cf. Guo & Finlay[11]) since there are no experimentally measured wavenumbers to compare with.

NONLINEAR SIMULATION OF SPLITTING AND MERGING OF VORTICES

The results on splitting and merging given above are based on linear stability theory. Nonlinearity sets in once a splitting or merging event develops past the initial linear stage. In order to study the nonlinear aspect of the problem and how it affects the validity of our linear theory results, we use the Galerkin spectral numerical method of Moser, Moin & Leonard (1983) to simulate the axisymmetric, time-dependent, incompressible Navier-Stokes equations in a curved channel. Periodic boundary conditions are used in the spanwise direction.

Outside the Eckhaus stable region, all wavenumbers are unstable. However, the previous section indicates that vortices with wavenumber in the Eckhaus valley are the least unstable to splitting and merging and are likely to be observed more often in experiments. In order to study the validity of this result in the presence of nonlinear splitting and merging processes and the interactions between vortices with different wavenumbers, we select the aspect ratio of our computational box to be $6\pi : 1$. Since an integer number of pairs of vortices must appear in the simulation region, the average wavenumber of the vortices is restricted to $n/3$, where n is an integer. We use 96 spanwise Fourier modes and 16 Chebyshev polynomials (in the radial direction) in this simulation.

We start the simulation using curved channel Poiseuille flow with low amplitude random noise ($< 0.1\%\overline{U}$) superimposed and use $Re = 2.2Re_c$ in a curved channel having $\eta = 0.975$. After $t = 500d/2\overline{U}$, finite amplitude $2D$ vortices develop rapidly. Fig. $6(a)$ shows contours of the Stokes stream function at $t = 650d/2\overline{U}$. Excluding the three pairs near the center of the box where a splitting event is underway, the average wavenumber of the vortices at this time is 2.79. This is very close to the wavenumber $\alpha = 2.82$ which has the maximum primary instability growth rate (see fig. 4). This demonstrates that during the early stage of the development of vortex flow, the primary instability does play an important role in wavenumber selection process, as suggested above.

As the solution proceeds in time, the flow goes through a sequence of splitting and merging processes and there is a decrease in the average wavenumber of the vortices until $t = 1300d/2\overline{U}$. For $t > 1300d/2\overline{U}$, the wavenumber of the vortices remains near the Eckhaus valley, which is rather flat and lies between $\alpha = 2.1$ and $\alpha = 2.35$, but the wavenumbers continue to fluctuate due to splitting and merging events. Figs. $6(b-g)$ show a typical cycle of the flow. In fig. $6(b)$, the average wavenumber of the first 3 vortex pairs from the left side of the box is 2.37, while the rest of the vortices have a wavenumber of 2.19, which is in the Eckhaus valley. There is a merging process just underway between the first two pairs from the left. This merging process becomes clearer in fig. $6(c)$. For the rest of the vortices, there is a very small decrease in wavenumber from 2.19 to 2.13. In fig. $6(d)$, one vortex pair has completely disappeared.

In fig. $6(c)$, there is a new pair beginning to appear near the left end of the box. This becomes clearer in figs. $6(d)$ and $6(e)$. Another new pair also begins to develop near the center of the box where the vortices have smaller α. The rest of vortices remain essentially unchanged. In fig. $6(f)$, the new pair near the left side is almost fully developed. The other one near the center continues to grow. The rest of vortices have been squeezed somewhat and there is an increase in their wavenumbers (from 2.17 in fig. $6d$ to 2.57 in fig. $6f$). It can be seen the second and third pairs from the left in fig. 6(f) have begun a merging process. In fig. 6(g), the new pair near the center has completed its development. At this time, the wavenumber is nearly the same for all vortices and the average wavenumber is 2.67. The merging of the second and third pairs from the left proceeds. As these two pairs continue to merge, the other vortices experience an adjustment in their wavenumbers which results in smaller wavenumbers.

The flow field produced at the start of a merging or splitting of two vortex pairs is similar to that given in fig. 2(b) and 3(b). We believe the merging and splitting events observed in experiments and in our nonlinear simulation are the nonlinear result of the Eckhaus instability discussed above.

From fig. 6 and similar results at other times in the simulation, we find that when the wavenumber of several pairs of neighboring vortices is close to the Eckhaus valley, the wavenumber of these pairs remains nearly constant for a long time and their adjustment due to splitting and merging events is very weak, as is the case in figs. 6($b-d$). But when the wavenumber of two or more neighboring pairs is not close to Eckhaus valley, there is a rapid and large change in wavenumber. Thus, when averaged over time, the average wavenumber is close to the Eckhaus valley obtained from our stability analysis.

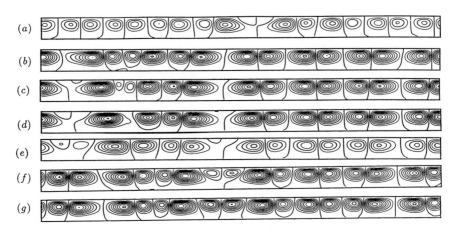

FIGURE 6. Contours of Stokes stream function for curved channel flow at $Re = 2.2Re_c$, $\eta = 0.975$ are shown in the (r,z)-plane at time (a) $650d/2\overline{U}$, (b) $1850d/2\overline{U}$, (c) $1949d/2\overline{U}$, (d) $1978d/2\overline{U}$, (e) $2009d/2\overline{U}$, (f) $2069d/2\overline{U}$ and (g) $2099d/2\overline{U}$. The aspect ratio is $\Gamma = 6\pi : 1$, with periodic spanwise boundary conditions.

REFERENCE

1. Ligrani, P. & Niver, R.D. 1988 Phys. Fluids **31**,3605–3618
2. Alfredsson, P.A. & and Persson, H. 1989 *J. Fluid Mech.* **202**, 543–557.
3. Matsson, J.O.E. & Alfredsson, P.H. 1990 J. Fluid Mech. **202**,543–557
4. Eckhaus, W. 1965 *Studies in nonlinear stability theory*. Springer, New York.
5. Riecke, H. & Paap, H. 1986 Phys. Rev. A **33**, 547–553
6. Jones, C.A. 1985 J. Compt. Phys. **61**,32–344
7. Finlay, W.H., Keller, J.B. & Ferziger, J.H. 1988 J. Fluid Mech. **194**,417–456
8. Finlay, W.H. 1990 J. Fluid Mech. **215**,209–227
9. Paap, H. & Riecke, H. 1990 Phys. Rev. A **41**, 1943–1951
10. Kelleher, M.D., Flentie, D.L. & McKee, R.J. 1980 J. Fluids Engng. **102**, 92–96
11. Guo, Y. & Finlay, W. H. 1991 J. Fluid. Mech. **228**:661-691.
12. Moser, R.D., Moin, P. & Leonard, A. 1983 J. Comp. Phys.**52**,524–544

TRANSIENT, OSCILLATORY AND STEADY CHARACTERISTICS OF DEAN VORTEX PAIRS

IN A CURVED RECTANGULAR CHANNEL

Phillip M. Ligrani

Department of Mechanical Engineering, Code ME/Li
Naval Postgraduate School
Monterey, California 93943-5000

INTRODUCTION

When Dean numbers are high enough, the flow in a curved channel is unstable to centrifugal instabilities, and secondary flows develop which eventually form into pairs of counter-rotating vortices. This flow is referred to as Dean vortex flow, and the accompanying pairs of streamwise-oriented vortices are referred to as Dean vortices. The present paper describes some transient, oscillatory, and steady characteristics of Dean vortex pairs in a curved rectangular channel with 40 to 1 aspect ratio and an inner to outer radius ratio of 0.979. In particular, attention is focussed on secondary instabilities observed in the form of oscillatory motions called twisting and undulating, and in the form of transient events consisting of splitting and merging of vortex pairs. Splitting and merging events are described from visualizations of flow in spanwise/radial planes at Dean numbers of 75 and 100 in the form of two distinctly different types of splitting events, and four distinctly different types of merging events. Even though these transient events produce fairly large amplitudes of unsteadiness, the Dean vortex pairs have preferred positions across the span of the channel about which the motion occurs. These preferred positions are evident from time-averaged distributions of streamwise velocity, which show a variety of steady vortex pair characteristics.

EXPERIMENTAL APPARATUS AND PROCEDURES

The present experimental results are obtained using the same 40 to 1 aspect ratio curved channel described by Ligrani and Niver (1988). A schematic drawing of the curved channel is shown in Figure 1. The facility is an open-circuit suction facility designed for low-speed transition studies. At the inlet, a honeycomb, screens and nozzle reduce spatial non-uniformities in the flow. These are followed by a 2.44 m long straight duct which provides fully developed laminar flow at the inlet of the curved section for Dean numbers up to 640 (Ligrani and Niver, 1988). The radii of the concentric convex and concave walls in the curved section are 596.9 mm and 609.6 mm, respectively. The 180 degree curved section is then followed by another 2.44 m long straight duct, additional flow management devices, and an outlet plenum. From the inlet of the first straight section to the outlet of the second straight section, the interior dimensions of the facility are 1.27 cm for the height and 50.80 cm for the width. The cylindrical coordinate system is aligned such that (r, x, z) are the normal, streamwise, and spanwise directions, respectively. y is measured from the concave wall, radially inwards. In the discussion which follows, the Dean number is defined as $De = 2 Re (d/r_i)^{1/2}$, where Reynolds number

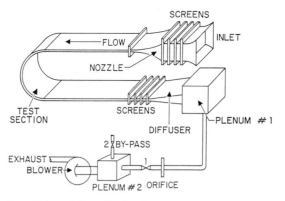

FIGURE 1. Schematic of the transparent curved channel.

Re equals $Ud/2\nu$, U is the mean (bulk) streamwise velocity, d is the channel width, ν is kinematic viscosity, and r_i is the radius of the convex (or inner) curved surface.

A slot, 3.2 mm wide and 76.2 mm long is located in the convex wall of the curved channel. It is lined with foam, and aligned in the spanwise direction 50.8 mm off of the centerline at a location 120 degrees from the start of curvature. It is used to allow probe insertion with no air leakage to the channel interior. A miniature five-hole pressure probe (tip diameter is 1.22 mm) is used in the slot to obtain surveys of the three mean velocity components (Ligrani et al., 1989a, 1989b).

In order to obtain information on secondary flow motions resulting from the wavy vortices, air in the channel was partially contaminated with smoke and then photographed. Ligrani and Niver (1988) and Ligrani et al. (1992b) describe the procedures used to obtain these flow visualization results. Because the smoke enters into the curved section in a layer near the convex surface, events eminating from the proximity of the convex surface are smoke rich, whereas secondary flows which begin near the concave side are relatively free of smoke.

EXPERIMENTAL RESULTS

At any streamwise location from 95 to 145 degrees from the start of curvature, a variety of different types of behavior develop as the Dean number increases. At low Dean numbers (<40-60), the fully is fully laminar. As the Dean number increases to any value between 40-60 and 64-75, flow visualizations show unsteady, wavy smoke layers evidencing small amplitude secondary flows. Then, at higher Dean numbers (>64-75), vortex pairs develop as the primary instability (Ligrani and Niver, 1988), where the streamwise location of the initial appearance of these pairs is dependent upon the Dean number. Secondary instabilities then develop at Dean numbers from about 75 to 200 in the form of vortex pair undulations, vortex pair twisting, and in the form of events where vortex pairs split and merge (Ligrani et al. 1992a, 1992b). Of these, twisting is particularly important because it leads to increases in longitudinal fluctuating intensities at Dean numbers above 160, particularly in upwash regions near the concave surface, which appear to be the most unstable part of the vortex pair structure. These fluctuating intensity increases are significant and strongly dependent upon the location within the vortex pair structure as the Dean number increases from 160 to about 400. Fully turbulent conditions then develop at Dean numbers greater than about 400.

Time-Averaged Results

Surveys of the mean streamwise velocity are presented in Figure 2 for Dean numbers from 35.0 to 187.7 at a location 120 degrees from the start of

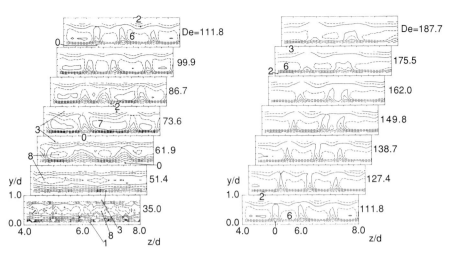

FIGURE 2. Time-averaged distributions of the streamwise velocity for Dean numbers from 35.0 to 187.7 as measured in a spanwise/radial plane located 120 degrees from the start of curvature.

curvature in the spanwise/radial plane located at z/d from 4.0 to 8.0. In these plots, $y/d = 0.0$ represents the concave surface, $y/d = 1.0$ repesents the convex surface, and z/d is measured with respect to the channel centerline. Streamwise velocities are normalized by U. Contour level magnitudes are tabulated in Table 1.

Results in Figure 2 for $De = 35.0$ and 51.4 show a spatially uniform, undisturbed flow. As Dean number increases to 61.9, areas of low pressure and low streamwise velocity develop near the concave surface. These velocity deficit regions are somewhat similar to each other, and correspond to secondary flow upwash regions directed from the concave surface to the convex surface. At higher Dean numbers (≥ 73.6), upwash regions are located between the two vortices in a pair, and correspond to areas of low pressure and low velocity which extend almost to the full channel height. Streamwise mean velocity surveys for $De \geq 162.0$ show qualitative differences compared to results at lower Dean numbers. In particular, three velocity deficits at $De=162.0$ are replaced by two deficits over the same channel span at $De=187.7$. Surveys at Dean numbers up to about 240 are then similar to the survey at $De=187.7$. At Dean numbers greater than about 240, time-averaged results show gradients only in the radial direction. Such spanwise uniform behavior probably results because spanwise lengthscales of unsteadiness amplitudes are greater than the spanwise spacings between vortex pairs.

Transient Dean Vortex Flows: Splitting and Merging Events

After the development of the Dean vortex pairs as the primary instability, secondary instabilities in the forms of splitting and merging events develop. These are transient events in themselves even though these events may occur repeatedly. These splitting and merging events are observed at De as large as 220 and at angular positions as small as 85 degrees from the start of curvature (Ligrani and Niver, 1988). Ligrani, Longest, Kendall, Fields, and Fuqua (1992b) provide additional documentation at Dean numbers of 75 and 100 from visualizations of the flow in spanwise/radial planes at angular positions from 95 degrees to 135 degrees from the start of curvature. These investigators describe two distinctly different types of splitting events and four distinctly different types of merging events. The two types of splitting events are: (A1) vortex pair emergence from flow near the concave surface (emergence), and (A2) abrupt development of two vortex pairs from a single pair (split). The four different types of merging events include: (D1) vortex pair disappearance into the flow near the concave surface (collapse), (D2) the sweeping of a vortex pair into the upwash of a neighboring pair (merge), (D3) engulfment

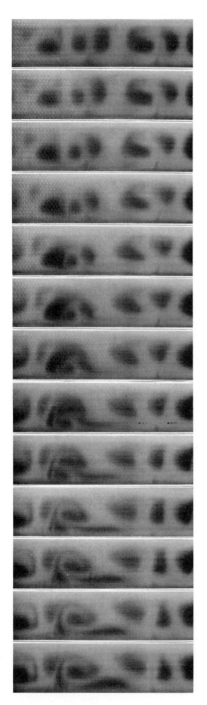 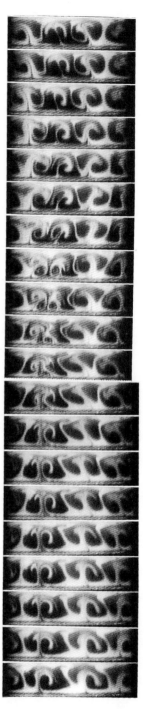

FIGURE 3. Time sequences of flow visualization photographs illustrating: (a) engulfment of a vortex pair for De=75 at a streamwise location 135^0 from the start of curvature (thirteen photographs), and (b) partial engulfment of a vortex pair for De=100 at a streamwise location 125^0 from the start of curvature (twenty photographs). The streamwise direction is into the plane of each photograph, the concave surface is on the bottom of each photograph, and the convex surface is on the top of each photograph. Photographs are spaced apart by 1/30 second (0.0333 sec.) intervals.

Table 1. Non-Dimensional Velocity Contour Levels of Figure 2.

0	0.45-0.60	4	1.05-1.20	7	1.50-1.65
1	0.60-0.75	5	1.20-1.35	8	1.65-1.80
2	0.75-0.90	6	1.35-1.50	9	1.80-1.95
3	0.90-1.05				

of a smaller vortex pair by a larger pair such that vortices having the same sign of vorticity merge together (engulfment), and (D4) cancellation of adjacent vortices having opposite signs of vorticity in two adjacent pairs (cancellation).

In the following discussion, examples of engulfment events are described along with information on the frequencies of splitting and merging events in general.

A time sequence of photographs which illustrates how one vortex pair may be engulfed by an adjacent vortex pair is presented in Figure 3a. These results are given for a Dean number of 75 at a streamwise location 135 degrees from the start of curvature. The photographs in the sequence are spaced 1/30 sec. apart and time increases as one proceeds down the page. The streamwise direction is into the plane of the paper. The convex surface is at the top of each photograph and the concave surface is at the bottom of each photograph. Three pairs of counter-rotating vortices are evidenced in the first (topmost) photograph of Figure 3a by mushroom-shaped smoke patterns. The upwash region (with respect to the concave surface) is a region of radially inward flow between the two vortices in a pair and corresponds to the dark thin "stem" of a mushroom-shaped smoke pattern (Ligrani and Niver, 1988). The downwash region (again, with respect to the concave surface) between adjacent pairs of vortices is seen as a bright area extending across the channel also approximately in the radial direction. Each vortex center then lies in the dark region between upwash and downwash regions.

The three vortex pairs initially in the top photograph of Figure 3a are present with about the same extent but with slightly different orientations. As time increases, the middle vortex pair becomes smaller, and starts to tilt such that the upwash region reaches an angle of about 60 degrees from the plane of the concave wall. This tilting and the motion of the smaller vortex pair towards the larger pair to its left occur because of pressures in fluid redistributed by secondary flows. In the sixth to the eighth photos, the right-hand vortex of the larger pair begins and continues to surround the entire smaller pair. In the ninth photo, the upwash region of the smaller pair appears to be severly tilted with repect to the concave surface, and the right-hand vortex of the larger pair is merging with what used to be the right-hand vortex of the smaller pair. Thus, the smaller pair is engulfed by the larger pair such that vortices (which make up one-half of each pair) with the same sign of vorticity merge together.

In some cases such as the one presented in Figure 3b, vortex pairs merge together in a different way. The data in this figure are for a Dean number of 100 at a streamwise position 125 degrees from the start of the curvature. In the initial photos of Figure 3b, three complete vortex pairs are evident with some rocking motion. From left to right, the three pairs in these photos are denoted B, C, and D, respectively. As the upwash region of pair B becomes normal to the concave surface (tenth photo), the upwash region of vortex pair C tilts to the left. The left-hand vortex of pair C then merges into near-wall flow, and the right-hand vortex merges with the right-hand vortex of pair B (photos 11-12). One vortex pair then replaces the two pairs in all subsequent photos. Spanwise locations of vortex pairs, presented as a function of time in Figure 4, also illustrate the collision of vortex pairs B and C which occurs 0.6-0.7 secs. or 80-93 channel time scales after the start of the photo sequence.

Simultaneous spanwise wavenumbers between vortex pairs A, B, C and D are given as a function of time in Figure 5. These values are determined from spanwise distances between pairs shown in Figure 4. As time becomes greater than 0.55 secs. (73. $d/(2.0\ U)$), the value of α between pairs B and C increases drastically just prior to the merging event. Prior to this increase, values of α between vortex pairs A and B, and between vortex pairs C and D range between 3.0 and 3.4. After the two vortex pairs have

285

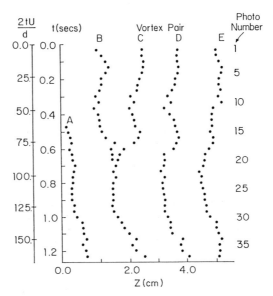

FIGURE 4. For the same experimental conditions as for Figure 3b, results are presented which illustrate spanwise locations of vortex pairs as a function of time for vortex pairs A, B, C, D, and E.

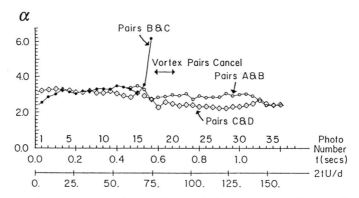

FIGURE 5. For the same experimental conditions as for Figure 3b, results are presented which illustrate the local spanwise wavenumber variations with time between vortex pairs A, B, C, and D.

merged, α values between the same pairs vary between 2.2 and 3.0. The spanwise-averaged spanwise wavenumber (determined from the spacings between pairs A, B, C and D) is about 3.2 over the time period prior to the merging event, and about equal to 2.6 afterwards. The overall average spanwise wavenumber (in space and time) at this experimental condition is 2.82-2.88 (when no perturbation exists in the flow). Values of α higher than this average, such as exist at times prior to 60.0 $d/(2.0\ U)$ secs., are prelude to a merging event and the disappearance of a vortex pair across the channel span. α values lower than the overall average, such as exist at times larger than 140.0 $d/(2.0\ U)$ secs., are prelude to the appearance of a new vortex pair across the channel span from a splitting event. After either a splitting event or a merging event, the reoccurance of the same type of event just afterwards is unlikely because of the alterations to the spanwise wavenumber which occur immediately following such events (Ligrani et al., 1992b).

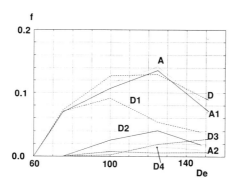

FIGURE 6. Non-dimensional frequencies of splitting and merging events as a function of Dean number as observed 95 degrees from the start of curvature.

Information on the frequencies of splitting and merging events is presented in Figure 6. These data were obtained from measurements in video sequences photographed in a spanwise/radial plane located 95 degrees from the start of curvature. On this plot, non-dimensional frequencies $f = 2\pi n(d/2)/U$ are presented as a function of Dean number, where n is the dimensional event frequency in Hz. The figure indicates that emergences are the most frequent splitting event, whereas collapses and merges are the most frequent types of merging events. Strong dependence on Dean number is evident. Overall frequency magnitudes of splitting events and merging events (A and D curves) are about equal for each Dean number providing evidence of a balance between these two types of phenomena. Such behavior provides additional evidence that vortex pairs have preferred spanwise locations and preferred spanwise wavenumbers across the span of the channel. The non-dimensional appearance and disappearance frequencies increase with Dean number for values up to around 125. At the Dean numbers increases further, decreases of these quantities with Dean number are evident.

Oscillating Dean Vortex Flows: Undulating and Twisting

Finlay et al. (1988) and Ligrani et al. (1992a) describe two different modes of oscillating waviness which manifest themselves as superimposed waves travelling in the streamwise direction at constant speeds. These modes are called undulating and twisting, and both produce vortices which are non-axisymmetric. Smoke visualization photographs of typical experimental undulating vortices show upwash (radially inward) flow regions which tilt from left to right as time increases. Pairs observed in numerical simulations also sideslip periodically in the spanwise direction. Twisting is different from undulating modes because it does not include spanwise motions in the form of sideslipping. In addition, the highest possible frequency of undulating vortices is an order of magnitude lower than fundamental frequencies observed during twisting. This change of frequencies is observed to occur experimentally when the Dean number exceeds about 130. The first photographic evidence of twisting at Dean numbers greater than about 130 is provided by Ligrani and Niver (1988) who indicate that initial De and θ of twisting are in rough agreement with values estimated by Finlay et al. (1988). Recent experimental evidence of undulations is provided by Ligrani et al. (1992a) at Dean numbers from 75 to about 130.

SUMMARY AND CONCLUSIONS

A transparent curved channel with mild curvature, an aspect ratio of 40 to 1, and an inner to outer radius ratio of 0.979 is employed to study the transient, oscillatory and steady characteristics of Dean vortex pairs at Dean numbers ranging from 40 to about 400. The transient modes of vortex motion consist of splittings and mergings of vortex pairs. The oscillating modes include twisting and undulating vortex motions. Even though these different modes of motion can produce fairly large amplitudes of unsteadiness, the Dean vortex pairs have preferred positions across the span of the channel about which the motion occurs. These preferred positions are evident from time-averaged flow field measurements, which show specific vortex pair characteristics and not the spanwise uniform behavior which would be present with random unsteadiness and without the preferred locations of vortex pairs.

Transient events, consisting of vortex pair splitting events and vortex pair merging events, are evident in visualizations of flows in cross-sectional planes at Dean numbers from 75 to 220 and at angular positions as small as 85 degrees from the start of curvature. These visualization results show two distinctly different types of splitting events, and four distinctly different types of merging events. Local spanwise wavenumber values higher than the overall average spanwise wavenumber (in space and time) for a particular Dean number are prelude to the merging of two vortex pairs across the channel span, and values lower than the overall average are prelude to the splitting of a vortex pair across the channel span.

ACKNOWLEDGEMENTS

This study was sponsored by the Propulsion Directorate, U. S. Army Aviation Research and Technology Activity-AVSCOM, NASA-Defense Purchase Requests C-80019-F and C-30030-P. Mr. Kestutis C. Civinskas was the program monitor. Technical contributions to the experimental portions of this study were made by Mr. L. R. Baun, Mr. W. A. Fields, Mr. J. M. Longest, Mr. M. R. Kendall, Mr. S. J. Fuqua, and Adjunct Research Professor C. S. Subramanian.

REFERENCES

Finlay, W. H., Keller, J. B., and Ferziger, J. H., 1988, Instability and transition in curved channel flow, J. Fluid Mech. 194:417.

Ligrani, P. M., Finlay, W. H., Fields, W. A., Fuqua, S. J., and Subramanian, C. S., 1992a, Features of wavy vortices in a curved channel from experimental and numerical studies, Physics of Fluids A, 4(4):695.

Ligrani, P. M., Longest, J. E., Kendall, M. R., Fields, W. A., and Fuqua, S. J., 1992b, Splitting, merging and spanwise wavenumber selection of Dean vortex pairs during their initial development in a curved rectangular channel, Physics of Fluids A, submitted.

Ligrani, P. M., and Niver, R. D., 1988, Flow visualization of Dean vortices in a curved channel with 40 to 1 aspect ratio, Phys. Fluids, 31:3605.

Ligrani, P. M., Singer, B. A., and Baun, L. R., 1989a, Miniature five-hole pressure probe for measurement of three mean velocity components in low-speed flows, J. Phys. E.-Sci. Instrum., 22:868.

Ligrani, P. M., Singer, B. A., and Baun, L. R., 1989b, Spatial resolution and downwash velocity corrections for multiple-hole pressure probes in complex flows, Exps. Fluids, 7:424.

ON THE SUBHARMONIC INSTABILITY OF FINITE-AMPLITUDE LONGITUDINAL VORTEX
ROLLS IN INCLINED FREE CONVECTION BOUNDARY LAYERS

C.C. Chen, A. Labhabi, H.-C. Chang[1] and R.E. Kelly[2]

[1] Department of Chemical Engineering
University of Notre Dame
Notre Dame, IN 46556

[2] Mechanical, Aerospace and Nuclear Engineering Department
University of California, Los Angeles
Los Angeles, CA 90024-1597

INTRODUCTION

More than twenty years ago, Sparrow and Husar (1969) investigated experimentally in water the buoyancy-induced instability of the free convection boundary layer that forms on the upper surface of a constant temperature, heated flat plate which is inclined at an angle θ relative to the vertical. The flow was visualized by an electrochemical technique using thymol blue, a pH indicator. Dark dye formed at the surface of the plate, which served as the negative electrode, and was swept towards regions of upwelling from the plate when Rayleigh-Bénard convection occurred.

The onset of Rayleigh-Bénard convection in a Boussinesq fluid with an imposed unidirectional shear is typically denoted by the formation of longitudinal convection rolls or vortices, at least if no side-wall constraints occur (for a review of results concerning onset, see, e.g., Kelly, 1977). The rolls are spanwise (z) periodic and, for the case of boundary layer flow, grow in amplitude in the streamwise (x) direction beyond the station at which the Reynolds number becomes critical ($Re = Re_c$) because the characteristic length scale (the boundary layer thickness) increases in that direction. The development of the rolls for the present case is shown in Fig. 1 of Sparrow and Husar (1969) and occurs initially with constant dimensional wavelength. Further downstream, however, two adjoining rolls appear to pinch off to give rise, at least locally, to a system of rolls with approximately twice the dimensional wavelength of the original rolls, as shown schematically in Fig. 1.

The station at which coalescence occurs is denoted by $Re_{1/2}$. In the experiment, this station varies somewhat in the spanwise direction but this variation is to be anticipated because controlled disturbances were not introduced upstream. A similar wavelength adjustment occurs for convection rolls in heated boundary layer flow of the Blasius type as is evident in Figs. 4(a) and (b) of the paper by Gilpin, Imura and Cheng (1978). However, it should be noted that such a pairing mechanism does not occur apparently in the somewhat similar problem of Görtler vortices found in the isothermal boundary layer forming along a concave wall; see Fig. 14 of the paper by Swearingen and Blackwelder (1987). Once pairing occurs in the free convection boundary layer, the larger wavelength rolls appear to break down rapidly due to a wavy instability.

The present paper is aimed at elucidating the wavelength adjustment mechanism described above. Our conclusion is that the mechanism arises from a subharmonic instability of the initial longitudinal rolls and occurs at a well-defined value of Reynolds number, say, $Re_{1/2} > Re_c$. The instability depends upon the growth of the boundary layer only parametrically, due to the fact that Re increases with x in a prescribed manner. It is not associated with the boundary layer thickness becoming equal to twice its value at Re_c or with any kind of Eckhaus instability. It is quite distinct from the gradual change in wavelength which occurs for transverse, two-dimensional waves arising due to a shear instability in boundary layer flows (which occurs for the present case when the angle of inclination is small). It is more akin to the subharmonic instability occurring in isothermal free shear layers (for an excellent overview of this phenomenon, see the review article by Ho and Huerre, 1984). However, the mechanism is perhaps more capable of being investigated on a reasonably sound basis for the present flow because a well-defined critical value for the control parameter occurs, the finite amplitude rolls are in quasi-equilibrium locally, and the flow is steady until after the subharmonic instability occurs. Comparison with experimental results will be made at the end.

ANALYSIS

Disturbance Equations

A detailed presentation of the analysis is contained in the forthcoming paper by Chen et. al. (1991), and so only a condensed version is given here. As far as the basic two-dimensional laminar flow is concerned, we assume that the Grashof number ($Gr_x = Ra_x/Pr$) based on distance (x) along the plate from the leading edge is large so that the flow can be described by boundary layer theory (e.g. see Sect. 4-12 of the book by White, 1974). A similarity solution then is obtainable for the dimensional stream function $\psi^*(x,y)$ and temperature $T^*(x,y)$ where y measures distance normal to the plate, of the form

$$\psi^* = 4\nu Gr_x^{1/4} F(\eta), \quad T^* = T_\infty + \Delta \overline{T} \Phi(\eta) \tag{1a,b}$$

where the similarity variable $\eta = y/h(x)$, $h(x)$ being the boundary layer thickness ($h = x/Gr_x^{1/4}$), and $\Delta \overline{T} = T^*(0) - T_\infty$.

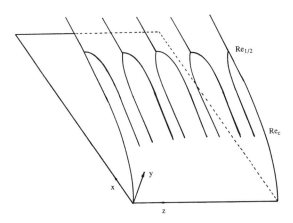

Figure 1. Schematic of longitudinal roll development and pairing in a free convection boundary layer. Boundaries of rolls correspond to upwelling of fluid from the plate.

The basic flow is now perturbed by a disturbance which depends on y, z and time t. The explicit dependence of the disturbance upon the streamwise variable is ignored although the implicit dependence is retained parametrically because the control parameter depends upon x and thereby allows the disturbance amplitude to vary with x. Essentially, we are ignoring any effect associated with the rate of change of the boundary layer thickness. Also, we ignore the details of how rolls with different wavelengths adjust locally to each other in the steady state. We are concerned primarily with determining the stability of the initial longitudinal rolls as x increases and, for this restricted purpose, establishing whether or not a disturbance grows or decays with time is sufficient.

If we define $z = h(x)\delta$, use ν and h to introduce a nondimensional time t, and use a characteristic basic flow velocity $\bar{U}(x)$ and $h(x)$ to introduce a nondimensional stream function $f(\eta,\delta)$ for the disturbed flow in the $y-z$ plane, then the following nondimensional equations can be established for the x-components of the disturbance velocity (u) and disturbance vorticity (ω), along with the disturbance temperature τ:

$$u_t = u_{\eta\eta} + u_{\delta\delta} - Vu_\eta + V'u + ReF''f_\delta + \tau \\ + Re[f_\delta u_\eta - f_\eta u_\delta] \tag{2a}$$

$$\omega_t = \omega_{\eta\eta} + \omega_{\delta\delta} - V\omega_\eta - V'\omega - \tan\theta\,\tau_\delta \\ + Re[f_\delta \omega_\eta - f_\eta \omega_\delta] \tag{2b}$$

$$\tau_t = P_r^{-1}\left(\tau_{\eta\eta} + \tau_{\delta\delta}\right) - V\tau_\eta + Re\phi' f_\delta + \eta\phi' u \\ + Re[f_\delta \tau_\eta - f_\eta \tau_\delta] \tag{2c}$$

where $V = \eta F' - 3F$, $\omega = f_{\eta\eta} + f_{\delta\delta}$, and Re is a Reynolds number based on the boundary layer thickness $(Re \sim Gr_x^{1/4})$. The boundary conditions are $u = \tau = f = f_\eta$ at $\eta = 0$ and ∞.

If the nonlinear and time-dependent terms of (2a,b,c) are ignored, these equations reduce to the equations used by Haaland and Sparrow (1973) to estimate Re_c for this flow. They pointed out the importance of the nonparallel flow term involving V in allowing a balance between diffusion and advection of disturbance vorticity at the edge of the boundary layer. We shall refer to the above system of equations as the modified parallel flow equations for the disturbance.

Computational Methods

For the linear problem the disturbance is assumed to be periodic in the δ (or z) direction so that for instance,

$$\tau(\eta,\delta,t) = \hat{\tau}(\eta)\exp(i\alpha\delta + \lambda t) \tag{3}$$

A mapping given by $\hat{\eta} = \exp(-\eta)$ is introduced to transform the semi-infinite interval $\eta \epsilon (0,\infty)$ to the finite interval $\hat{\eta} \epsilon (0,1)$. A Chebyshev-Tau spectral expansion is now applied with "extra" terms added so as to satisfy the boundary conditions, e.g.,

$$\hat{\tau} = \sum_{i=1}^{n+1} c_i P_i(\hat{\eta}) \tag{4}$$

where the $P_i(\hat{\eta})$ are the odd Chebyshev polynomials of degree $2i - 1$. By substituting (3), (4) and similar representations for $\hat{u}$ and $\hat{\omega}$ into the linearized form of (2a,b,c), taking the inner product with n bases and accounting for the boundary conditions, a $(3n) \times (3n)$ eigenvalue problem is obtained.

In order to investigate the stability of a finite-amplitude state of convection with regard to a subharmonic instability, a more general expression for τ is taken, namely,

$$\tau(\eta,\delta,t) = \sum_{m=-\infty}^{m=\infty} A_{m/2}(t) \hat{\tau}_{m/2}(\eta) \exp(im\alpha\delta/2) \tag{5}$$

where $A_{-m/2}$ is the complex conjugate of $A_{m/2}$. In the actual computations, the $m = 0$ mode is deleted because this simplification does not affect the prediction of subharmonic instability to leading order. By again utilizing the Chebyshev-Tau spectral method and taking inner products, a system of coupled nonlinear ordinary differential equations is obtained with the form

$$\frac{dA_{m/2}}{dt} = \lambda_{m/2} A_{m/2} + (\alpha Re/\alpha_c) \sum_{n=-\infty}^{n=\infty} P[n/2, (m-n)/2] A_{n/2} A_{(m-n)/2} \tag{6}$$

where $n \neq 0$ again and $\lambda_{m/2}$ is the growth or decay rate for $Re \neq Re_c$. In the actual computations, the value of Prandtl number was usually set at $Pr = 5.5$, and a six mode truncation was found to ensure accuracy for $|A_1|$ to the third decimal place for the values of Re of interest.

RESULTS

For the linear problem, the neutral stability curve is shown in Fig. 2 and yields $\tilde{Re}_c = Re_c \tan\theta = 24.26$ and $\alpha_c = 1.292$. Calculations were also done for a value of $Pr = 6.7$ in order to check the current results with the earlier numerical results of Haaland and Sparrow (1973). Satisfactory agreement was obtained although the small wavenumber asymptote occurs for a smaller value of α in their calculation.

The dashed line in Fig. 2 corresponds to a maximum growth rate trajectory, which occurs for $\alpha \cong \alpha_c$. However, this is not the trajectory corresponding to the experimental results of Sparrow and Husar (1969), for which the dimensional wavelength is constant, and so α must increase with $Re(x)$. If we assume that $\alpha = \alpha_c$ at $Re = Re_c$, then the path in Fig. 2 corresponding to the experimental results follows the formula

$$\alpha = \alpha_* = \alpha_c (Re/Re_c)^{1/3} \tag{7}$$

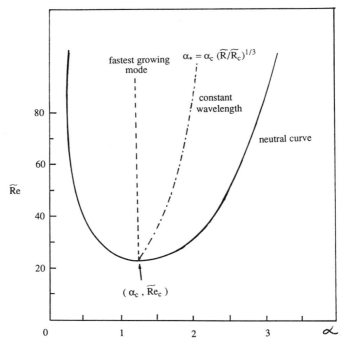

Figure 2. Neutral curve for linear stability and trajectories for the fastest growing disturbance and for a constant dimensional wavelength disturbance, $Pr = 5.5$.

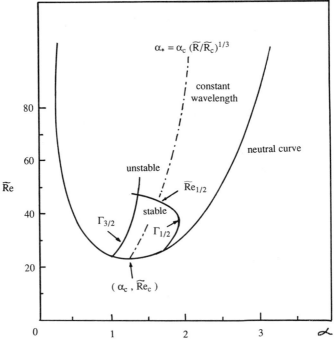

Figure 3. Neutral curves for stability of finite amplitude, single wave number states with respect to subharmonic disturbances, $Pr = 5.5$.

based on the variation of h and Re with x. This path is shown by the dash-dot curve in Fig. 2. It is estimated to be contained well within the Eckhaus stability boundary for the values of Re of interest here.

The stability boundaries for finite amplitude states ($Re > Re_c$, various α) with respect to a subharmonic of order 2 ($m = 1$) or with respect to the harmonic of this mode corresponding to $m = 3$ are shown in Fig. 3 and are labeled $\Gamma_{1/2}$ and $\Gamma_{3/2}$, respectively. If we follow the path defined by (7), it is clear that the original rolls are destabilized first by a subharmonic of order 2 when $\tilde{Re} = \tilde{Re}_{1/2} \cong 41.0$ or equivalently when $\tilde{Re}_{1/2}/\tilde{Re}_c \cong 1.7$. From the experiments of Sparrow and Husar (1969) it is estimated that $\tilde{Re}_{1/2}/\tilde{Re}_c \cong 2.8$ although there is considerable uncertainty over the experimental value of $\tilde{Re}_c$ due to the fact that controlled disturbances were not introduced into the flow. In view of this uncertainty as well as the approximations made in the analysis, the comparison seems to be reasonable.

An alternative estimate of $\tilde{Re}_{1/2}$ can be obtained by considering the stability of the subharmonic to sideband disturbances and making use of the Eckhaus stability boundaries

$$(\alpha_- - \alpha_c)/\sqrt{3} < (\alpha - \alpha_c) < (\alpha_+ - \alpha_c)/\sqrt{3} \qquad (8)$$

for stability with regard to sideband disturbances. The trajectory of a subharmonic of order 2 would be similar to the trajectory given by α_* but would be shifted to the left in Fig. 2. The question then is where along this trajectory does the subharmonic begin to extract energy from the fundamental. It was argued that the subharmonic itself should be stable on the Eckhaus basis. Thus, once the subharmonic crosses the Eckhaus stability boundary, it can presumably begin to destabilize the fundamental. Using the boundaries given by (8), this criterion yields $\tilde{Re}_{1/2} = 42$, which is very close to the numerical result. However, the agreement, though remarkable, might be fortuitous because the Eckhaus boundary (8) pertains really to the stability of a single convection state, not two coexisting ones. Also, the bounds given by (8) hold only when $Re \cong Re_c$ (see Cheng and Chang, 1990).

ACKNOWLEDGEMENT

The work conducted at the University of Notre Dame was supported by the NSF under Grant No. ENG-8451116 and by the Center of Applied Mathematics at the University.

REFERENCES

Chen, C.C., Labhabi, A., Chang, H.-C. and Kelly, R.E., 1991, Spanwise pairing of finite-amplitude longitudinal vortex rolls in inclined free convection boundary layers, accepted for publication in J. Fluid Mech.

Cheng, M. and Chang, H.-C., 1990, A generalized sideband stability theory via center manifold projection, Phys. Fluids A, 2:1364.

Gilpin, R.R., Imura, H. and Cheng, K.C., 1978, Experiments on the onset of longitudinal vortices in horizontal Blasius flow heated from below, J. Heat Transfer, 100:71.

Haaland, S.E. and Sparrow, E.M., 1973, Vortex instability of natural convection flow on inclined surfaces, Int. J. Heat Mass Transfer, 16:2355.

Ho., C.-M. and Huerre, P., 1984, Perturbed free shear layers, Annual Rev. of Fluid Mech., 16:365.

Kelly, R.E., 1977, The onset and development of Rayleigh-Bénard convection in shear flows: a review, in "Physicochemical Hydrodynamics," D.B. Spalding, ed. Advance Publications, London.

Sparrow, E.M. and Husar, R.B., 1969, Longitudinal vortices in natural convection flow on inclined plates, J. Fluid Mech., 37:251

Swearingen, J.D. and Blackwelder, R.F., 1987, The growth and breakdown of streamwise vortices in the presence of a wall, J. Fluid Mech., 182:255.

White, F.M., 1974, "Viscous Fluid Flow," McGraw-Hill, Inc., New York.

CENTRIFUGAL INSTABILITIES IN ROTATING FRAME ABOUT FLOW AXIS

Innocent Mutabazi* and Christiane Normand
Service de Physique Théorique du CEA, Centre d'Etudes de Saclay,
F-91191, Gif sur Yvette Cedex, France

Michel Martin and José Eduardo Wesfreid
Laboratoire de Physique et Mécanique de la Matière Hétérogène, URA
CNRS 857, Ecole Supérieure de Physique et de Chimie Industrielles de
Paris, 10 rue Vauquelin, F-75231 Paris Cédex 05

INTRODUCTION

Flows with curved streamlines play an important role in everyday life and industrial systems such as rotating machines and chemical reactors. They also represent a prototype of investigation of transition to chaos in hydrodynamic systems. Under certain circumstances, flows with curved streamlines lose their stability and longitudinal rolls are generated as a result of centrifugal instability. The simplest flow systems with curved streamlines are the Taylor-Couette flow between two coaxial independently rotating cylinders and the curved Poiseuille channel flow [1].

Another source of longitudinal rolls is the rotation of the flow system about its own axis. Indeed, the plane Poiseuille channel flow, which is stable to spanwise infinitesimal perturbations, becomes unstable under slow rotation about its own axis and longitudinal rolls appear in the flow [2]. The rotation generates a Coriolis force which has a stabilizing or destabilizing effect depending on its interaction with the shear force. Therefore, there is a similarity between streamline curvature generating centrifugal force and flow rotation generating Coriolis force.

The interaction between rotation and streamline curvature may generate new patterns, in particular, oscillatory modes, as a result of two different competing instability mechanisms: the Coriolis and the centrifugal forces. In the following, we will analyze the linear stability results of the curved, finite gap channel flow, rotating around its own axis. Previous study of such a problem has been performed in the small gap approximation with a first order correction [3]. The results are understood using a generalized Rayleigh criterion [4].

---

*Present address: Physics Department, the Ohio State University, 174 W. 18th Avenue, Columbus, OH 43210, USA

GOVERNING EQUATIONS

In a rotating frame of reference, the Navier-Stokes equations for incompressible fluid of density ρ and kinematic viscosity v read [1]:

$$\frac{\partial \mathbf{v}}{\partial t} + (\mathbf{v}.\nabla)\mathbf{v} - 2\mathbf{v} \wedge \Omega = -\nabla(P/\rho) + v\Delta\mathbf{v} \qquad (1\text{-a})$$

$$\text{div } \mathbf{v} = 0 \qquad (1\text{-b})$$

where P is the dynamical pressure containing the centrifugal energy term ($-1/2\ \Omega^2 r^2$) associated with the rotation of the frame of reference. Flow equations in curved geometries can be written in cylindrical coordinates (r, θ, z). The perturbed flow is then described by the base flow $(V(r), P)$ and the perturbations characteristics $(\mathbf{v}', p')$: $\mathbf{v} = V(r)\mathbf{e}_\theta + \mathbf{v}'$, $p = P + p'$, where $\mathbf{e}_\theta$ is the unit vector along the base flow direction. If we scale the variables and the perturbative velocity as follows:

$$t = t'd^2/v, r = R(1+\delta x), \vartheta = \delta^{1/2}y, z = dz', \mathbf{v}' = (u, \delta^{-1/2}v, w)(d/v), \delta = d/R$$

we obtain after dropping primes off t', z' and p', the following linear stability equations:

$$(L_0 + c^2\delta^2)u + 2c^2\delta\frac{\partial v}{\partial y} - 2\text{Ta}(cV + \text{Ro})v = -\frac{\partial p}{\partial x} \qquad (2\text{-a})$$

$$(L_0 + c^2\delta^2)v + 2c^2\delta^2\frac{\partial u}{\partial y} - \text{Ta}[DV + \delta(cV + 2\text{Ro})] = -c\delta\frac{\partial p}{\partial y} \qquad (2\text{-b})$$

$$L_0 w = -\frac{\partial p}{\partial z} \qquad (2\text{-c})$$

$$\frac{\partial u}{\partial x} + c\delta u + \frac{\partial v}{\partial y} + \frac{\partial w}{\partial z} = 0 \qquad (2\text{-d})$$

where

$$D = \frac{d}{dx}, L_0 = \frac{\partial}{\partial t} + c\text{Ta}V\frac{\partial}{\partial y} - \frac{\partial^2}{\partial x^2} - \frac{\partial^2}{\partial z^2} - c\delta\frac{\partial}{\partial x} - c^2\delta^2\frac{\partial^2}{\partial y^2}, c = \frac{1}{2+\delta x}$$

The flow control parameters are defined as $\text{Ta} = (V_m d/v)\, \delta^{1/2}$, $\text{Ro} = \Omega R/V_m$. The last number is the inverse of the Rossby number and will be referred as "rotation number" in the remainder of the paper. The characteristic velocity V_m is the mean velocity across the channel.

The boundary conditions are:

$$u = Du = v = 0 \qquad \text{at} \qquad x = 0 \text{ and } x = 1 \qquad (3)$$

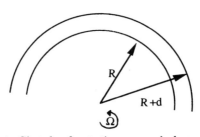

Figure 1. Sketch of rotating curved channel flow

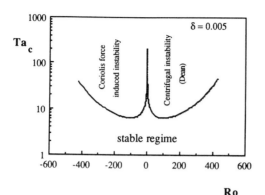

Figure 2. Phase diagram for $\delta = 0.005$

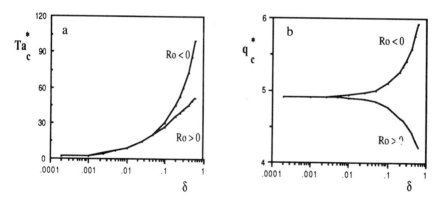

Figure 3. Critical parameters at the point where restabilization by rotation begins

RESULTS

The equations (2) together with the boundary conditions (3) are solved numerically using fourth order Runge-Kutta procedure. Here are the main results:

1. In the phase diagram (Ro, Ta$_c$), there are two branches of stability corresponding one to stationary modes induced by centrifugal force, the other to stationary modes induced by rotation (Fig.2). The two branches intersect in a codimension-two point, in the neighbourhood of which critical modes are oscillatory. The two branches of stability intersect near the point Ro = -1 where the net flux induced by rotation is equal and opposite to that induced by external pressure gradient.

2. For small gap size channels, the point at which restabilization begins, is δRo* = 1/2 and the critical parameters (q$_c$, Ta$_c$) are independent of the gap size. For wide gap size channels, the point where restabilization by rotation begins and the correspoinding critical parameters depend upon the gap size δ (Fig.3).

3. For small gap size channels and low rotation regime, the system exhibits similarities with the Taylor-Dean system.

4. For small gap size channels and high rotation regime, rotation effects dominate centrifugal force and lead to restabilization of flow (relaminarization).

DISCUSSION

Some of these results may be understood in the framework of the Rayleigh circulation criterion generalized in such a way that rotation effects are included. This is achieved by using the so called "displaced particle argument" [5]. One gets the Rayleigh discriminant in non-dimensional form:

$$\Phi(x) = 2\delta\{(\text{Ro} + cV)[DV + \delta(cV + 2\text{Ro})]\} \quad (4)$$

The flow is unstable when $\Phi(x) < 0$ and stable when $\Phi(x) > 0$: therefore, it consists of alternating stable and unstable layers, the width of which depend on Ro and δ (Fig.4). One unstable layer corresponds to Coriolis force driven perturbations, the other corresponds to centrifugal instability. The competition between different destabilizing mechanisms in two potentially unstable layers gives rise to critical oscillatory modes.

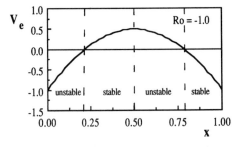

Figure 4. Generalized Rayleigh criterion applied to the profile
$V_e(x) = V(x) + \text{Ro}$ in the small gap size channels
and for low rotation regime

The relaminarization of the flow by rotation can be understood using the asymptotic solution of the stability equations and their dynamical similarity with those describing the Taylor-Couette system [1]. In the case of small gap size channels, the critical parameters have the following asymptotic behaviour:

$$q_c = \frac{12.08}{2 - \delta|\text{Ro}|} \qquad \text{Ta}_c = \frac{39.7}{[|\text{Ro}|(1 - \delta|\text{Ro}|/3)^5]^{1/2}} \qquad (5)$$

From relations (4) and (5), for $\delta|\text{Ro}| > 3$, it appears that the flow has no potentially unstable layer in the gap and no perturbations can evolve in it, it remains two-dimensional.

Application of the generalized Rayleigh discriminant (4) to the boundary layer flow over curved surfaces may be done if one approximates the boundary layer flow by a piece-wise function [5]:

$$V(x) = \begin{cases} x & \text{if } 0 < x < 1 \\ 1 & \text{if } x > 1 \end{cases}$$

For boundary layer flow of small thickness, rotation of a concave plate will enhance the centrifugal instability if Ro > 0 and act against it for Ro < 0; rotation of a convex plate which is centrifugally stable should induce instability for $-1 < \text{Ro} < 0$. These results are under investigation and will be developed in a forthcoming publication [6].

CONCLUDING REMARKS

The effect of rotation on Dean instability in curved Poiseuille channel flow of finite size gap is investigated within linear stability theory and generalized Rayleigh criterion. Rotating flows are met in technical problems such as rotating machines, chromatography, and in geophysical proglems. In particular, the occurrence of longitudinal rolls in chromatographic systems should change sensitively the efficiency of species separation based on Field Flow Fragmentation in highly rotating channels.

ACKNOWLEDGEMENT

We would like to thank A. Aouidef and V. Vladimirov for interesting discussions. One of us (I.M.) very much appreciates the financial support from the Theoretical Physics Service of the Commissariat à l'Energie Atomique (France) where a part of this work was performed.

REFERENCES

[1] S. Chandrasekhar, Hydrodynamic and Hydromagnetic Stability, Oxford University Press, Clarendon 1961
[2] D.K. Lezius and J.P. Johnston, J. Fluid Mech. **77**, 153 (1976)
[3] O.J.E. Matsson and P.H. Alfredsson, J. Fluid Mech. **210**, 537 (1990)
[4] I. Mutabazi, C. Normand J.E. Wesfreid, preprint 1991
[5] P.G. Drazin and W.H. Reid, Hydrodynamic Stability, Cambridge University Press, Cambridge (1981)
[6] A. Aouidef, I. Mutabazi and J.E. Wesfreid, preprint 1991

A GUIDE TO LITERATURE RELATED TO THE TAYLOR-COUETTE PROBLEM

Randall Tagg

Department of Physics - Campus Box 172
University of Colorado, P.O. Box 173364
Denver, CO 80217-3364

There are nearly 1500 references in the following list. The size of this list might have two consequences. First, a student undertaking research in this subject could feel daunted by the task of making an original contribution. Second, a grant officer might conclude that all the important work has been done and further funding would have diminishing returns. A survey of the literature, and especially an attempt to classify it, shows that these consequences would be wrong. We have come to the laboratory and we have seen many enticing patterns of flow in this simple geometry, but we are far from conquering the subject. Much work remains before we gain a firm understanding of the connection between transitional flows and turbulence. Also, we have by no means exploited all the practical possibilities of what we have learned.

Consider the following:

1.) a large proportion of the literature deals with the primary instability alone, and it is only here where we might claim to have a "deep" understanding of the subtle interplay of nonlinearity, symmetry, and boundary conditions in establishing patterns of secondary flow;

2.) analytical techniques continue to unfold in organizing our knowledge of such a wide variety of flow patterns;

3.) many new experimental and data-analysis techniques are tried and tested in this system before moving on to more complex geometries;

4.) while there are significant contributions, the literature is remarkably sparse on the subject of subsequent flow patterns (wavy vortex flow, "twists", modulated wavy vortices, turbulent vortices, spiral turbulence, etc.);

5.) powerful, well-resolved numerical computations have only recently been applied to the more complex flows and the results are very promising. However, care should be taken to recognize the constraints imposed by modeling assumptions, e.g. periodic boundary conditions. Experiments often show subtle but important differences when compared to numerical simulations and these can lead to improved modeling. Thus there is a very profitable interplay between laboratory results and computer simulations;

6.) many interesting variants of the fundamental problem of flow between coaxial rotating cylinders are being investigated. These variants (e.g. eccentric cylinders, azimuthal pressure gradients, axial flow, radial temperature gradients, or the influence of Coriolis forces) may provide many of the opportunities for practical applications of the subject;

7.) finally, the Taylor-Couette system, like the hydrogen atom, is a fundamental "toy" of physics; as new ideas of general import arise in the subjects of turbulence and chaotic flows, we need such a "simple" system for providing the first tests.

For those approaching the literature for the first time, some suggestions are:

General fluid dynamics textbooks that discuss the Taylor-Couette system:
Acheson (1990); Kundu (1990);Landau and Lifshitz (1987); Panton (1984); Tritton (1988);White [1991]

Texts on stability theory:
Chandrasekhar (1961) (Dover reprint 1981); Drazin & Reid (1981); Joseph (1976); Lin (1955); Shivamoggi (1986)

General reviews:
Cognet (1984); DiPrima(1981);DiPrima & Swinney (1985); Iooss (1984); Kataoka (1986); Koschmieder (1975b)[1992]; Monin (1986); Stuart (1971)(1986); Swinney & Gollub (1978)(1985)

Tutorials on theoretical methods:
Golubitsky et al. (1988); Haken (1983a)(1983b); Kirchgässner (1975); Manneville (1990); Richtmeyer (1981);

Experimental methods:
Donnelly (these proceedings); Weidman (1989)

Numerical methods:
We are unaware of a review, but see Hussaini & Voigt (1990) for a recent overview of numerics applied to instability problems;

Pattern formation, fronts, etc.:
Ahlers (1989); Collet & Eckmann (1990); Hohenberg & Cross (1987); Newell (1989); Manneville (1990);

Wave interactions:
Craik (1985); Infeld & Rowlands (1990); Whitham [1974];

Chaos:
Bergé et al. (1986); Thompson & Stewart (1986) (see the chapter contributed by H. L. Swinney);

Some classic papers and flow surveys:
Taylor (1923); Coles (1965); Snyder (1970); Andereck et al. (1986)

As an exercise, the reader should look for the following variants of the problem in the list of references:
- A. Temperature gradients
- B. Coriolis forces
- C. Axial and/or radial through-flow
- D. Magnetic fields; ferrofluids
- E. Electric fields; electrorheological fluids
- F. Eccentric cylinders and lubrication theory
- G. Spheres and ellipsoids
- H. Non-cylindrical walls
- I. Corrugated walls
- J. Gas dynamics
- K. Non-Newtonian fluids
- L. Liquid crystals
- M. Immiscible fluids or mixtures
- N. Porous media
- O. Critical phenomena with shear
- P. Taylor-Dean problem
- Q. Görtler vortices

Finally, a few notes. First, sincere apologies are given for omissions (for example, it has not yet been possible to compile the references to most theses). Even so, we have tried to err on the side of inclusiveness. Thus many papers on rheology are in the list. Such papers provide many clues for problems that are both of fundamental theoretical importance and of immediate practical use, e.g. concerning the behavior of all sorts of non-Newtonian fluids. A similar statement applies to papers on Görtler vortices and the Dean problem (although here the list is far from complete). The list has been "spiced" with references to papers relevant to, but not specifically dealing with, the Taylor-Couette problem. This hopefully provides a link to a much broader set of literature. No effort has been made, however, to represent the vast and important literature on Rayleigh-Bénard convection.

Second, in performing computer database searches for papers, one becomes aware of certain keywords which can be recommended to authors in choosing titles so their papers will be more easily found. Here is a rough synopsis of the search strategy used to prepare this list (XOR means "exclusive-or"; * is a wild-card allowing arbitrary characters to be added); the percentages of titles which match the criteria are shown:

Couette	28%
XOR ((Taylor OR wavy) AND (vort*))	+13%
XOR ((rotating OR flow) AND (cylinder*))	+13%
XOR (Taylor AND (experiment OR stabilit* OR problem))	+ 3%
	57%

The resulting list must in general be filtered, e.g., to eliminate papers on plane Couette flow or on the Rayleigh-Taylor instability of heavy fluid over light fluid. Many of the titles not captured by this search are either variants (e.g. flow between spheres) or related topics (e.g. general treatments on bifurcation theory that mention the Taylor-Couette experiment). The success rate of the above search will be higher, then, if abstracts are included in the search. However, the selectivity of the above search has not been quantified.

Acknowledgement is given to several authors who have contributed more of their own titles to this list. Special thanks goes to Dr. Dieter Roth for sending many references that were omitted from an early version. The assistance of Tim Jorgenson and other staff members of the University of Texas libraries is gratefully acknowledged. Finally, this project would not have been possible without the support of the research groups of Professors Harry Swinney and Russell Donnelly. This list of references will grow and further versions (including electronic versions) will be made available. Inquiries and notice of omissions (please!) may be sent via e-mail to: RTAGG@CUDNVR.DENVER.COLORADO.EDU (or RTAGG@CUDENVER.BITNET).

- (1978), " XIII Symposium on advanced problems and methods in fluid mechanics," Arch. Mech. **30**.
- (1980), "Study of blood suspension aggregation in a Couette flow," Rev. Phys. Appl. **15**, 1357-66.
Abbott, J. R., N. Tetlow, and A. L. Graham (1991), "Experimental observations of particle migration in concentrated suspensions: Couette flow," J. Rheol. **35**, 773.
Abbott, T. N. G. and K. Walters (1970), "Rheometrical flow systems. Part 3. Flow between rotating eccentric cylinders," J. Fluid Mech. **43**, 257-267.
Abboud, M. (1988), "Ein Beitrag zur theoretischen Untersuchung von Taylor-Wirblen im Spalt zwischen Zylinder/Kegel-Konfigurationen," Z. Angew. Math. Mech. **68**, 275-7.
Abboud, M. (1990), "Numerical research into Taylor vortices in the cleft between a cone and a cylinder," Z. Angew. Math. Mech. **70**, T441-442.
Abdallah, Y. A. G. and J. E. R. Coney (1988), "Adiabatic and diabatic flow studies by shear stress measurements in annuli with inner cylinder rotation," J. Fluids Eng. **110**, 399-405.
Abraham, N. B., J. P. Gollub, and H. L. Swinney (1984), "Testing nonlinear dynamics," Physica D **11D**, 252-264.
Acheson, D. J. (1990), "Ch. 9 Instability," in Elementary Fluid Dynamics (Oxford University Press, Oxford).
Adjizian, J. C., C. Droulle, G. Osterman, B. Pignon, and Gg. Potron (1984), "Comparative interest of two coaxial viscometers: ecktacytometer and low shear 30," Biorheology, suppl. 1, 95-7.
Afendikov, A. L. and K. I. Babenko (1985), "On the loss of stability of Couette flow at different Rossby numbers," Dok. Akad. Nauk SSSR **28**, 548-551 [Sov. Phys. Dokl. **30**, 202-4 (1985)].
Agarwal, R. S. (1988), "Numerical solution of micropolar fluid flow and heat transfer between two co-axial porous circular cylinders," Int. J. Eng. Sci. **26**, 1133.
Ahlers, G. (1989), "Experiments on bifurcation and one-dimensional patterns in nonlinear systems far from equilibrium," in Lectures in the Sciences of Complexity, edited by D. L. Stein (Addison-Wesley), pp. 175-224.
Ahlers, G. and D. S. Cannell (1983), "Vortex-front propagation in rotating Couette-Taylor flow," Phys. Rev. Lett. **50**, 1583-6.
Ahlers, G., D. S. Cannell, and M. A. Dominguez-Lerma (1982), "Fractional mode numbers in wavy Taylor vortex flow," Phys. Rev. Lett. **49**, 368-371.
Ahlers, G., D. S. Cannell, and M. A. Dominguez-Lerma (1983), "Possible mechanism for transitions in wavy Taylor-vortex flow," Phys. Rev. A **27**, 1225-1227.
Ahlers, G., D. S. Cannell, M. A. Dominguez-Lerma, and R. Heinrichs (1986), "Wavenumber selection and Eckhaus instability in Couette-Taylor flow," Physica D **23D**, 202-19.
Ait Aider, A. (1983), "Sequences transitoires en ecoulement de Couette-Taylor. Influence du rapport d'aspect," Thesis: Doctorat de 3eme Cycle, INPL de Lorraine.
Aitta, A. (1986), "Quantitative Landau model for bifurcations near a tricritical point in Couette-Taylor flow," Phys. Rev. A **34**, 2086-92.
Aitta, A. (1989), "Dynamics near a tricritical point in Couette-Taylor flow," Phys. Rev. Lett. **62**, 2116-19.
Aitta, A., G. Ahlers, and D. S. Cannell (1985), "Tricritical phenomena in rotating Couette-Taylor flow," Phys. Rev. Lett. **54**, 673-6.
Akbay, U., E. Becker, S. Krozer, and S. Sponagel (1980), "Instability of slow viscometric flow," Mech. Res. Commun. **7**, 199-204.
Aksel, N. (1991), " A new instability mechanism applied to the compressible Couette flow. ," Continuum Mechanics Thermodynamics **3**, 147.
Ali, M. and P. D. Weidman (1990), "On the stability of circular Couette flow with radial heating," J. Fluid Mech. **220**, 53-84.
Alziary de Roquefort, T. and G. Grillaud (1978), "Computation of Taylor vortex flow by a transient implicit method," Comput. Fluids **6**, 259-69.
Andereck, C. D. and G. W. Baxter (1988), "An overview of the flow regimes in a circular Couette system," in Propagation in Systems Far from Equilibrium, edited by J. Wesfreid, H. Brand, P. Manneville, G. Albinet, and N. Boccara (Springer-Verlag, Berlin), pp. 315-324.
Andereck, C. D., R. Dickman, and H. L. Swinney (1983), "New flows in a circular couette system with co-rotating cylinders," Phys. Fluids **26**, 1394-401.
Andereck, C. D., S. S. Liu, and H. L. Swinney (1986), "Flow regimes in a circular Couette system with independently rotating cylinders," J. Fluid Mech. **164**, 155-83.
Andreichikov, I. P. (1977), "Branching of the secondary modes in the flow between cylinders," Izv. Akad. Nauk SSSR, Mekh. Zhidk. Gaza, 47 [Fluid Dyn. (USSR) **12**, 38 (1977)].
Anson, D. K. and K. A. Cliffe (1989), "A numerical investigation of the Schaeffer homotopy in the problem of Taylor-Couette flows," Proc. R. Soc. London, Ser. A **426**, 331-342.
Anson, D. K., T. Mullin, and K. A. Cliffe (1989), "A numerical and experimental investigation of a new solution in the Taylor vortex problem," J. Fluid Mech. **207**, 475.

Antimirov, M. Ya. and A. A. Kolyshkin (1980), "Stability of Couette flow in a radial magnetic field," Magn. Gidrodin. **16**, 45-51 [Magnetohydrodynamics **16**, 145-50 (1980)].

Aoki, H., H. Nohira, and H. Arai (1967), Bull. JSME **10**, 523.

Apter, D. M., P. K. Vlasov, A. A. Karpukhin, A. I. Ovchinnikov, and Z. F. Chukhanov (1987), "Power requirements of pressureless hydraulic-pipeline container-carrying systems," Fluid Mech. - Sov. Res. **16**, 112-16.

Astaf'eva, N. M. (1985), "Numerical simulation of spherical Couette flow asymmetric relative to the plane of the equator," Izv. Akad. Nauk SSSR, Mekh. Zhidk. Gaza **20**, 56-62 [Fluid Dyn. (USSR) **20**, 383-9 (1985)].

Astill, K. N. (1961), Ph.D. Thesis, Massachusetts Inst. of Technology.

Astill, K. N. (1964), J. Heat Transfer **86**, 383.

Astill, K. N. and K. C. Chung (1976), "A numerical study of instability of the flow between rotating cylinders," Am. Soc. Mech. Eng. Publ. 76-FE-27.

Astill, K. N., J. T. Ganley, and B. W. Martin (1968), "The developing tangential velocity profile for axial flow in an annulus with a rotating inner cylinder," Proc. R. Soc. London, Ser. A **307**, 55-69.

Atten, P. and T. Honda (1982), "The electroviscous effect and its explanation. I. The electrohydrodynamic origin study under unipolar DC injection," J. Electrostat. **11**, 225-245.

Atten, P., R. Baron, and M. Goniche (1980), "Superposition of a cylindrical Couette flow and of an electrically induced secondary flow," J. Phys. (Paris), Lett. **41**, L1-4.

Atten, P., R. Baron, and M. Goniche (1979), "Superposition of a cylindrical flow and of a secondary movement electrically induced," C. R. Acad. Sci., Ser. B **289**, 135-7.

Aubert, J. H. and M. Tirrell (1980), "Macromolecules in nonhomogeneous velocity gradient fields," J. Chem. Phys. **72**, 2694-701.

Aubry, N., P. Holmes, J. L. Lumley, and E. Stone (1987), "Models for coherent structures in the wall layer," in *Advances in Turbulence. Proceedings of the First European Turbulence Conference*, edited by G. Comte-Bellot and J. Mathieu (Springer-Verlag, Berlin, Germany).

Aydin, M. and H. J. Leutheusser (1980), "Very low velocity calibration and application of hot-wire probes," Disa Inf., 17-18.

Babcock, K. L., G. Ahlers, and D. S. Cannell (1991), "Noise sustained structure in Taylor-Couette flow with through-flow," Phys. Rev. Lett. **67**, 3388.

Babenko, K. I., A. L. Afendikov, and S. P. Yur'ev (1982), "Bifurcation of Couette flow between rotating cylinders in the case of a double eigenvalue," Dok. Akad. Nauk SSSR **266**, 73-8 [Sov. Phys. Dokl. **27**, 706-9 (1982)].

Babenko, K. I. and A. L. Afendikov (1984), "Stability of Taylor vortices," Dok. Akad. Nauk SSSR **278**, 828-33 [Sov. Phys. Dokl. **29**, 784-7 (1984)].

Babkin, V. A. (1987), "Plug Couette flow of fiber suspension between coaxial cylinders," Izv. Akad. Nauk SSSR, Mekh. Zhidk. Gaza [Fluid Dyn. (USSR) **22**, 849-56 (1987)].

Babkin, V. A. (1988), "Anisotropic turbulence in a flow of incompressible fluid between rotating coaxial cylinders," J. Appl. Math. Mech. (USSR) **52**, 176.

Ball, K. S. (1989), "An experimental study of heat transfer in a vertical annulus with a rotating inner cylinder," Int. J. Heat Mass Transf. **32**, 1517.

Ball, K. S. and B. Farouk (1987), "On the development of Taylor vortices in a vertical annulus with a heated rotating inner cylinder," Int. J. Numer. Methods Fluids **7**, 857-67.

Ball, K. S. and B. Farouk (1988), "Bifurcation phenomena in Taylor-Couette flow with buoyancy effects," J. Fluid Mech. **197**, 479-501.

Ball, K. S. and B. Farouk (1989), "A flow visualization study of the effects of buoyancy on Taylor vortices," Phys. Fluids A **1**, 1502-1507.

Banerjee, M. B., R. G. Shandil, J. R. Gupta, and R. L. Behl (1979), "On Landau's conjecture in hydrodynamics. IV. Determination of amplitude of steady secondary flow past marginal state in rotatory Couette flow problem," Natl. Acad. Sci. Lett. (India) **2**, 67-9.

Banerjee, M. B., R. G. Shandil, and A. K. Gupta (1985), "Instability of revolving fluids of variable viscosity," J. Math. Phys. Sci. **19**, 349-64.

Bar-Yoseph, P., S. Seelig, A. Solan, and K. G. Roesner (1987), "Vortex breakdown in spherical gap," Phys. Fluids **30**, 1581-1583.

Bar-Yoseph, P., A. Solan, R. Hillen, and K. G. Roesner (1990), "Taylor vortex flow between eccentric coaxial rotating spheres," Phys. Fluids A **2**, 1564-1573.

Bar-Yoseph, P., A. Solan, and K. G. Roesner (1991), "Polar vortex motion between rotating spherical shells: vortex breakdown, steady and time-periodic," Eur. J. Mech. B/Fluids **10 (suppl.)**, 325.

Barcilon, V. and H. C. Berg (1971), "Forced axial flow between rotating concentric cylinders," J. Fluid Mech. **47**, 469-479.

Barcilon, A. and J. Brindley (1984), "Organized structures in turbulent Taylor-Couette flow," J. Fluid Mech. **143**, 429-49.

Barcilon, A., J. Brindley, M. Lessen, and F. R. Mobbs (1979), "Marginal instability of Taylor-Couette flows at a very high Taylor number," J. Fluid Mech. **94**, 453-63.

Barenghi, C. F. (1990), "A spectral method for time modulated Taylor-Couette flow," Comput. Meth. Appl. Mech. Eng. **80**, 223.

Barenghi, C. F. (1991a), "Computations of transitions and Taylor vortices in temporally modulated Taylor-Couette flow," J. Comput. Phys. **95**, 175.

Barenghi, C. F. (1991b), "Vortices and the Couette flow of Helium-II," (preprint).

Barenghi, C. F. and C. A. Jones (1987), "On the stability of superfluid helium between rotating concentric cylinders," Phys. Lett. A **122**, 425-30.

Barenghi, C. F. and C. A. Jones (1988), "The stability of the Couette flow of helium II," J. Fluid Mech. **197**, 551-69.

Barenghi, C. F. and C. A. Jones (1989), "Modulated Taylor-Couette flow," J. Fluid Mech. **208**, 127-160.

Bark, F. H., A. V. Johansson, and C. -G. Carlsson (1984), "Axisymmetric stratified two-layer flow in a rotating conical channel," J. Mec. Theor. Appl. **3**, 861-78.

Barkley, D., G. E. Karniadakis, I. G. Kevrekidis, A. J. Smits, and Z. -H. Shen, "Chaotic advection in a complex annular geometry," in IUTAM Symposium on Fluid Mechanics of Stirring and Mixing.

Barkovskii, yu. S. and V. I. Yudovich (1978), "Taylor vortex formation in the case of variously rotating cylinders and the spectral properties of one class of boundary-value problems," Dok. Akad. Nauk SSSR **242**, 784-7 [Sov. Phys. Dokl. **23**, 716-17 (1978)].

Barratt, P. J., J. M. Manley, and V. A. Nye (1987), "A linear analysis of instabilities In Couette flow of nematic liquid crystals," Rheol. Acta **26**, 34-39.

Barratt, P. J. and I. Zuniga (1982), "A theoretical investigation of the Pieranski-Guyon instability In Couette flow of nematic liquid crystals," J. Non-Newtonian Fluid Mech. **11**, 23-36.

Barratt, P. J. and I. Zuniga (1984), "A theoretical investigation of Benard-Couette instabilities In nematic liquid crystals," J. Phys. D **17**, 775-86.

Bartels, F. (1982), "Taylor vortices between two concentric rotating spheres," J. Fluid Mech. **119**, 1-25.

Bashtovoi, V. G. and V. A. Chernobai (1988), "Fluid friction in the Couette flow of a magnetic fluid," Magn. Gidrodin. **24**, 63-6 [Magnetohydrodynamics **24**, 189-92 (1988)].

Bassom, A. P. (1989), "On the effect of crossflow on nonlinear Görtler vortices in curved channel flows," Quart. J. Mech. Appl. Math. **42**, 495.

Bassom, A. P. and P. Hall (1989), "On The Generation Of Mean Flows By The Interaction Of Görtler Vortices And Tollmien-Schlichting Waves In Curved Channel Flows," Stud. Appl. Math. **24**, 185.

Bassom, A. P. and S. D. Seddougui (1990), "The onset of three-dimensionality and time-dependence in Görtler vortices: neutrally stable wavy modes," J Fluid Mech. **220**, 661.

Batchelor, G. K. (1960), "A theoretical model of the flow at speeds far above the critical. Appendix to R. J. Donnelly and N. J. Simon," J. Fluid Mech. **7**, 416-418.

Bau, H. H. and K. E. Torrance (1981), "Onset of convection in a permeable medium between vertical coaxial cylinders," Phys. Fluids **24**, 382-5.

Baxter, G. W. and C. D. Andereck (1986), "Formation of dynamical domains in a circular Couette system," Phys. Rev. Lett. **57**, 3046-9.

Bayly, B. J. (1988), "Three-dimensional centrifugal-type instabilities in inviscid two-dimensional flows," Phys. Fluids **31**, 56-64.

Beard, D. W., M. H. Davies, and K. Walters (1966), "The stability of elastico-viscous flow between rotating cylinders. Part 3. Overstability in viscous and Maxwell fluids," J. Fluid Mech. **23**, 321-334.

Bearden, J. A. (1939), Phys. Rev. **56**, 1023.

Becker, K. M. and J. Kaye (1962), J. Heat Transfer **84**, 97.

Beer, T., H. -W. Suss, and H. -D. Tscheuschner (1988), "A process rheometer for non-Newtonian liquids," Wiss. Z. Tech. Univ. Dresd. **37**, 9-12.

Belyaev, Yu. N., A. A. Monakhov, S. A. Shcherbakov, and I. M. Yavorskaya (1984), "Nonuniqueness of the sequence of transitions to turbulence in rotating layers," Dok. Akad. Nauk SSSR **279**, 51-4 [Sov. Phys. Dokl. **29**, 872-4 (1984)].

Belyaev, Yu. N., A. A. Monakov, S. A. Scherbakov, and I. M. Yavorskaya (1984), "Some routes to stochasticity in spherical Couette flow and characteristics of its attractors," Preprint.

Belyaev, Yu. N. and I. M. Yavorskaya (1983), "Transition to stochasticity of viscous flow between rotating spheres," in Nonlinear Dynamics and Turbulence, edited by G. I. Barenblatt, G. Iooss, and D. D. Joseph (Pitman, Boston, London, Melbourne), pp. 61-70.

Belyaev, Yu. N. and I. M. Yavorskaya (1991), "Spherical Couette flow: transitions and onset of chaos," Fluid Dyn. (USSR) **26**, 7.

Ben-Jacob, E., H. Brand, G. Dee, L. Kramer, and J. S. Langer (1985), "Pattern propagation in nonlinear dissipative systems," Physica D **14**, 348-364.

Benjamin, T. B. (1976), "Applications of Leray-Schauder degree theory to problems of hydrodynamic instability," Math. Proc. Camb. Phil. Soc. **79**, 373-392.

Benjamin, T. B. (1977), "Quelques nouveaux resultats concernant des phenomenes de bifurcation en mecanique des fluides," in Computing Methods in Applied Sciences and Engineering, Lect. Notes in Physics 91 (Springer).

Benjamin, T. B. (1978a), "Applications of generic bifurcation theory in fluid mechanics," in Contemporary Developments in Continuum Mechanics and Partial Differential Equations, edited by G. M. de la Penha and L. A. Medeiros (North Holland).

Benjamin, T. B. (1978b), "Bifurcation phenomena in steady flows of a viscous fluid. I. Theory," Proc. R. Soc. London, Ser. A **359**, 1-26.

Benjamin, T. B. (1978c), "Bifurcation phenomena in steady flows of a viscous fluid. II. Experiments," Proc. R. Soc. London, Ser. A **359**, 27-43.

Benjamin, T. B. (1981), "New observations in the Taylor experiment," in Transition and Turbulence: Proc. of a Symposium, edited by R. E. Meyer (Academic Press, New York), pp. 25-41.

Benjamin, T. B. (1980), "Mathematical aspects of the Taylor problem," in Dynamical Systems, Stability, and Turbulence, Lecture Notes in Math.

Benjamin, T. B. and T. Mullin (1982), "Notes on the multiplicity of flows in the Taylor experiment," J. Fluid Mech. **121**, 219-230.

Benjamin, T. B. and T. Mullin (1981), "Anomalous modes in the Taylor experiment," Proc. R. Soc. London, Ser. A **377**, 221-49.

Benjamin, T. B. and S. K. Pathak (1987), "Cellular flows of a viscous liquid that partly fills a horizontal rotating cylinder," J. Fluid Mech. **183**, 399-420.

Bennett, J. and P. Hall (1988), "On the secondary instability of Taylor-Görtler vortices to Tollmien-Schlichting waves in fully developed flows," J. Fluid Mech. **186**, 445-69.

Bennett, J., P. Hall, and F. T. Smith (1991), "The strong nonlinear interaction of Tollmien-Schlichting waves and Taylor-Görtler vortices in curved channel flow," J. Fluid Mech. **223**, 475.

Berberian, J. G. (1980), "A method for measuring small rotational velocities for a Couette-type viscometer," Rev. Sci. Instrum. **51**, 1136-7.

Bergé, P., Y. Pomeau, and C. Vidal (1986), Order within Chaos: towards a deterministic approach to turbulence (Wiley, New York).

Berger, M. H. (1987), "Finite element analysis of flow in a gas-filled rotating annulus," Int. J. Numer. Methods Fluids **7**, 215-31.

Berkovskii, B. M., S. V. Isaev, and B. E. Kashevskii (1980), "One effect of internal rotational degrees of freedom in the hydrodynamics of microstructural fluids," Dok. Akad. Nauk SSSR **253**, 62-5 [Sov. Phys. Dokl. **25**, 519-21 (1980)].

Berkovskii, B. M., V. Yu. Veretenov, and S. M. Malyavin, "Numerical investigation of the stability of Couette flow of a ferrofluid in an homogeneous magnetic field relative to axisymmetric excitations," in Odinnadtsatoe Rizhskoe Soveshchanie Po Magnitnoi Gidrodinamike (11th Riga Conference On Magnetohydrodynamics), vol. 3 (Institut Fiziki an Latviiskoi SSR, Riga, USSR), pp. 115-18.

Berland, T., T. Jossang, and J. Feder (1986), "An experimental study of the connection between the hydrodynamic and phase-transition descriptions of the Couette-Taylor instability," Phys. Scr. **34**, 427-31.

Bestehorn, M., R. Friedrich, and H. Haken (1988), "The oscillatory instability of a spatially homogeneous state in large aspect ratio systems of fluid dynamics," Z. Phys. B **72**, 265-275.

Bhat, G. S., R. Narasimha, and S. Wiggins, "A simple dynamical system that mimics open flow turbulence," Phys. Fluids A **2**, 1983-2001.

Bhattacharjee, J. K. (1987), Convection and Chaos in Fluids (World Scientific, Singapore), pp. 203-215, 234-239.

Bhattacharjee, J. K., K. Banerjee, and K. Kumar (1986), "Modulated Taylor-Couette flow as a dynamical system," J. Phys. A **19**, L835-9.

Bhattacharya, D. K. and B. C. Eu (1987), "Nonlinear transport processes and fluid dynamics: effects of thermoviscous coupling and nonlinear transport coefficients on plane Couette flow of Lennard-Jones fluids," Phys. Rev. A **35**, 821-836.

Bielek, C. A. and E. L. Koschmieder (1990), "Taylor vortices in short fluid columns with large radius ratio," Phys. Fluids A **2**, 1557-1563.

Bjorklund, I. S. and W. M. Kays (1959), J. Heat Transfer **81**, 175.

Bland, S. B. and W. H. Finlay (1991), "Transitions toward turbulence in a curved channel," Phys. Fluids A **3**, 106-114.

Blaszcyk, J. and R. Petela (1986), "Application of a modified rotary rheometer to the investigation of slurries," Rheol. Acta **25**, 521-6.

Blennerhassett, P. J. and P. Hall (1979), "Centrifugal instabilities of circumferential flows in finite cylinders: linear theory," Proc. R. Soc. London, Ser. A **365**, 191-207.

Block, H., E. M. Gregson, W. D. Ions, G. Powell, R. P. Singh, and S. M. Walker (1978), "The measurement of birefringent, viscous and dielectric properties of liquids under shear," J. Phys. E **11**, 251-5.

Block, H., E. M. Gregson, A. Qin, G. Tsangaris, and S. M. Walker (1983), "A Couette cell with fixed stator alignment for the measurement of flow modified permittivity and electroviscosity," J. Phys. E **16**, 896-902.

Bocharov, YU. V. and A. D. Vuzhva (1989), "Hydrodynamics of a nematic liquid crystal near the threshold for the Freedericksz transition," Pis'ma Zh. Tekh. Fiz. **14**, 1460-2 [Sov. Tech. Phys. Lett. **14**, 635-6 (1989)].

Bogatyrev, T. P., V. G. Gilev, and V. D. Zimin (1980), "Spatial-time spectra of stochastic oscillations in convective cells," Pis'ma Zh. Eksp. Teor. Fiz. **32**, 229-32 [JETP Lett. **32**, 210-213 (1980)].

Bohlin, L. and K. Fontell (1978), "Flow properties of lamellar liquid crystalline lipid-water systems," J. Colloid Interface Sci. **67**, 272-83.

Bolstad, J. H. and H. B. Keller (1987), "Computation of anomalous modes in the Taylor experiment," J. Comput. Phys. **69**, 230-51.

Bonnet, J. and T. Alziary de Roquefort (1976), "Ecoulement entre deux spheres concentriques en rotation," J. Mecanique **15**, 373-397.

Booz, O. and H. Fasel (1978), "Numerical calculation of rotation-symmetrical flows between rotating coaxial cylinders," Z. Angew. Math. Mech. **58**, T252-4.

Borisevich, V. D., E. V. Levin, and V. V. Naumochkin (1989), "Numerical Investigation of Viscous Gas Secondary Flows in a Rotating Cylinder with Sources and Sinks," Izv. Akad. Nauk SSSR, Mekh. Zhidk. Gaza **24** [Fluid Dyn. (USSR) **24**, 520 (1989)].

Bouabdallah, A. (1980), "Transitions et turbulence dans l'ecoulement de Taylor-Couette," Thesis: Doctorat es Sciences, INP de Lorraine (France).

Bouabdallah, A. (1981), "Effect of the geometric factor on the laminar-turbulent transition in Taylor-couette flow," in <u>Symmetries And Broken Symmetries In Condensed Matter Physics. Proc. of the Colloque Pierre Curie</u>, edited by N. Boccara (IDSET, Paris, France), pp. 407-17.

Bouabdallah, A. and G. Cognet (1980), "Laminar-turbulent transition in Taylor-Couette flow," in <u>Laminar Turbulent Transition (IUTAM Conference)</u>, edited by R. Eppler and H. Fasel (Springer-Verlag, Berlin), pp. 368-377.

Boudourides, M. A. (1990), "Adiabatic circular Couette flow with temperature-dependent viscosity," J. Math. Anal. Appl. **152**, 461.

Bowen, R. (1977), "A model for Couette flow data," in <u>Turbulence Seminar</u>, edited by A. Dold and B. Eckmann (Springer-Verlag, Berlin), pp. 117-34.

Boyarevich, V. V. and R. P. Millere (1984), "Increasing the rotation process in convergent flows," in <u>Odinnadtsatoe Rizhskoe Soveshchanie Po Magnitnoi Gidrodinamike (11th Riga Conference On Magnetohydrodynamics)</u>, vol. 1 (Institut Fiziki an Latviiskoi SSR, Riga, USSR), pp. 75-8.

Brachet, M. E., P. Coullet, and S. Fauve (1987), "Propagative phase dynamics in temporally intermittent systems," Europhys. Lett. **4**, 1017-1022.

Brand, H. R. (1984), "Nonlinear phasedynamics for the spatially periodic states of the Taylor instability," Prog. Theor. Phys. **71**, 1096-1099.

Brand, H. R. (1985a), "Phase dynamics for incommensurate nonequilibrium systems," Phys. Rev. A **32**, 3551-3553.

Brand, H. R. (1985b), "Phase dynamics for spiraling Taylor vortices," Phys. Rev. A **31**, 3454-6.

Brand, H. R. (1985c), "Phase dynamics - the analogue of hydrodynamics for large aspect-ratio pattern forming nonequilibrium systems," unpublished lecture notes, Univ. of Texas, Austin.

Brand, H. R. (1987), "Phase dynamics with a material derivative due to a flow field," Phys. Rev. A **35**, 4461-4463.

Brand, H. R. (1988), "Phase dynamics - a review and perspective," in <u>Propagation in Systems far from Equilibrium</u>, edited by J. Wesfreid, H. Brand, P. Manneville, G. Albinet, and N. Boccara (Springer-Verlag, Berlin), pp. 206-224.

Brand, H. and M. C. Cross (1983), "Phase dynamics for the wavy vortex state of the Taylor instability," Phys. Rev. A **27**, 1237-1239.

Brand, H. R. and R. J. Deissler (1989), "Confined states in phase dynamics," Phys. Rev. Lett. **63**, 508-11.

Brand, H. R. and K. Kawasaki (1984), "Macroscopic dynamics of systems with a small number of topological defects and non-equilibrium systems," J. Phys. A **17**, L905-L910.

Brand, H. R., J. E. Wesfried, M. A. Azouni, and S. Kai, "Spirals with a continuously changing pitch in nonequilibrium systems," (preprint).

Brandstater, A., U. Gerdts, A. Lorenzen, and G. Pfister (1981), in <u>Nonlinear Phenomenon at Instabilities and Phase Transitions</u>, edited by T. Riste (Plenum, New York).

Brandstater, A., G. Pfister, and E. O. Schulz-Dubois (1982), "Excitation of a Taylor vortex mode having resonant frequency dependence of coherence length," Phys. Lett. A **88A**, 407-9.

Brandstater, A., J. Swift, H. L. Swinney, and A. Wolf (1984), "A strange attractor in a Couette-Taylor experiment," in <u>Turbulence and Chaotic Phenomena in Fluids (IUTAM Symposium)</u>, edited by T. Tatsumi (North Holland, Amsterdam), pp. 179-84.

Brandstater, A., J. Swift, H. L. Swinney, A. Wolf, J. Doyne Farmer, E. Jen, and P. J. Crutchfield (1983), "Low-dimensional chaos in a hydrodynamic system," Phys. Rev. Lett. **51**, 1442-5.

Brandstater, A. and H. L. Swinney (1987), "Strange attractors in weakly turbulent Couette-Taylor flow," Phys. Rev. A **35**, 2207-20.

Brandstater, A. and H. L. Swinney (1984), "Distinguishing low-dimensional chaos from random noise in a hydrodynamic experiment," in <u>Fluctuations and Sensitivity in Nonequilibrium Systems</u>, edited by W. Horsthemke and D. Kondepudi (Springer-Verlag, Berlin), pp. 166-171.

Braun, H. and Chr. Friedrich (1989), "Transient processes in Couette flow of a Leonov fluid influenced by dissipation," J. Non-Newtonian Fluid Mech. **33**, 39.

Braun, M. J. and R. C. Hendricks (1991), "Non-intrusive laser-based, full-field quantitative flow measurements aided by digital image processing. Part 1: Eccentric cylinders," Tribology international **24**, 195.

Brewster, D. B., P. Grosberg, and A. H. Nissan (1959), "The stability of viscous flow between horizontal concentric cylinders," Proc. R. Soc. London, Ser. A **251**, 76-91.

Brewster, D. B. and A. H. Nissan (1958), "The hydrodynamics of flow between horizontal concentric cylinders - I.," Chem. Eng. Sci. **7**, 215-21.

Brice, J. C. and P. A. C. Whiffin (1977), "Changes in fluid flow during Czochralski growth," J. Cryst. Growth **38**, 245-8.

Brindley, J. (1986), "Spatio-temporal behavior of flows with circular constraints," Physica D **23D**, 240-245.

Brindley, J. (1991), "Nonlinear mode competition and co-existence and approach to turbulence in closed rotating flow," Eur. J. Mech. B/Fluids **10 (suppl.)**, 326.

Brindley, J., C. Kaas-Petersen, and A. Spence (1989), "Path-following methods in bifurcation problems," Physica D **34D**, 456-461.

Brindley, J. and F. R. Mobbs (1987), "An experimental investigation of transition to turbulencein Taylor-Couette flow using digital image processing," in <u>Advances in Turbulence. Proceedings of the First European Turbulence Conference</u> , edited by G. Comte-Bellot and J. Mathieu (Springer-Verlag, Berlin), pp. 7-15.

Brindley, J., F. R. Mobbs, and I. Moroz (1982), "Two-mode instability in Taylor-Couette flow," Preprint.

Brodskii, I. I., V. A. Kozlachkov, I. I. Korshever, V. S. L'vov, S. L. Musher, yu. E. Nesterikhin, S. A. Pavlov, A. A. Predtechenskii, I. G. Remel, and A. V. Shafarenko (1984), "High productivity real time computing system for processing hydrophysical information," Avtometriya, 3-12 [Autom. Monit. Meas.].

Broomhead, D. S. and S. C. Ryrie (1988), "Particle paths in wavy vortices," Nonlinearity **1**, 409-34.

Brown, C. S., P. J. Barham, and C. G. Cannon (1981), "Oriented polymers from solution. IV. Polyethylene/polypropylene blend films," J. Polym. Sci., Polym. Phys. Ed. **19**, 1047-53.

Brunn, P. (1982), "The general solution of the equations of creeping motion of a micropolar fluid and its application," Int. J. Eng. Sci. **20**, 575-85.

Brunn, P. O. (1983), "Hydrodynamically induced cross stream migration of dissolved macromolecules (modelled as nonlinearly elastic dumbbells)," Int. J. Multiphase Flow **9**, 187-202.

Bucherer, C., J. C. Lelievre, and C. Lacombe (1988), "Theoretical and experimental study of the time dependent flow of red blood cell suspension through narrow pores," Biorheology **25**, 639-49.

Buggisch, H. and G. Eitelberg (1984), "Marangoni effect in the Couette flow," J. Non-Newtonian Fluid Mech. **15**, 1-12.

Bühler, K. (1982), "Ein Beitrag zum Stabilitatsverhalten der Zylinderspaltstromung mit Rotation und Durchfluss," Strömungsmechanik und Strömungsmaschinen **32**, 35-44.

Bühler, K., "Visualization of flow instabilities in spherical gaps," in <u>Symposium on Flow Visualization</u>.

Bühler, K. (1983a), "Instabilitäten im Spalt konzentrischer Kugeln bei Rotation und Durchfluss," Report No. IB 221-83 A 08 (DFVLR, Göttingen), pp. 82-86.

Bühler, K. (1983b), "Stromungen im Spalt konznetrischer Kugeln mit Durchfluss," Report No. AZ: Bu 533/1-1 (DFG).

Bühler, K. (1983c), "Instabilitaten spiralformiger Strömungen zwischen konzentrischen Kugeln," Z. Angew. Math. Mech. **63**, T235-T239.

Bühler, K. (1984a), "Instabilitäten spiralforminger Strömungen im Zylinderspalt," Z. Angew. Math. Mech. **64**, T180-4.

Bühler, K. (1984b), "The influence of a base flow on the onset of thermal instabilities in horizontal fluid layers and the analogy to the Taylor problem," Strömungsmechanik und Strömungsmaschinen **34**, 67-76.

Bühler, K. (1986), "Hydrodynamic instabilities in a spherical gap," Strömungsmechanik und Strömungsmaschinen **38**, 11-24.

Bühler, K. (1987), "Stationary and time-dependent vortex streams in a gap between concentric spheres," Z. Angew. Math. Mech. **67**, T268-70.

Bühler, K. (1988), "Wirbelströmungen im Kugelspalt bei hohen Reynoldszahlen," Report No. Az: Bu 533/1-5 (DFG).

Bühler, K. (1989a), "Asymmetrische Stromungsformen im Kugelspalt," Z. Angew. Math. Mech. **69**, T611-614.

Bühler, K. (1989b), "Scherschichtströmungen im Kugelspalt," Z. Flugwiss. Weltraumforsch. **13**, 8-15.

Bühler, K. (1990), "Symmetric and asymmetric Taylor vortex flow in spherical gaps," Acta Mech. **81**, 3-38.

Bühler, K. (1991a), "Wirbelströmungen mit Scherschichten," Z. Angew. Math. Mech. **71**, T485-488.

Bühler, K. (1991b), "Dynamical behavior of instabilities in spherical gap flows: theory and experiment," Eur. J. Mech. B/Fluids **10 (suppl.)**, 187-192.

Bühler, K., "Shear layer instabilities in spherical gap flows," Eur. J. Mech. B/Fluids (to appear).

Bühler, K., J. E. R. Coney, M. Wimmer, and J. Zierep (1986), "Advances in Taylor Vortex Flow: a report on the fourth Taylor Vortex Flow Working Party Meeting," Acta Mech. **62**, 47-61.

Bühler, K., K. R. Kirchartz, and M. Wimmer (1989), "Strömungsmechanische Instabilitäten," Strömungsmechanik und Strömungsmaschinen **40**, 99-126.

Bühler, K. and N. Polifke (1990), "Dynamical behavior of Taylor vortices with superimposed axial flow," in Nonlinear Evolution of Spatio-temporal Structures in Dissipative Continuous Systems, edited by F. H. Busse and L. Kramer (Plenum, New York), pp. 21-29.

Bühler, K. and A. Schmitz (1989), "A shear flow instability in cyclindrical geometry," Exp. Fluids **7**, 568-570.

Bühler, K., M. Wimmer, and J. Zierep (1983), "Taylor vortex flow between two rotating spheres and cones with and without mass flux," in Summary of the Proceedings of the 60th Anniversary Taylor Vortex Flow, edited by K. J. Park and R. J. Donnelly, pp. 34-36.

Bühler, K. and J. Zierep (1982), "Die Stromung im Spalt zwischen zwei rotierenden Kugeln mit Durchfluss," Z. Angew. Math. Mech. **62**, T198-T201.

Bühler, K. and J. Zierep (1983), "Transition to turbulence in a spherical gap," in Proceedings of the 4th International Symposium on Turbulent Shear Flows, pp. 16.1-16.6.

Bühler, K. and J. Zierep (1984), "New secondary instabilities for high Re-number flow between two rotating spheres," in Second IUTAM-Symposium on Laminar-Turbulent Transition.

Bühler, K. and J. Zierep (1987), "Dynamical instabilites and transition to turbulence in spherical gap flows," in Advances in Turbulence: Proceedings of the First European Turbulence Conference, edited by G. Comte-Bellot and J. Mathieu (Springer-Verlag, Berlin).

Bureau, M., J. C. Healy, D. Bourgoin, and M. Joly (1978), "Experimental study in vitro of the rheological behaviour of blood in the transient regime to low velocity shear flow," Rheol. Acta **17**, 612-25.

Bureau, M., J. C. Healy, D. Bourgoin, and M. Joly (1980), "Rheological hysteresis of blood at low shear rate," Biorheology **17**, 191-203.

Burkhalter, J. E. and E. L. Koschmieder (1973), "Steady supercritical Taylor vortex flow," J. Fluid Mech. **58**, 547-560.

Burkhalter, J. E. and E. L. Koschmieder (1974), "Steady supercritical Taylor vortices after sudden starts," Phys. Fluids **17**, 1929-1935.

Bust, G. S., B. C. Dornblaser, and E. L. Koschmieder (1985), "Amplitudes and wavelegnths of wavy Taylor vortices," Phys. Fluids **28**, 1243-1247.

Buzug, Th., T. Riemers, and G. Pfister, "Optimal reconstruction of strange attractors from purely geometrical arguments," (preprint).

Buzug, Th., T. Reimers, and G. Pfister (1990), "Dimensions and Lyapunov spectra from measured time series of Taylor-Couette flow," in Nonlinear Evolution of Spatio-temporal Structures in Dissipative Continuous Systems: Proceedings of a NATO Advanced Research Workshop, edited by F. H. Busse and L. Kramer (Plenum, New York).

Caldwell, D. R. and R. J. Donnelly (1962), "On the reversibility of the transition past instability in Couette flow," Proc. R. Soc. London, Ser. A **267**, 197-205.

Cameron, J. R. (1989), "Viscometry of nonhomogeneous flows and the behaviour of a titanium-crosslinked hydroxypropyl guar gel in Couette flow," J. Rheol. **33**, 15-46.

Cannell, D. S., M. A. Dominguez-lerma, and G. Ahlers (1983), "Experiments on wave number selection in rotating Couette-Taylor flow," Phys. Rev. Lett. **50**, 1365-8.

Capper, P., J. C. Brice, C. L. Jones, W. G. Coates, J. J. G. Gosney, C. Ard, and I. Kenworthy (1988), "Interfaces and flow regimes in ACRT grown Cd/sub x/Hg/sub 1-x/Te crystals," J. Cryst. Growth **89**, 171-6.

Capper, P., J. J. G. Gosney, and C. L. Jones (1984), "Application of the accelerated crucible rotation technique to the Bridgman growth of Cd/sub x/Hg/sub 1-x/Te: simulations and crystal growth," J. Cryst. Growth **70**, 356-64.

Capper, P., J. J. Gosney, C. L. Jones, and I. Kenworthy (1986), "Bridgman growth of Cd/sub x/Hg/sub 1-x/Te using ACRT," J. Electron. Mater. **15**, 371-6.

Capper, P., J. J. Gosney, C. L. Jones, and E. J. Pearce (1986), "Fluid flows induced in tall narrow containers by ACRT," J. Electron. Mater. **15**, 361-70.

Carmi, S. and J. I. Tustaniwskyj (1981), "Stability of modulated finite-gap cylindrical Couette flow: linear theory," J. Fluid Mech. **108**, 19-42.

Castle, P. and F. R. Mobbs, "Hydrodynamic stability of the flow between eccentric rotating cylinders: visual observations and torque measuements," Proc. Inst. Mech. Eng. (London) **182**, 41-52.

Castle, P., F. R. Mobbs, and P. H. Markho (1971), "Visual observations and torque measurements in the Taylor vortex regime between eccentric rotating cylinders," J. Lubr. Technol. **93**, 121-129.

Chan, P. C. -H. and L. G. Leal (1981), "An experimental study of drop migration in shear flow between concentric cylinders," Int. J. Multiphase Flow **7**, 83-99.

Chandrasekhar, S. (1953), "The stability of viscous flow between rotating cylinders in the presence of a magnetic field," Proc. R. Soc. London, Ser. A **216**, 293-309.

Chandrasekhar, S. (1954a), "The stability of viscous flow between rotating cylinders," Mathematika **1**, 5-13.

Chandrasekhar, S. (1954b), "The stability of viscous flow between rotating cylinders in the presence of a radial temperature gradient," J. Rat. Mech. Anal. **3**, 181-207.

Chandrasekhar, S. (1954c), "On characteristic value problems in high order differential equations which arise in studies on hydrodynamics and hydromagnetic stability," Am. Math. Monthly **61**, 32-45.

Chandrasekhar, S. (1958), "The stability of viscous flow between rotating cylinders," Proc. R. Soc. London, Ser. A **246**, 301-311.

Chandrasekhar, S. (1960), Proc. Natl. Acad. Sci. USA **46**, 137, 141.

Chandrasekhar, S. (1961a), <u>Hydrodynamic and Hydromagnetic Instability</u> (Clarendon Press, Oxford).

Chandrasekhar, S. (1961b), "Adjoint differential systems in the theory of hydrodynamic stability," J. Math. Mech. **10**, 683-690.

Chandrasekhar, S. (1962), "The stability of spiral flow between rotating cylinders," Proc. R. Soc. London, Ser. A **265**, 188-197.

Chandrasekhar, S. and R. J. Donnelly (1957), "The hydrodynamic stability of helium II between rotating cylinders. I," Proc. R. Soc. London, Ser. A **241**, 9-28.

Chandrasekhar, S. and D. D. Elbert (1962), "The stability of viscous flow between rotating cylinders. II.," Proc. R. Soc. London, Ser. A **268**, 145-52.

Chandrasekharan, E. and G. Ramanaiah (1984), "Couette flow of a dusty fluid between two infinite coaxial cylinders with one of the moving boundaries suddenly stopped," Indian J. Technol. **22**, 401-5.

Chang, T. S. and W. K. Sartory (1967a), "Hydromagnetic stability of dissipative flow between rotating cylinders. Part 1. Stationary critical modes," J. Fluid Mech. **27**, 65-79.

Chang, T. S. and W. K. Sartory (1967b), "Corrigendum to: Hydromagnetic stability of dissipative flow between rotating cylinders. Part 1. Stationary critical modes," J. Fluid Mech. **28**, 821.

Chang, T. S. and W. K. Sartory (1969), "Hydromagnetic stability of dissipative flow between rotating cylinders. Part 2. Oscillatory critical modes and asymptotic results," J. Fluid Mech. **36**, 193-206.

Channabasappa, M. N., G. Ranganna, and B. Rajappa (1983), "Stability of Couette flow between rotating cylinders lined with porous material. I," Indian J. Pure Appl. Math. **14**, 741-56.

Channabasappa, M. N., G. Ranganna, and B. Rajappa (1984), "Stability of viscous flow in a rotating porous medium in the form of an annulus: the small-gap problem," Int. J. Numer. Methods Fluids **4**, 803-11.

Chelyshkov, V. S. (1985), "An autooscillation regime branching off from circular Couette flow for viscous isothermal gas," Gidromekhanika, 35-40.

Chen, C. F. and D. K. Christensen (1967), "Stability of flow induced by an impulsively started rotating cylinder," Phys. Fluids **10**, 1845-1846.

Chen, C. F. and R. P. Kirchner (1971), "Stability of time-dependent rotational Couette flow Part 2. Stability analysis," J. Fluid Mech. **48**, 365-384.

Chen, J. and J. Kuo (1990), "The linear stability of steady circular Couette flow with a small radial temperature gradient," Phys. Fluids A **2**, 1585-1591.

Chen, J. -C. and G. P. Neitzel (1982), "Strong stability of impulsively initiated Couette flow for both axisymmetric and nonaxisymmetric disturbances," J. Appl. Mech. **49**, 691-6.

Chen, J. -C., G. P. Neitzel, and D. F. Jankowski (1985), "The influence of initial condition on the linear stability of time-dependent circular Couette flow," Phys. Fluids **28**, 749-51.

Chen, J. -C., G. P. Neitzel, and D. F. Jankowski (1987), "Numerical experiments on the stability of unsteady circular Couette flow with random forcing," Phys. Fluids **30**, 1250-8.

Chen, K. (1991), "Elastic instability of the interface in Couette flow of viscoelastic liquids," J. Non-Newtonian Fluid Mech. **40**, 261.

Chen, K. S., A. C. Ku, and T. M. Chan (1990), "Flow in the half-filled annulus between horizontal concentric cylinders in relative rotation," J. Fluid Mech. **213**, 149.

Chen, S., N. V. Thakor, and J. W. Wagner (1986), "A microprocessor-based two-channel thromboelastograph," IEEE Trans. Biomed. Eng. **BME-33**, 887-90.

Cheng, M. and H. Chang (1990), "A generalized sideband stability theory via center manifold projection," Phys. Fluids A **2**, 1364.

Cheng-Han, T., S. Iacobellis, and W. Lick (1987), "Flocculation of fine-grained lake sediments due to a uniform shear stress," J. Great Lakes Res. **13**, 135-46.

Chevray, R. (1989), "Chaos and the onset of turbulence," in <u>Advances in Turbulence</u>, edited by W. K. George and R. E. A. Arndt (Hemisphere, New York), pp. 127-158.

Choo, Y. K. (1976), "Influence of heat transfer on the stability of parallel flow between concnetric cylinders," Phys. Fluids **19**, 1676-9.

Chossat, P. (1984), "Bifurcation in the presence of symmetry in classical problems of hydrodynamics," J. Mec. Theor. Appl., spec. suppl., 157-192.

Chossat, P. (1985a), "Bifurcation d'ondes rotatives superposees," C. R. Acad. Sci., Ser. A **300**, 209.

Chossat, P. (1985b), "Interaction d'ondes rotatives dans le probleme de Couette-Taylor," C. R. Acad. Sci., Ser. A **300**, 251-4.

Chossat, P., Y. Demay, and G. Iooss (1984), "Recent results about secondary bifurcations in the Couette-Taylor problem," in <u>Proceedings of the International Conference on Nonlinear Mechanics</u>, edited by C. Wei-Zang (Science Press, Beijing), pp. 1001-1004.

Chossat, P., Y. Demay, and G. Iooss (1987), "Interaction of azimuthal modes in the Couette-Taylor problem," Arch. Ration. Mech. Anal. **99**, 213-48.

Chossat, P. and M. Golubitsky, "Hopf bifurcation in the presence of symmetry, center manifold and Liapunov-Schmidt reduction," in <u>Oscillation, Bifurcation, and Chaos</u>, edited by F. V. Atkinson, W. F. Langford, and A. B. Mingarelli, CMS Conf. Proc., Vol. 8 (American Mathematical Society, Providence), pp. 343-352.

Chossat, P. and M. Golubitsky (1988a), "Iterates of maps with symmetry," SIAM J. Math. Anal. **19**, 1259.

Chossat, P. and M. Golubitsky (1988b), "Symmetry increasing bifurcation of chaotic attractors," Physica D **32D**, 423-436.

Chossat, P., M. Golubitsky, and B. L. Keyfitz (1987), "Hopf-Hopf Mode Interactions with O(2) Symmetry," Dyn. Stab. Sys. **1**, 255-292.

Chossat, P. and G. Iooss (1985), "Primary and secondary bifurcations in the Couette-Taylor problem," Jpn. J. Appl. Mech. **2**, 37-68.

Chow, A. W. and G. G. Fuller (1985), "Some experimental results on the development of Couette flow for non-Newtonian fluids," J. Non-Newtonian Fluid Mech. **17**, 233-43.

Christmann, L. and W. Knappe (1976), "Rheological measurements on plastic melts with a new rotational rheometer," Rheol. Acta **15**, 296-304.

Chung, K. (1976), "Stability study of a viscous flow between rotating coaxial cylinders," Ph.D. Thesis, Tufts University.

Chung, K. C. and K. N. Astill (1977), "Hydrodynamic instability of viscous flow between rotating coaxial cylinders with fully developed axial flow," J. Fluid Mech. **81**, 641-55.

Cladis, P. E. (1991), in <u>Nematics, Mathematical and Physical Aspects</u>, edited by J. -M. Coron, J. -M. Ghidaglia, and F. Helein (Kluwer Academic, Dordrecht), pp. 65-91.

Cladis, P. E., Y. Couder, and H. R. Brand (1985), Phys. Rev. Lett. **55**, 2945-.

Cladis, P. E. and S. Torza (1975), Phys. Rev. Lett. **35**, 1283-.

Cladis, P. E. and S. Torza (1976), in <u>Colloid and Interface Science</u>, vol. 4, edited by M. Kelker (Academic, New York), pp. 487-499.

Claussen, M. (1984), "Surface-layer similarity in turbulent circular Couette flow," J. Fluid Mech. **144**, 123-31.

Cliffe, K. A. (1983), "Numerical calculations of two-cell and single-cell Taylor flows," J. Fluid Mech. **135**, 219-233.

Cliffe, K. A. (1988), "Numerical calculations of the primary-flow exchange process in the Taylor problem," J. Fluid Mech. **197**, 57-79.

Cliffe, K. A., A. D. Jepson, and A. Spence (1985), "The numerical solution of bifurcation problems with symmetry with application to the finite Taylor problem," in Numerical Methods for Fluid Dynamics II, edited by K. W. Morton and M. J. Baines (Oxford University Press, Oxford), pp. 155-176.

Cliffe, K. A. and T. Mullin (1985), "A numerical and experimental study of anomalous modes in the Taylor experiment," J. Fluid Mech. **153**, 243-58.

Cliffe, K. A. and T. Mullin (1986), "A numerical and experimental study of the Taylor problem with asymmetric end conditions," in Sixth International Symposium on Finite Element Methods in Flow Problems (Inst. Nat. Recherche Inf. & Autom., Le Chesnay, France), pp. 377-81.

Cliffe, K. A. and A. Spence (1983), "The calculation of high order singularities in the finite Taylor problem," in Numerical Methods for Bifurcation Problems, edited by T. Kupper, H. D. Mittelmann, and H. Weber (Birkhauser, Basel), pp. 129-144.

Cliffe, K. A. and A. Spence (1985), "Numerical calculations of bifurcations in the finite Taylor problem," in Numerical Methods for Fluid Dynamics II, edited by K. W. Morton and M. J. Baines (Oxford University Press, Oxford), pp. 177-197.

Cognet, G. (1968), "Contribution a l'etude de l'ecoulement de Couette par la methode polarographique," Thesis: Doctorat es Sciences, Universite de Nancy (France).

Cognet, G. (1971), "Utilisation de la polarographie pour l'etude de l'ecoulement de Couette," J. Mecanique **10**, 65-90.

Cognet, G. (1984), "The ways to turbulence in Couette flow between coaxial cylinders," J. Mec. Theor. Appl., spec. suppl., 7-44.

Cognet, G., A. Bouabdallah, and A. A. Aider (1982), "Laminar turbulent transition in Taylor-Couette flow, influence of geometrical parameters," in Stability in the Mechanics of Continua, 2nd Symposium, edited by F. H. Schroeder (Springer, Berlin), pp. 330-340.

Cohen, S. and D. M. Maron (1991), "Analysis of a rotating annular reactor in the vortex flow regime," Chemical engineering science **46**, 123.

Cole, J. A. (1966), Proc. of the Second Australasian Conference Hydrodynamics and FLuid Mechanics (Univ. of Aukland, New Zealand), pp. B313.

Cole, J. A. (1969), "Taylor vortices with eccentric rotating cylinders," Nature (London) **221**, 253-254.

Cole, J. A. (1974a), "Taylor vortices with short rotating cylinders," J. Fluids Eng. **96**, 69-70.

Cole, J. A. (1974b), "Taylor vortex behavior in annular clearances of limited length," in Proc. of the Fifth Australasian Conf. on Hydraulics and Fluid Mechanics, pp. 514-521.

Cole, J. A. (1976), "Taylor-vortex instability and annulus-length effects," J. Fluid Mech. **75**, 1-15.

Cole, J. A. (1977), "Cell cize and flow history effects in the Taylor vortex regime," in Proc. 6th Australian Conf. on Hydraulics and Fluid Mechanics, pp. 331-4.

Cole, J. A., "The effect of cylinder radius ratio on wavy vortex onset," in Third Taylor Vortex Flow Working Party Meeting, pp. 1.a1.

Coles, D. (1965), "Transition in circular Couette flow," J. Fluid Mech. **21**, 385-425.

Coles, D. (1967), "A note on Taylor instability in circular Couette flow," J. Appl. Mech. **89**, 529-534.

Coles, D. (1981), "Prospects for useful research on coherent structure in turbulent shear flow," Proc. Indian Acad. Sci. Eng. Sci. **4**, 111-27.

Coles, D. and C. Van Atta (1966a), "Progress report on a digital experiment in spiral turbulence," AIAA J. **4**, 1969-1971.

Coles, D. and C. Van Atta (1966b), "Measured distortion of a laminar circular Couette flow by end effects," J. Fluid Mech. **25**, 513-521.

Coles, D. and C. Van Atta (1967), "Digital experiment in spiral turbulence," Phys. Fluids Suppl. (Boundary Layers and Turbulence), S120-S121.

Coney, J. E. R. and J. Atkinson (1978), "The effect of Taylor vortex flow on the radial forces in an annulus having variable eccentricity and axial flow," J. Fluids Eng. **100**, 210-14.

Coney, J. E. R. and D. A. Simmers (1979a), "A study of fully-developed, laminar, axial flow and Taylor vortex flow by means of shear stress measurements," J. Mech. Eng. Sci. **21**, 19-24.

Coney, J. E. R. and D. A. Simmers (1979b), "The determination of shear stress in fully developed laminar axial flow and Taylor vortex flow, using a flush-mounted hot film probe," Disa Inf., 9-14.

Conrad, P. W. and W. O. Criminale (1965), "The stability of time-dependent laminar flow: Flow with curved streamlines," Z. Angew. Math. Phys. **16**, 569-581.

Constantin, P., C. Foias, B. Nicolaenko, and R. Temam (1989), Integral Manifolds and Inertial Manifolds for Dissipative Partial Differential Equations (Springer, New York).

Coombes, A. and A. Keller (1979), "Oriented polymers from solution. I. A novel method for producing polyethylene films," J. Polym. Sci., Polym. Phys. Ed. **17**, 1637-47.

Cooper, E. R., D. F. Jankowski, G. P. Neitzel, and T. H. Squire (1985), "Experiments on the onset of instability in unsteady circular Couette flow," J. Fluid Mech. **161**, 97-113.

Copper, A. L. and D. L. Book (1979), "Formation of a viscous boundary layer on the free surface of an imploding rotating liquid cylinder," J. Fluid Mech. **93**, 305-17.

Coriell, S. R., G. B. Mcfadden, R. F. Boisvert, and R. F. Sekerka (1984), "Effect of a forced Couette flow on coupled convective and morphological instabilities during unidirectional solidification," J. Cryst. Growth **69**, 15-22.

Cornish, R. J. (1933), "Flow of water through fine clearances with relative motion of the boundaries," Proc. R. Soc. London, Ser. A **140**, 227-40.

Couette, M. M. (1888), Comptes Rendus **107**, 388.

Couette, M. M. (1988), Bull. Sci. Phys., 4, 262.

Couette, M. M. (1890), "Etudes sur le frottement des liquides," Ann. Chim. Phys. **21**, 433-510.

Coughlin, K. (1990), "Quasiperiodic Taylor-Couette flow," Ph.D. Thesis, Harvard University.

Coughlin, K. T., P. S. Marcus, R. P. Tagg, and H. L. Swinney (1991), "Distinct quasiperiodic modes with like symmetry in a rotating fluid," Phys. Rev. Lett. **66**, 1161-1164.

Coughlin, K. T. and P. S. Marcus (1990a), "Modulated waves in Taylor-Couette flow. Part 1. Analysis," Preprint.

Coughlin, K. T. and P. S. Marcus (1990b), "Modulated waves in Taylor-Couette flow. Part 2. Numerical simulation," Preprint.

Coullet, P., C. Elphick, L. Gil, and J. Lega (1987), "Topological defects of wave patterns," Phys. Rev. Lett. **59**, 884-887.

Coullet, P., C. Elphick, and D. Repaux (1987), "Nature of spatial chaos," Phys. Rev. Lett. **58**, 431-434.

Coullet, P. and S. Fauve (1985), "Propogative phasedynamics for systems with Galilean invariance," Phys. Rev. Lett. **55**, 2857-2859.

Coullet, P., S. Fauve, and E. Tirapegui (1985), "Large scale instability of nonlinear standing waves," J. Phys. (Paris), Lett. **46**, 787-791.

Coullet, P., L. Gil, and J. Lega (1989), "Defect-mediated turbulence," Phys. Rev. Lett. **62**, 1619.

Coullet, P., L. Gil, and D. Repaux (1989), "Defects and subcritical bifurcations," Phys. Rev. Lett. **62**, 2957-60.

Coullet, P., R. E. Goldstein, and H. Gunaratne (1989), "Parity-breaking transitions of modulated patterns in hydrodynamic systems," Phys. Rev. Lett. **63**, 1954.

Coullet, P. and G. Iooss (1990), "Instabilities of one-dimensional cellular patterns," Phys. Rev. Lett. **64**, 866-869.

Coullet, P. and D. Repaux (1987), "Models of pattern formation from a singularity theory point of view," in Instabilities and Nonequilibrium Structures, edited by E. Tirapegui and D. Villarroel (Reidel, Dordrecht), pp. 179-195.

Coullet, P. H. and E. A. Spiegel (1983), "Amplitude equations for systems with competing instabilities," SIAM J. Appl. Math. **43**, 776-821.

Coyle, D. J., C. W. Macosko, and L. E. Scriven (1990), "Stability of symmetric film-splitting between counter-rotating cylinders," J. Fluid Mech. **216**, 437.

Craik, A. D. D. (1985), "Mode interactions in Taylor-Couette flow," in Wave Interactions and Fluid Flows (Cambridge Univ. Press, Cambridge, England), pp. 244-257.

Crawford, G. (1983), "Transitions in Taylor wavy-vortex flow," Ph.D. Thesis, Univ. of Oregon.

Crawford, G. L., K. Park, and R. J. Donnelly (1985), "Vortex pair annihilation in Taylor wavy-vortex flow," Phys. Fluids **28**, 7-9.

Crawford, J. D. (1991), "Introduction to bifurcation theory," Rev. Mod. Phys. **63**, 991.

Crawford, J. D., M. Golubitsky, and W. F. Langford, "Modulated rotating waves in O(2) interactions," Dyn. Stab. Sys. **3**, 159-175.

Crawford, J. D., M. Golubitsky, M. G. M. Gomes, E. Knobloch, and I. N. Stewart (1989), "Boundary conditions as symmetry constraints," (Preprint).

Crawford, J. D. and E. Knobloch (1988), "On degenerate Hopf bifurcation with broken O(2) symmetry," Nonlinearity **1**, 617.

Crawford, J. D. and E. Knobloch (1991), "Symmetry and symmetry-breaking bifurcations in fluid dynamics," Ann. Rev. Fluid Mech. **23**, 341-387.

Crawford, J. D., E. Knobloch, and H. Riecke (1990), "Period-doubling mode interactions with circular symmetry," Physica D **44**, 340.

Cressely, R., R. Hocquart, T. Wydro, and J. P. Decruppe (1985), "Numerical evaluation of extinction angle and birefringence in various directions as a function of velocity gradient," Rheol. Acta **24**, 419-26.

Cross, M. C. (1984), "Wave-number selection by soft boundaries near threshold," Phys. Rev. A **29**, 391-2.

Cruickshank, J. O. (1987), "A new method for predicting the critical Taylor number in rotating cylindrical flows," J. Appl. Mech. **54**, 713-19.

Crutchfield, J., D. Farmer, N. Packard, R. Shaw, G. Jones, and R. J. Donnelly (1980), "Power spectral analysis of a dynamical system," Phys. Lett. A **76A**, 1-4.

Cummins, P. G., E. Staples, and B. Millen (1990), "A Couette shear flow cell for small-angle neutron scattering studies," Meas. Sci. Technol. **1**, 179.
Curl, M. L. and W. P. Graebel (1972), SIAM J. Appl. Math. **23**, 380.
Dagan, A. (1989), "Pseudo-spectral and asymptotic sensitivity investigation of coutner-rotating vortices," Comput. Fluids **17**, 509.
Dai, R. X. and A. Z. Szeri (1990), "A numerical study of finite Taylor flows," Int. J. Non-Linear Mech. **25**, 45-60.
Dang, H. T. (1983), "Cylindrical Couette problem of rarefied binary mixture," C. R. Acad. Sci., Ser. B **296**, 1015-18.
Dangelmayr, G. and E. Knobloch (1987), "On the Hopf bifurcation with broken O(2) symmetry," in The Physics of Structure Formation: Theory and Simulation, edited by W. Guttinger and G. Dangelmayr (Springer-Verlag, Berlin), pp. 387-393.
Darby, R. (1985), "Couette viscometer data reduction for materials with a yield stress," J. Rheol. **29**, 369-78.
Das Gupta, S., D. V. Khakhar, and S. K. Bhatia (1991), "Axial transport of granular solids in horizontal rotating cylinders. Part 1: Theory," Powder technology **67**, 145.
Daskalakis, J. (1990), "Couette Flow Through a Porous Medium of a High Prandtl Number Fluid with Temperature Dependent Viscosity," Int. J. Energy Res. **14**, 21.
Daskopoulos, P. and A. M. Lenhoff (1989), "Flow in curved ducts: bifurcation structure for stationary ducts," J. Fluid Mech. **203**, 125.
Datta, S. K. (1965), "Stability of spiral flow between concentric circular cylinders at low axial Reynolds numbers," J. Fluid Mech. **21**, 635-640.
Davey, A. (1962), "The growth of Taylor vortices in flow between rotating cylinders," J. Fluid Mech. **14**, 336-368.
Davey, A., R. C. Diprima, and J. T. Stuart (1968), "On the instability of Taylor vortices," J. Fluid Mech. **31**, 17-52.
Davis, M. W. and E. J. Weber (1960), "Liquid-liquid extraction between rotating cylinders," Ind. Eng. Chem. **52**, 929-934.
Davis, S. H. (1976), "The stability of time-periodic flows," Ann. Rév. Fluid Mech. **8**, 57-74.
de Araujo, J. H. C., V. Raus, and A. S. Vargas (1990), "Finite Element Solution of Flow Between Eccentric Cylinders with Viscous Dissipation," Int. J. Numer. Methods Fluids **11**, 849.
Dean, W. R. (1928), "Fluid motion in a curved channel," Proc. R. Soc. London, Ser. A **121**, 402-420.
Debler, W. R. (1966), "On the analogy between thermal and rotational hydrodynamic stability," J. Fluid Mech. **24**, 165-176.
Debler, W., E. Funer, and B. Schaaf (1969), "Torque and flow patterns in supercritical circular Couette flow," in Proc. of the Twelfth Intl. Cong. of Appl. Mechanics, edited by M. Hetenyi and W. G. Vincenti (Springer-Verlag, Berlin).
Decruppe, J. P., R. Hocquart, T. Wydro, and R. Cressley (1989), "Flow birefringence study at the transition from laminar to Taylor vortex flow," J. Phys. (Paris) **50**, 3371-3394.
Decruppe, J. P., R. Hocquart, and R. Cressley (1991), "Experimental study of the induced flow birefringence of a suspension of rigid particles at the transition from Couette flow to Taylor vortex flow," Rheol. Acta **30**, 575.
Dee, G. (1985a), "Propagation into an unstable state," J. Stat. Phys. **39**, 705-17.
Dee, G. (1985b), "Dynamical properties of propagating front solutions of the amplitude equation," Physica D **15**, 295-304.
Dee, G. and J. S. Langer (1983), "Propagating pattern selection," Phys. Rev. Lett. **50**, 383-386.
Deissler, R. J. (1985), "Noise-sustained structure, intermittency, and the Ginzburg-Landau equation," J. Stat. Phys. **40**, 371.
Deissler, R. J. (1987a), "Turbulent bursts, spots and slugs in a generalized Ginzburg-Landau equation," Phys. Lett. A **120**, 334-340.
Deissler, R. J. (1987b), "Spatially growing waves, intermittency, and convective chaos in an open-flow system," Physica D **25**, 233-260.
Deissler, R. J., "External noise and the origin and dynamics of structure on convectively unstable systems," J. Stat. Phys. (to appear).
Deissler, R. J. and H. R. Brand (1988), "Generation of counterpropagating nonlinear interacting traveling waves by localized noise," Phys. Lett. A **130**, 293-298.
Deissler, R. J., R. E. Ecke, and H. Haucke (1987), "Universal scaling and transient behavior of temporal modes near a Hopf bifurcation: theory and experiment," Phys. Rev. A **36**, 4390-4401.
Demay, Y. and G. Iooss (1984), "Computation of bifurcated solutions for the Couette-Taylor problem, both cylinders rotating," J. Mec. Theor. Appl., spec. suppl., 193-216.
Denier, J. P. and P. Hall (1991), "Nonlinear short wavelength Taylor vortices," Eur. J. Mech. B/Fluids **10 (suppl.)**, 277-282.
Denier, J. P. and P. Hall (1991), "The effect of wall compliance on the Goertler vortex instability," Phys. Fluids A **3**, 2000-2002.

Denier, J. P., S. O. Seddougui, and P. Hall (1991), "On the receptivity problem for Görtler vortices: vortex motions induced by wall roughness," Phil. Trans. Roy. Soc. London, Ser. A **335**, 51.

Deutsch, S. and W. M. Phillips (1979), "The stability of blood cell suspensions to small disturbances in circular Couette flow: experimental results for the Taylor problem," J. Biomech. Eng. **101**, 289-92.

di Meglio, J. -M., D. A. Weitz, and P. M. Chaikin (1987), "Competition between shear-melting and Taylor instabilities in colloidal crystals," Phys. Rev. Lett. **58**, 136-9.

Dinar, N. and H. B. Keller (1985), "Computations of Taylor vortex flows using multigrid continuation methods," Preprint, Caltech.

DiPrima, R. C. (1955), "Application of the Galerkin method to the calculation of the stability of curved flows," Quart. Appl. Math. **13**, 55-62.

DiPrima, R. C. (1959), "The stability of viscous flow between rotating concentric cylinders with a pressure gradient acting round the cylinders," J. Fluid Mech. **6**, 462.

DiPrima, R. C. (1960), "The stability of a viscous fluid between rotating cylinders with an axial flow," J. Fluid Mech. **9**, 621-631.

DiPrima, R. C. (1961a), "Stability of nonrotationally symmetric disturbances for viscous flow between rotating cylinders," Phys. Fluids **4**, 751-755.

DiPrima, R. C. (1961b), "Some variational principles for problems in hydrodynamic and hydromagnetic stability," Quart. Appl. Math. **18**, 375-385.

DiPrima, R. C. (1963a), "A note on the stability of flow in loaded journal bearings," Trans. Amer. Soc. Lub. Engrs. **6**, 249-253.

DiPrima, R. C. (1963b), "Stability of curved flows," J. Appl. Mech. **30**, 486-492.

DiPrima, R. C. (1967), "Vector eigenfunction expansions for the growth of Taylor vortices in the flow between rotating cylinders," in Nonlinear Partial Differential Equations, edited by W. F. Ames (Academic Press, New York), pp. 19-42.

DiPrima, R. C. (1979), "Nonlinear hydrodynamic stability," in Proc. 8th U.S.National Cong. of Applied Mechanics (Western Periodicals), pp. 39-60.

DiPrima, R. C. (1981), "Transition in flow between rotating concentric cylinders," in Transition and Turbulence, edited by R. E. Meyer (Academic Press, New York), pp. 1-24.

DiPrima, R. C. and D. W. Dunn (1956), "The effect of heating and cooling on the stability of the boundary-layer flow of a liquid over a curved surface," J. Aero. Sci. **23**, 913-16.

DiPrima, R. C. and P. M. Eagles (1977), "Amplification rates and torques for Taylor-vortex flows between rotating cylinders," Phys. Fluids **20**, 171-5.

DiPrima, R. C., P. M. Eagles, and B. S. Ng (1984), "The effect of radius ratio on the stability of Couette flow and Taylor vortex flow," Phys. Fluids **27**, 2403-11.

DiPrima, R. C., P. M. Eagles, and J. Sijbrand, "Interaction of axisymmetric and nonaxisymmetric disturbances in the flow between concentric counterrotating cylinder: bifurcations near multiple eigenvalues," Preprint, Rensselaer Poltechnic Institute.

DiPrima, R. C., W. Eckhaus, and L. A. Segel (1971), "Non-linear wave-number interaction in near-critical two-dimensional flows," J. Fluid Mech. **49**, 705-744.

DiPrima, R. C. and R. N. Grannick (1971), "A nonlinear investigation of the stability of flow between counterrotating cylinders," in Instability of Continuous Systems (IUTAM Symposium), edited by H. Leipholz (Springer-Verlag, Berlin, New York), pp. 55-60.

DiPrima, R. C. and G. J. Habetler (1969), "A completeness theorem for non-selfadjoint eigenvalue problems in hydrodynamic stability," Arch. Ration. Mech. Anal. **34**, 218-227.

DiPrima, R. C. and P. Hall (1984), "Complex eigenvalues for the stability of Couette flow," Proc. R. Soc. London, Ser. A **396**, 75-94.

DiPrima, R. C. and C. H. T. Pan (1964), "The stability of flow between concentric cylindrical surfaces with a circular magnetic field," J. Appl. Math. Phys. **15**, 560-567.

DiPrima, R. C. and A. Pridor (1979), "The stability of viscous flow between rotating concentric cylinders with an axial flow," Proc. R. Soc. London, Ser. A **366**, 555-573.

DiPrima, R. C. and E. H. Rogers (1969), "Computing problems in nonlinear hydrodynamic stability," Phys. Fluids Suppl. **11**, 155-165.

DiPrima, R. C. and R. Sani (1965), "The convergence of the Galerkin method for the Taylor-Dean stability problem," Quart. Appl. Math. **23**, 183-187.

DiPrima, R. C. and J. Sijbrand (1982), "Interactions of axisymmetric and nonaxisymmetric disturbances in the flow between concentric rotating cylinders: bifurcations near multiple eigenvalues," in Stability in the Mechanics of Continua, edited by F. H. Schroeder (Springer, Berlin), pp. 383-386.

DiPrima, R. C. and J. T. Stuart (1964), "Nonlinear aspects of instability in flow between rotating cylinders," in Proc. of the 11th International Congress of Applied Mechanics, edited by H. Gortler (Springer-Verlag, Berlin, Heidelberg, New York), pp. 1037-1044.

DiPrima, R. C. and J. T. Stuart (1972a), "Non-local effects in the stability of flow between eccentric rotating cylinders," J. Fluid Mech. **54**, 393-415.

DiPrima, R. C. and J. T. Stuart (1972b), "Flow between eccentric rotating cylinders," J. Lubr. Technol. **94**, 266-274.

DiPrima, R. C. and J. T. Stuart (1974), "Development and effects of super-critical Taylor-vortex flow in a lightly loaded journal bearing," J. Lubr. Technol. **96**, 28-35.

DiPrima, R. C. and J. T. Stuart (1975), "The nonlinear calculation of Taylor-vortex flow between eccentric rotating cylinders," J. Fluid Mech. **67**, 85-111.

DiPrima, R. C. and J. T. Stuart (1983), "Hydrodynamic stability," J. Appl. Mech. **50**, 983-91.

DiPrima, R. C. and H. L. Swinney (1985), "Instabilities and transition in flow between concentric rotating cylinders," in Hydrodynamic Instabilities and the Transition to Turbulence, 2nd ed., edited by H. L. Swinney and J. P. Gollub, Topics in Applied Physics, vol. 45 (Springer-Verlag, Berlin), pp. 139-80.

Doering, C. R., J. D. Gibbon, D. D. Holm, and B. Nicolaenko (1987), "Exact Lyapunov dimension of the universal attractor for the complex Ginzburg-Landau equation," Phys. Rev. Lett. **59**, 2911-2914.

Dominguez-Lerma, M. A., G. Ahlers, and D. S. Cannell (1985), "Effects of 'Kalliroscope' flow visualization particles on rotating Couette-Taylor flow," Phys. Fluids **28**, 1204-6.

Dominquez-Lerma, M. A., G. Ahlers, and D. S. Cannell (1984), "Marginal stability curve and linear growth rate for rotating Couette-Taylor flow and Rayleigh-Benard convection," Phys. Fluids **27**, 856-60.

Dominguez-Lerma, M. A., D. S. Cannell, and G. Ahlers (1986), "Eckhaus boundary and wave-number selection in rotating Couette-Taylor flow," Phys. Rev. A **34**, 4956-70.

Dong-Ha Kim, Tong-Kun Lim, and Unyob Shim (1986), "The characteristics of the spatial modes in Couette flow," New Phys. (Korean Phys. Soc.) **26**, 20-5.

Donnelly, R. J. (1958), "Experiments on the stability of viscous flow between rotating cylinders I. Torque measurements," Proc. R. Soc. London, Ser. A **246**, 312-325.

Donnelly, R. J. (1959), "Experiments on the hydrodynamic instability of helium II between rotating cylinders," Phys. Rev. Lett. **3**, 507-508.

Donnelly, R. J. (1962), "Experimental determination of stability limits," in Proc. Symposia in Applied Mathematics, v.13 Hydrodynamic instability (American Math. Soc.), pp. 41-53.

Donnelly, R. J. (1963), "Experimental confirmation of the Landau law in Couette flow," Phys. Rev. Lett. **10**, 282-284.

Donnelly, R. J. (1964), "Experiments on the stability of viscous flow between rotating cylinders III. Enhancement of stability by modulation," Proc. R. Soc. London, Ser. A **281**, 130-139.

Donnelly, R. J. (1965), "Experiments on the stability of viscous flow between rotating cylinders IV. The ion technique," Proc. R. Soc. London, Ser. A **283**, 509-519.

Donnelly, R. J. (1991), "Taylor-Couette flow: the early days," Physics Today **44**, 32-39.

Donnelly, R. J. (1990), in Nonlinear evolution of spatio-temporal structures in dissipative systems, edited by F. Busse and L. Kramer (Plenum, New York).

Donnelly, R. J. and D. R. Caldwell (1964), "Experiments on the stability of hydromagnetic Couette flow," J. Fluid Mech. **19**, 257-263.

Donnelly, R. J. and D. Fultz (1960a), "Experiments on the stability of viscous flow between rotating cylinders II. Visual observations," Proc. R. Soc. London, Ser. A **258**, 101-123.

Donnelly, R. J. and D. Fultz (1960b), "Experiments on the stability of spiral flow between rotating cylinders," Proc. Natl. Acad. Sci. USA **46**, 1150-1154.

Donnelly, R. J. and M. M. LaMar (1987), "Absolute measurement of the viscosity of classical and quantum fluids by rotating-cylinder viscometers," Phys. Rev. A **36**, 4507-4510.

Donnelly, R. J. and M. M. LaMar (1988), "Flow and stability of helium II between concentric cylinders," J. Fluid Mech. **186**, 163-198.

Donnelly, R. J. and M. Ozima (1960), "Hydromagnetic stability of flow between rotating cylinders," Phys. Rev. Lett. **4**, 497-498.

Donnelly, R. J. and M. Ozima (1962), "Experiments on the stability of flow between rotating cylinders in the presence of a magnetic field," Proc. R. Soc. London, Ser. A **266**, 272-286.

Donnelly, R. J., K. Park, R. Shaw, and R. W. Walden (1980), "Early nonperiodic transitions in Couette flow," Phys. Rev. Lett. **44**, 987-9.

Donnelly, R. J., F. Reif, and H. Suhl (1962), "Enhancement of hydrodynamic stability by modulation," Phys. Rev. Lett. **9**, 363-365.

Donnelly, R. J. and K. Schwarz (1963), "The approach to equilibrium in nonlinear hydrodynamics," Phys. Lett. **5**, 322-324.

Donnelly, R. J. and K. W. Schwarz (1965), "Experiments on the stability of viscous flow between rotating cylinders VI. Finite-amplitude experiments (with an appendix by P. H. Roberts)," Proc. R. Soc. London, Ser. A **283**, 531-546.

Donnelly, R. J. and N. J. Simon (1960), "An empirical torque relation for supercritical flow between rotating cylinders (with an appendix by G.K. Batchelor)," J. Fluid Mech. **7**, 401-418.

Donnelly, R. J. and C. E. Swanson (1986), "Quantum turbulence," J. Fluid Mech. **173**, 387-429.

Donnelly, R. J. and D. T. Tanner (1965), "Experiments on the stability of viscous flow between rotating cylinders V. The theory of the ion technique," Proc. R. Soc. London, Ser. A **283**, 520-530.

Donner, M., M. Siadat, and J. F. Stoltz (1988), "Erythrocyte aggregation: approach by light scattering determination," Biorheology **25**, 367-75.

Doppke, H. and W. Heller (1982), "Analysis by means of light scattering of laminar, vortical, and turbulent flow," J. Rheol. **26**, 199-211.

Drazin, P. G. and W. H. Reid (1981), Hydrodynamic Stability (Cambridge Univ. Press, Cambridge).

Drouin, D. and G. Cognet, "Reduction of skin friction for dilute polymer solutions in circular Couette flow, measured with polarographic probes," in Super Laminar Flow in Journal Bearings: Lyon Symposium:Tribology Series, pp. 61-65.

Drouot, R. (1982), "Diffusion of macromolecules in non-uniform velocity gradients," Rheol. Acta **21**, 635-6.

Dubois-Violette, E. and P. Manneville (1978), "Stability of Couette flow in nematic liquid crystals," J. Fluid Mech. **89**, 273-303.

Duck, P. W. (1979), "Flow induced by a torsionally oscillating wavy cylinder," Quart. J. Mech. Appl. Math. **32**, 73-91.

Dufaux, J., D. Quemada, and P. Mills (1980), "Determination of rheological properties of red blood cells by Couette viscometry," Rev. Phys. Appl. **15**, 1367-74.

Duffy, B. R. (1980), "Flow of a liquid with an anisotropic viscosity tensor: some axisymmetric flows," J. Non-Newtonian Fluid Mech. **7**, 359-67.

Dungan, S. R. and H. Brenner (1988), "Sedimentation and dispersion of non-neutrally buoyant Brownian particles in cellular circulatory flows simulating local fluid agitation," Phys. Rev. A **38**, 3601-8.

Dunn, F. (1985), "Cellular inactivation by heat and shear," Radiat. Environ. Biophys. **24**, 131-9.

Duty, R. L. and W. H. Reid (1964), "On the stability of viscous flow between rotating cylinders. Part 1. Asymptotic analysis," J. Fluid Mech. **20**, 81-94.

Duvivier, C., J. Didelon, J. P. Arnould, J. M. Zahm, E. Puchelle, C. Kopp, and B. Obrecht (1984), "A new viscoelastometer for studying the rheological properties of bronchial mucus in clinical practice," Biorheology, suppl. 1, 119-22.

Eagles, P. M. (1971), "On the stability of Taylor vortices by fifth-order amplitude expansions," J. Fluid Mech. **49**, 529-550.

Eagles, P. M. (1972), "On the stability of slowly varying flow between concentric cylinders," Proc. R. Soc. London, Ser. A **355**, 209-24.

Eagles, P. M. (1974), "On the torque of wavy vortices," J. Fluid Mech. **62**, 1-9.

Eagles, P. M. (1985a), "Development of Taylor-Couette flow on an intermediate timescale," Proc. R. Soc. London, Ser. A **398**, 289-305.

Eagles, P. M. (1985b), "Ramped Taylor-Couette flow," Phys. Rev. A **31**, 1955-6.

Eagles, P. M. and K. Eames (1983), "Taylor vortices between almost cylindrical boundaries," J. Eng. Math. **17**, 263-80.

Eagles, P. M., J. T. Stuart, and R. C. Diprima (1978), "The effects of eccentricity on torque and load in Taylor-vortex flow," J. Fluid Mech. **87**, 209-31.

Easthope, P. L. and D. E. Brooks (1980), "A comparison of rheological constitutive functions for whole human blood," Biorheology **17**, 235-47.

Ebert, F., S. U. Schoffel, and G. D. Catalano (1986), "Comment on 'A prediction of particle behavior via the Basset-Boussinesq-Oseen equation'," AIAA J. **24**, 1403-5.

Eckhaus, W. (1962a), "Problemes non lineaires dans las theorie de la stabilite," J. Mecanique **1**, 49-77.

Eckhaus, W. (1962b), "Problemes non lineaires de stabilite dans un espace a deux dimensions. Premiere Partie: solutions periodiques," J. Mecanique **1**, 413-438.

Eckhaus, W. (1963), "Problemes non lineaires de stabilite dans un espace a deux dimensions. Deuxieme partie: stabilite des solutions periodiques," J. Mecanique **2**, 153-172.

Eckhaus, W. (1965), Studies in Nonlinear Stability Theory (Springer-Verlag, Berlin).

Eckmann, J. -P. and C. E. Wayne (1991), "Propagating Fronts and the Center Manifold Theorem," Commun. Math. Phys. **136**, 285.

Eckstein, E. C., D. G. Bailey, and A. H. Shapiro (1977), "Self-diffusion of particles in shear flow of a suspension," J. Fluid Mech. **79**, 191-208.

Economides, D. G. and G. Moir (1981), "Taylor vortices and the Goldreich-Schubert instability," Geophys. Astrophys. Fluid Dyn. **16**, 299-317.

Edwards, W. S. (1990a), "Linear spirals in the finite Couette-Taylor problem," in Instability and Transition, vol. II, edited by M. Y. Hussaini and R. G. Voigt (Springer, New York), pp. 408-425.

Edwards, W. S. (1990b), "New stability analyses for the Couette-Taylor problem," Ph.D. Thesis, Univ. of Texas at Austin.

Edwards, W. S., S. R. Beane, and S. Varma (1991), "Onset of wavy vortices in the finite-length Couette-Taylor problem," Phys. Fluids A **3**, 1510.

Edwards, W. S., R. P. Tagg, B. C. Dornblaser, H. L. Swinney, and L. S. Tuckerman (1991), "Periodic traveling waves with nonperiodic pressure," Eur. J. Mech. B/Fluids **10 (suppl.)**, 205-210.

Edwards, W. S., R. P. Tagg, B. C. Dornblaser, H. L. Swinney, and L. S. Tuckerman, "Erratum: Periodic traveling waves with nonperiodic pressure," European Journal of Mechanics/B Fluids (to appear).

Eisenberg, E., C. W. Tobias, and C. R. Wilke (1955), Chem. Eng. Prog., Symp. Ser. **51**, 1.

El-Dujaily, M. J. and F. R. Mobbs (1990), "The effect of end walls on subcritical flow between concentric and eccentric rotating cylinders," Int. J. Heat Fluid Flow **11**, 72.

Eldabe, N. T. and A. A. Hassan (1991), "Non-Newtonian-flow formation in Couette motion in magnetohydrodynamics with time-varying suction," Can. J. Phys. **69**, 75.

Elezgaray, J. and A. Arneodo (1991), "Modeling reaction-diffusion pattern formation in the Couette flow reactor," J. Chem. Phys. **95**, 323.

Elleaume, P., J. P. Hulin, and B. Perrin (1978), "Hydrodynamic instabilities in the rotating Couette flow of superfluid helium," J. Phys. (Paris) Colloq. **39**, C6/163-4.

Elliott, L. (1973), "Stability of a viscous fluid between rotating cylinders with an axial flow and pressure gradient round the cylinders," Phys. Fluids **16**, 577-580.

Erenberg, V. B. and V. N. Pokrovskii (1981), "Inhomogeneous shear flows of linear polymers," Inzh.-Fiz. Zh. **41**, 449-456 [J. Eng. Phys. **41**, 966-71 (1981)].

Espurz, A., A. Espurz Nieto, and M. V. Fabian (1982), "Magnetoviscous behaviour of a magnetite ferrofluid," An. Fis. Ser. B **78**, 239-44.

Essabbah, H. and C. Lacombe (1980), "Rheological study in transient flow of polycythaemic blood," J. Biophys. Med. Nucl. **4**, 239-42.

Evans, D. J. and H. J. M. Hanley (1981), "Thermodynamic fluctuation theory for shear flow," Physica A **108A**, 567-74.

Evans, M. W. (1989), "Group theoretical statistical mechanics applied to Couette flow," Chem. Phys. **132**, 1.

Evans, M. W. and D. M. Heyes (1988), "Correlation functions in Couette flow from group theory and molecular dynamics," Mol. Phys. **65**, 1441.

Evans, M. W. and D. M. Heyes (1990), "Correlation functions in non-Newtonian Couette flow. A group theory and molecular dynamics approach," J. Chem. Soc., Faraday Trans. **86**, 1041.

Fabisiak, W. and C. R. Huang (1980), "Mathematical analysis of the hysteresis rheogram of human blood," Biorheology **17**, 391-6.

Fage, A. (1938), "The influence of wall oscillations, wall rotation and entry eddies on the breakdown of laminar flow in an annular pipe," Proc. R. Soc. London, Ser. A **165**, 513-17.

Farcy, A. and T. Alziary de Roquefort (1988), "Chebyshev pseudospectral solution of the incompressible Navier-Stokes equations in curvilinear domains," Comput. Fluids **16**, 459-73.

Farmer, J. D. and J. J. Sidorowich (1987), "Predicting chaotic time series," Phys. Rev. Lett. **59**, 845-8.

Fasel, H. and O. Booz (1984), "Numerical investigation of supercritical Taylor-vortex flow for a wide gap," J. Fluid Mech. **138**, 21-52.

Fauve, S. (1987), "Large scale instabilities of cellular flows," in <u>Instabilities and Nonequilibrium Structures</u>, edited by E. Tirapegui and D. Villarroel (Reidel, Dordrecht), pp. 63-88.

Fauve, S., E. W. Bolton, and M. E. Brachet (1987), "Nonlinear oscillatory convection: a quantitative phase dynamics approach," Physica D **29D**, 202-14.

Fauve, S., S. Douady, and O. Thual (1991), "Drift instabilities of cellular patterns," J. Phys. (Paris) II. **1**, 311.

Fauve, S. and O. Thual (1990a), "Solitary waves generated by subcritical instabilities in dissipative systems," Phys. Rev. Lett. **64**, 282.

Fauve, S. and O. Thual (1990b), "Localized structures in cellular flows," Journal of Physics: Condensed Matter **2 (suppl.)**, 465.

Fearn, D. R. and W. S. Weiglhofer (1991), "Magnetic instabilities in rapidly rotating spherical geometries I. from cylinders to spheres," Geophys. Astrophys. Fluid Dyn. **56**, 159.

Feke, D. L. and W. R. Schowalter (1985), "The influence of Brownian diffusion on binary flow-induced collision rates in colloidal dispersions," J. Colloid Interface Sci. **106**, 203-4.

Fenstermacher, P. R. (1979), "Laser doppler velocimetry of the onset of chaos in Taylor vortex flow," Ph.D. Thesis, City College of the City University of New York.

Fenstermacher, P. R., H. L. Swinney, and J. P. Gollub (1979), "Dynamical instabilities and the transition to chaotic Taylor vortex flow," J. Fluid Mech. **94**, 103-28.

Fenstermacher, P. R., H. L. Swinney, S. V. Benson, and J. P. Gollub (1979), "Bifurcations to periodic, quasiperiodic, and chaotic regimes in rotating and convecting fluids," Ann. N. Y. Acad. Sci. **316**, 652-666.

Finlay, W. H. (1989), "Perturbation expansion and weakly nonlinear analysis for two-dimensional vortices in curved or rotating channels," Phys. Fluids A **1**, 854-860.

Finaly, W. H. (1990), "Transition to oscillatory motion in rotating channel flow," J. Fluid Mech. **215**, 209-227.

Finlay, W. H., J. B. Keller, and J. H. Ferziger (1988), "Instability and transition in curved channel flow," J. Fluid Mech. **194**, 417-56.

Finlay, W. H. and K. Nandakumar (1990), "Onset of two-dimensional cellular flow in finite curved channels of large aspect ratio," Phys. Fluids A **2**, 1163-1174.

Fletcher, D. F., S. J. Maskell, and M. A. Patrick (1985), "Heat and mass transfer computations for laminar flow in a axisymmetric sudden expansion," Comput. Fluids **13**, 207-21.

Fourtune, L., I. Mutabazi, and C. D. Andereck (1992), "A model of the time-dependence disappearance in the flow pattern in the Taylor-Dean system,," in Ordered and Turbulent Patterns in Taylor-Couette flow, edited by C. D. Andereck and F. H. F.Hayot (Plenum Publishing Corporation).

Frank, G. and R. Meyer-Spasche (1981), "Computation of transitions in Taylor vortex flows," Z. Angew. Math. Phys. **32**, 710-20.

Fraser, A. (. (1984), "Low dimensional chaos in a hydrodynamic system," Ph.D. Thesis, Univ. of Texas, Austin.

Frederking, T. H. K., S. C. Soloski, and Y. I. Kim (1985), "Forced convection influence on cryostatic stability deduced from transverse fluid motion," in Proceedings of the 9th International Conference on Magnet Technology. MT-9 1985, edited by C. Marinucci and P. Weymuth (Swiss Inst. Nucl. Res., Villigen, Switzerland), pp. 790-3.

Freitas, C. J. and R. L. Street (1988), "Non-linear transient phenomena in a complex recirculating flow: a numerical investigation," Int. J. Numer. Methods Fluids **8**, 769-802.

Frene, J. and M. Godet (1971), Tribology **4**, 216.

Friebe, H. W. (1976), "The stability of very dilute long chain polymers in Couette flow," Rheol. Acta **15**, 329-55.

Frisch, U. and S. A. Orszag (1990), "Turbulence: challenges for theory and experiment," Phys. Today, 24-32.

Fu, Y. B. and P. Hall (1991), "Nonlinear development and scondary instability of large-amplitude Görtler vortices in hypersonic boundary layers," Eur. J. Mech. B/Fluids **10 (suppl.)**, 283-288.

Fujimura, K. (1990), "Linear stability of a rotating film flow attached inside a circular cylinder," Phys. Fluids A **2**, 1182.

Fujisawa, N. and H. Shirai (1986), "On the stability of turbulent wall jets along concave surfaces," Bull. JSME **29**, 3761-6.

Fuller, G. G., J. M. Rallison, R. L. Schmidt, and L. G. Leal (1980), "The measurement of velocity gradients in laminar flow by homodyne light-scattering spectroscopy," J. Fluid Mech. **100**, 555-75.

Fung, L., K. Nandakumar, and J. H. Masliyah (1987), "Bifurcation phenomena and cellular-pattern evolution in mixed-convection heat transfer," J. Fluid Mech. **177**, 339-57.

Gadala-maria, F. and A. Acrvos (1980), "Shear-induced structure in a concentrated suspension of solid spheres," J. Rheol. **24**, 799-814.

Galkin, V. S. and V. I. Nosik (1987), "The principle of material frame-indifference and cylindrical Couette flow of a rarefied gas," Prikl. Mat. Mekh. **51**, 957-61 [Appl. Math. Mech. **51**, 736-9 (1987)].

Gardiner, S. R. M. and R. H. Sabersky (1978), "Heat transfer in an annular gap," Int. J. Heat Mass Transf. **21**, 1459-66.

Gaster, M., "On transition to turbulence in boundary layers," in Turbulence And Chaotic Phenomena In Fluids. Proceedings Of The International Symposium, edited by T. Tatsumi (North Holland, Amsterdam, Netherlands), pp. 99-106.

Gazley, C., Jr. (1958), "Heat-transfer characteristics of the rotational and axial flow between rotating cylinders," Trans. ASME **80**, 79-90.

Gebhardt, T. and S. Grossmann (1991), "Stokes modes in Taylor-Couette geometry," Z. Naturforschung., Teil A **46**, 669.

Georgescu, A. and I. Oprea (1988), "Bifurcation (catastrophe) surfaces in multiparametric eigenvalue problems in hydromagnetic stability theory," Bul. Inst. Politeh. Bucur. Constr. Mas. **50**, 9-12.

Gerdts, U. (1985), Ph.D. Thesis, Universität Kiel, Kiel, Germany.

Glasgow, L. A. and R. H. Luecke (1977), "Stability of centrifugally stratified helical Couette flow," Ind. Eng. Chem., Fundam. **16**, 366-71.

Goh, C. J., N. Phan-thien, and J. D. Atkinson (1985), "On migration effects in circular Couette flow," J. Chem. Phys. **81**, 6259-65.

Gol'dshtik, M. A., E. M. Zhdanova, and V. N. Shtern (1984), "Spontaneous twisting of a submerged jet," Dok. Akad. Nauk SSSR **277**, 815-18 [Sov. Phys. Dokl. **29**, 615-17 (1984)].

Gollub, J. P. and M. H. Freilich (1974a), in Fluctuations, Instabilities, and Phase Transitions (NATO Advanced Study Institute), edited by T. Riste (Plenum, New York).

Gollub, J. P. and M. H. Freilich (1974b), "Optical heterodyne study of the Taylor instability in a rotating fluid," Phys. Rev. Lett. **33**, 1465-1468.

Gollub, J. P. and M. H. Freilich (1976), "Optical heterodyne test of perturbation expansions for the Taylor instability," Phys. Fluids **19**, 618-626.

Gollub, J. P. and H. L. Swinney (1975), "Onset of turbulence in a rotating fluid," Phys. Rev. Lett. **35**, 927-930.

Golovin, A. M. (1978), "A phenomenological theory of turbulence," Fluid Mech. - Sov. Res. **7**, 161-7.

Golubitsky, M. and W. F. Langford (1988), "Pattern formation and bistability in flow between counterrotating cylinders," Physica D **32D**, 362-92.

Golubitsky, M., I. Stewart, D. G. Schaeffer, and W. F. Langford (1988), "Case study 6: the Taylor-Couette system," in Singularities and Groups in Birfurcation Theory, vol. 2 (Springer-Verlag, New York), pp. 485-512.

Golubitsky, M. and I. Stewart (1986), "Symmetry and stability in Taylor-Couette flow," SIAM J. Math. Anal. **17**, 249-88.

Gorman, M., L. A. Reith, and H. L. Swinney (1980), "Modulation patterns, multiple frequencies, and other phenomena in cirular Couette flow," Ann. N. Y. Acad. Sci. **357**, 10-21.

Gorman, M. and H. L. Swinney (1979), "Visual observation of the second characteristic mode in a quasiperiodic flow," Phys. Rev. Lett. **43**, 1871-5.

Gorman, M. and H. L. Swinney (1981), "Recent results on instabilities and turbulence in Couette flow," Physica A **106A**, 123-7.

Gorman, M. and H. L. Swinney (1982), "Spatial and temporal characteristics of modulated waves in the circular Couette system," J. Fluid Mech. **117**, 123-42.

Gorman, M., H. L. Swinney, and D. A. Rand (1981), "Doubly periodic circular Couette flow: experiments compared with predictions from dynamics and symmetry," Phys. Rev. Lett. **46**, 992-5.

Görtler, H. (1940), "Über eine dreidimensionale Instabilität laminarer Grenzschichten an konkaven Wänden," Nach. Ges. Wiss. Göttingen, Math.-phys. **1**, 1-26 [Tech. Memor. Nat. Adv. Com. Aero., Wash., No. 1375].

Görtler, H. (1959), "Über eine Analogie zwischen den Instabilitäten laminarer Grenzschichtströmungen an konkaven Wänden und an erwärmten Wänden," Ingen.-Arch. **28**, 71-78.

Gotoh, T. and S. Kuwabara (1982), "A nonlinear analysis on stability of the Taylor-Couette flow," J. Phys. Soc. Jpn. **51**, 1647-54.

Graham, R. and J. A. Domaradzki (1982), "Local amplitude equation of Taylor vortices and its boundary condition," Phys. Rev. A **26**, 1572-9.

Gravas, N. and B. W. Martin (1978), "Instability of viscous axial flow in annuli having a rotating inner cylinder," J. Fluid Mech. **86**, 385-94.

Graziani, G. (1990), "Green's function method for axisymmetric flows: analysis of the Taylor-Couette flow," Computational Mechanics **7**, 77.

Greave, P. L., R. I. Grosvenor, and B. W. Martin (1983), "Factors affecting the stability of viscous axial flow in annuli with a rotating inner cylinder," Int. J. Heat Fluid Flow **4**, 187-197.

Green, J. and W. M. Jones (1982), "Couette flow of dilute solutions of macromolecules: embryo cells and overstability," J. Fluid Mech. **119**, 491-505.

Greenspan, H. P. and L. N. Howard (1963), "On a time-dependent motion of a rotating fluid," J. Fluid Mech. **17**, 385-404.

Greenstein, T. and T. J. Som (1976), "Frictional force exerted on a slowly rotating eccentrically positioned sphere inside a circular cylinder," Phys. Fluids **19**, 161-162.

Griffiths, R. W. (1987), "Effects of Earth's rotation on convection in magma chambers," Earth Planet. Sci. Lett. **85**, 525-36.

Gregory, N. and W. S. Walker (1951), "The effect on transition of isolated surface excrescences in the boundary layer," Rep. Memor. Aero. Res. Coun., Lond., Report No. 2779.

Grifoll, J., X. Farriol, and F. Giralt (1986), "Mass transfer at smooth and rough surfaces in a circular Couette flow," Int. J. Heat Mass Transf. **29**, 1911-18.

Gu, Z. H. and T. Z. Fahidy (1985a), "Visualization of flow patterns in axial flow between horizontal coaxial rotating cylinders," Can. J. Chem. Eng. **63**, 14-21.

Gu, Z. H. and T. Z. Fahidy (1985b), "Characteristics of Taylor vortex structure in combined axial and rotating flow," Can. J. Chem. Eng. **63**, 710-15.

Gu, Z. H. and T. Z. Fahidy (1986), "The effect of geometric parameters on the structure of combined axial and Taylor-vortex flow," Can. J. Chem. Eng. **64**, 185-9.

Guckenheimer, J. (1986), "Strange attractors in fluids: another view," Ann. Rev. Fluid Mech. **18**, 15-31.

Guesbaoui, H. (1978), "Contribution a l'etude de la stabilite de l'ecoulement de Couette entre cylindres coaxiaux," Thesis: Doctorat de 3eme Cycle, INP Lorraine (France).
Guillope, C., D. Joseph, K. Nguyen, and F. Rosso (1987), "Nonlinear stability of rotating flow of two fluids," J. Mec. Theor. Appl. **6**, 619-45.
Guo, Y. and W. H. Finlay (1991), "Splitting, merging and wavelength selection of vortices in curved and/or rotating channel flow due to Eckhaus instability," J. Fluid Mech. **228**, 661.
Gupta, R. K. and S. C. Gupta (1977), "Couette flow of a dusty gas between two infinite coaxial cylinders," Proc. Indian Natl. Sci. Acad., Part A **43**, 56-67.
Gupta, S. C. and P. C. Jain (1981), "Couette flow of a viscous electrically conducting fluid in a porous annulus," Def. Sci. **31**, 53-61.
Haas, P. A. (1987), "Turbulent dispersion of aqueous drops in organic liquids," AIChE J. **33**, 987-95.
Haas, R. and K. Bühler (1989), "Einfluss nichtnewtonischer Stoffeigenschaften auf die Taylor-Wirbelströmung," Rheol. Acta **28**, 402-413.
Hacisslamoglu, M. and J. Langlinais (1990), "Discussion of flow of a power-law fluid in an eccentric annulus," SPE drilling engineering **5**, 95.
Hagerty, W. (1950), "Use of an optical property of glycerine-water solutions to study viscous fluid flow problems," J. Appl. Mech. **17**, 54-58.
Haken, H. (1983a), <u>Synergetics - An Introduction: Nonequilibrium Phase Transitions and Self-Organization in Physics, Chemistry and Biology</u>, 3rd ed. (Springer-Verlag, Berlin).
Haken, H. (1983b), <u>Advanced Synergetics: Instability Hierarchies of Self-Organizing Systems and Devices</u> (Springer-Verlag, Berlin).
Hakim, V., P. Jacobsen, and Y. Pomeau (1990), "Fronts vs. solitary waves in nonequilibrium system," Europhys. Lett. **11**, 19.
Hakim, V. and Y. Pomeau (1991), "On stable localized structures and subcritical instabilities," Eur. J. Mech B/Fluids **10 (suppl.)**, 137-143.
Hall, P. (1975), "The stability of unsteady cylinder flows," J. Fluid Mech. **67**, 29-63.
Hall, P. (1980a), "Centrifugal instabilities of circumferential flows in finite cylinders: nonlinear theory," Proc. R. Soc. London, Ser. A **372**, 317-56.
Hall, P. (1980b), "Centrifugal instabilities in finite containers: a periodic model," J. Fluid Mech. **99**, 575-96.
Hall, P. (1981), "Centrifugal instability of a Stokes layer: subharmonic destabilization of the Taylor vortex mode," J. Fluid Mech. **105**, 523-30.
Hall, P. (1982a), "Taylor-Görtler vortices in fully developed or boundary-layer flows: linear theory," J. Fluid Mech. **124**, 475-94.
Hall, P. (1982b), "Centrifugal instabilities of circumferential flows in finite cylinders: the wide gap problem," Proc. R. Soc. London, Ser. A **384**, 359-79.
Hall, P. (1983a), "On the nonlinear stability of slowly varying time-dependent viscous flows," J. Fluid Mech. **126**, 357-68.
Hall, P. (1983b), "The linear development of Görtler vortices in growing boundary layers," J. Fluid Mech. **137**, 363-384.
Hall, P. (1983c), "On the stability of the unsteady boundary layer on a cylinder oscillating transversely in a viscous fluid," Report No. ICASE 83-45 (ICASE, NASA Langley Research Ctr., Hampton, VA).
Hall, P. (1983d), "The evolution equations for Taylor vortices in the small gap limit," Report No. ICASE 83-55 (ICASE, NASA Langley Research Ctr., Hampton, VA).
Hall, P. (1984), "Evolution equations for Taylor vortices in the small-gap limit," Phys. Rev. A **29**, 2921-3.
Hall, P. (1985), "Instability of time-periodic flows," Report No. NASA CR-178009, ICASE Rep. No. 85-46 (ICASE, NASA Langley Research Ctr., Hampton, VA).
Hall, P. (1990), "Gortler vortices in gowing boundary layers: The leading edge receptivity problem, linear growth adn the nonlinear breakdown stage," Mathematika **37**, 151.
Hall, P. and W. D. Lakin (1988), "The fully nonlinear development of Görtler vortices in growing boundary layers," Proc. R. Soc. London, Ser. A **415**, 421-444.
Hall, P. and M. Malik (1989), "The growth of Gortler vortices in compressible boundary layers," J. Eng. Math. **23**, 239.
Hall, P. and S. Seddougui (1989), "On the onset of three-dimensionality and time-dependence in Görtler vortices," J. Fluid Mech. **204**, 405-20.
Hall, P. and F. T. Smith (1988), "The nonlinear interaction of Tollmien-Schlichting waves and Taylor-Görtler vortices in curved channel flows," Proc. R. Soc. London, Ser. A **417**, 255-82.
Hamersma, P. J., J. ellenberger, and J. M. H. Fortuin (1982), "A three-parameter model describing the behaviour of a viscoelastic liquid in a tangential annular flow," Rheol. Acta **21**, 705-12.
Hamersma, P. J., J. Ellenberger, and J. M. H. Fortuin (1983), "Derivation of a three-parameter model describing the results of steady-state shear stress-shear rate measurements of

viscoelastic polymer solutions, with different types of equipment and in a seven-decade shear-rate range," Chem. Eng. Sci. **38**, 819-25.

Hammer, P. W. (1991), "Bifurcations of Taylor-Couette flow subject to a nonaxisymmetric Coriolis force," Ph.D. Thesis, Univ. of Oregon.

Hammer, P. W., R. J. Wiener, and R. J. Donnelly, "Bifurcation phenomena in Taylor-Couette flow subject to a Coriolis force," in Ordered and Turbulent Patterns in Taylor-Couette Flow, edited by C. D. Andereck and F. Hayot, Proceedings of a NATO Advanced Research Workshop (Plenum).

Hammer, P. W., R. J. Wiener, C. E. Swanson, and R. J. Donnelly (1991), "Bifurcations of Taylor-Couette flow subject to an external Coriolis force," (preprint).

Hämmerlin, G. (1955), "Über das Eigenwertproblem der dreidimensionalen Instabilität laminarer Grenzschichten an konkaven Wänden," J. Rat. Mech. Anal. **4**, 279-321.

Hämmerlin, G. (1956), "Zur Theorie der dreidimensionalen Instabilität laminarer Grenzschichten," Z. Angew. Math. Phys. **7**, 156-64.

Hanks, R. W. (1983), "Couette viscometry of Casson fluids," J. Rheol. **27**, 1-6.

Harada, I. (1980), "A numerical study of weakly compressible rotating flows in a gas centrifuge," Nucl. Sci. Eng. **73**, 225-41.

Harris, D. L. and W. H. Reid (1964), "On the stability of viscous flow between rotating cylinders. Part. 2. Numerical analysis," J. Fluid Mech. **20**, 95-101.

Hasoon, M. A. and B. W. Martin (1977), "The stability of viscous axial flow of an annulus with a rotating inner cylinder," Proc. R. Soc. London, Ser. A **352**, 351-380.

Hayafuji, H. and K. Wada (1986), "Some characteristics of axial laminar Couette flow contained between circular double tubes. II. Theoretical analysis for concentric circular double tubes model," Rep. Fac. Sci. Technol. Meijo Univ., 54-8.

Hayafuji, H. and K. Wada (1987), "Some characteristics of axial Couette flow contained between circular crown section double tubes. II. Results of sample calculation," Rep. Fac. Sci. Technol. Meijo Univ., 46-52.

Hayashi, O., K. Jinda, T. Takahashi, and H. Ueno (1979), "The viscosity of liquid butadiene-acrylonitrile copolymers," Kobunshi Ronbunshu **36**, 567-73.

Hegseth, J. (1990), "Spatiotemporal patterns in flow between two independently rotating cylinders," Ph.D. Thesis, Ohio State University.

Hegseth, J. J., C. D. Andereck, F. Hayot, and Y. Pomeau (1989), "Spiral turbulence and phase dynamics," Phys. Rev. Lett. **62**, 257-60.

Hegseth, J. J., C. D. Andereck, F. Hayot, and Y. Pomeau (1991), "Spiral turbulence: development and steady state properties," Eur. J. Mech. B/Fluids **10 (suppl.)**, 221-226.

Heinrichs, R., D. S. Cannell, and G. Ahlers (1986), "Effects of finite geometry on the wavenumber of Taylor vortex flow," Phys. Rev. Lett. **56**, 1794-7.

Heinrichs, R. M., D. S. Cannell, G. Ahlers, and M. Jefferson (1988), "Experimental test of the perturbation expansion for the Taylor instability at various wavenumbers," Phys. Fluids **31**, 250-5.

Hendriks, F. and A. Aviram (1982), "Use of zinc iodide solutions in flow research," Rev. Sci. Instrum. **53**, 75-8.

Herbert, T. (1979), "Higher eigenstates of Görtler vortices," in Recent Developments In Theoretical And Experimental Fluid Mechanics, edited by U. Muller, K. G. Roesner, and B. Schmidt (Springer-Verlag, Berlin), pp. 322-30.

Herbert, T. (1980), "Numerical studies on nonlinear hydrodynamic stability by computer-extended perturbation series," in Proc. 7th International Conf. on Numerical Methods in Fluid Dynamics.

Herbert, T. (1983), "On perturbation methods in nonlinear stability theory," J. Fluid Mech. **126**, 167-186.

Herbert, T., "Nonlinear effects in hydrodynamic stability," in Special Course On Stability and Transition of Laminar Flow (Agard-R-709) (Agard, NEUILLY-SUR-SEINE, FRANCE).

Herbst, L., H. Hoffmann, J. Kalus, H. Thurn, and K. Ibel (1985), "Relaxation of aligned rod-like micelles," in Neutron Scattering in the 'Nineties. Proceedings of a Conference (IAEA, Vienna, Austria), pp. 501-6.

Herron, I. H. (1985a), "Exchange of stabilities for Görtler flow," SIAM J. Appl. Math. **45**, 775-9.

Herron, I. H. (1985b), "Linear vs energy stability for time-periodic flows," Phys. Fluids **28**, 2298-9.

Herron, I. H. (1991), "Stability criteria for flow along a convex wall," Phys. Fluids A **3**, 1825-1827.

Hess, S. (1984), "Influence of an orienting field on the viscosity of a molecular liquid," Z. Naturforsch., Teil A **39A**, 22-6.

Heuser, G. and R. Opitz (1980), "A Couette viscometer for short time shearing of blood," Biorheology **17**, 17-24.

Hill, N. A. (1988), "Numerical studies of 'side-by-side' and other modes for the Taylor problem in a finite annulus," Comput. Fluids **16**, 445-58.

Hille, P., R. Vehrenkamp, and E. O. Schulz-DuBois (1985), "The development of primary and secondary flow in a curved square duct," J. Fluid Mech. **151**, 219-241.

Hiremath, P. S. and B. G. Kittur (1989), "A note on flow between longitudinally corrugated cylinders," Appl. Sci. Res. **46**, 379.

Hirst, D., "The aspect ratio dependence of the attractor dimension in Taylor-Couette flow," Ph.D. Thesis, Univ. of Texas, Austin.

Ho, B. P. and L. G. Leal (1976), "Migration of rigid spheres in a two-dimensional unidirectional shear flow of a second-order fluid," J. Fluid Mech. **76**, 783-99.

Ho, C. Y., J. L. Nardacci, and A. H. Nissan (1964), AIChE J. **10**, 194.

Hoare, M., T. J. Narendranathan, J. R. Flint, D. Heywood-Waddington, D. J. Bell, and P. Dunnill (1982), "Disruption of protein precipitates during shear in Couette flow and in pumps," Ind. Eng. Chem., Fundam. **21**, 402-6.

Hocking, L. M. (1981), "The instability of flow in the narrow gap between two prolate spheroids. Ii. Arbitrary axis ratio," Quart. J. Mech. Appl. Math. **34**, 475-88.

Hocking, L. M. and J. Skiepko (1981), "The instability of flow in the narrow gap between two prolate spheroids. I. Small axis ratio," Quart. J. Mech. Appl. Math. **34**, 57-68.

Hohenberg, P. C. (1979), "Hydrodynamic instabilities and turbulence (and measurement techniques)," in <u>Light Scattering in Solids</u>, edited by J. L. Birman, H. Z. Cummins, and K. K. Rebane (Plenum, New York), pp. 23-7.

Hohenberg, P. C. (1985), "Nonequilibrium steady states with spatial patterns," Phys. Scr. **T9**, 93-4.

Hohenberg, P. C. and M. C. Cross (1987), "An introduction to pattern formation in nonequilibrium systems," in <u>Fluctuations and Stochastic Phenomena in Condensed Matter. Proceedings of the Sitges Conference on Statistical Mechanics</u>, edited by L. Garrido (Springer-Verlag, Berlin), pp. 55-92.

Hohenberg, P. C. and J. S. Langer (1982), "Nonequilibrium phenomena: outlines and bibliographies if a workshop," J. Stat. Phys. **28**, 193-226.

Holderied, M., L. Schwab, and K. Stierstadt (1988), "Rotational viscosity of ferrofluids and the Taylor instability in a magnetic field," Z. Phys. B **70**, 431-3.

Hollis Hallet, A. C. (1953), "Experiments with a rotating cylinder viscometer in liquid helium II," Proc. Cambridge Philos. Soc. **49**, 717-727.

Hong, S. H. (1976), "Transformation of the perturbation equations for the analysis of magnetohydrodynamic Couette flow stability," Z. Angew. Math. Phys. **27**, 483-5.

Horseman, N. J. and P. Bassom (1990), "Long-wave/short-wave interactions in flow between concentric cylinders," J. Fluid Mech. **215**, 525.

Hsieh, D. Y. and F. Chen (1984), "On model study of Couette flow," Phys. Fluids **27**, 321-2.

Huggins, E. R. and D. P. Bacon (1980), "Vortex currents and hydrodynamic instability in Taylor cells," Phys. Rev. A **21**, 1327-30.

Hughes, T. H. and W. H. Reid (1964), "The effect of a transverse pressure gradient on the stability of Couette flow," J. Appl. Math. Phys. **15**, 573-581.

Hughes, T. H. and W. H. Reid (1968), "The stability of spiral flow between rotating cylinders," Phil. Trans. Roy. Soc. London, Ser. A **263**, 57-91.

Hung, W. L. (1978), "Stability of Couette flow by the method of energy," M.S. Thesis, Univ. of Minnesota.

Hung, W. L., D. D. Joseph, and B. R. Munson (1972), "Global stability of spiral flow. Part 2," J. Fluid Mech. **51**, 593-612.

Hunter, R. J. and J. Frayne (1979), "Couette flow behavior of coagulated colloidal suspensions. IV. Effect of viscosity of the suspension medium," J. Colloid Interface Sci. **71**, 30-8.

Husband, D. M. and F. Gadala-maria (1987), "Anisotropic particle distribution in dilute suspensions of solid spheres in cylindrical Couette flow," J. Rheol. **31**, 95-110.

Hussain, A. K. M. F. (1986), "Coherent structures and turbulence," J. Fluid Mech. **173**, 303-56.

Hussaini, M. Y. and R. G. Voigt (1990), Eds., <u>Instability and Transition</u>, vols. I and II (Springer, New York).

Hyun, J. M. and H. S. Kwak (1989), "Flow of a Double-Diffusive Stratified Fluid in a Differentially-Rotating Cylinder," Geophys. Astrophys. Fluid Dyn. **46**, 203.

Infeld, E. and G. Rowlands (1990), <u>Nonlinear Waves, Solitons and Chaos</u> (Cambridge Univ. Press, Cambridge, England).

Iooss, G. (1984), "Bifurcation and transition to turbulence in hydrodynamics," in <u>Bifurcation Theory and Apllications: lectures given at the 2nd 1983 Session of the Centro Internationale Matematico Estivo (C.I.M.E.)</u>, edited by L. Salvadori (Springer-Verlag, Berlin), pp. 152-201.

Iooss, G. (1986a), "Secondary bifurcations of Taylor vortices into wavy inflow or outflow boundaries," J. Fluid Mech. **173**, 273-88.

Iooss, G. (1986b), "Recent progresses in the Couette-Taylor problem," Preprint, Universite de Nice.

Iooss, G. (1987), "Reduction of the dynamics of a bifurcation problem using normal forms and symmetries," in Instabilities and Nonequilibrium Structures, edited by E. Tirapegui and D. Villarroel (Reidel, Dordrecht), pp. 3-40.

Iooss, G., P. Coullet, and Y. Demay (1986), "Large scale modulations in the Taylor-Couette problem with counterrotating cylinders," Preprint no. 89, Universite de Nice.

Iooss, G. and D. D. Joseph (1990), Elementary Stability and Bifurcation Theory, 2nd ed. (Springer, New York).

Iooss, G. and A. Mielke (1991), "Bifurcating time-periodic solutions of Navier-Stokes equations in infinite cylinders," J. Nonlinear Sci. **1**, 107-146.

Iooss, G., A. Mielke, and Y. Demay (1989), "Theory of steady Ginzburg-Landau equation, in hydrodynamic stability problems," Eur. J. Mech. B/Fluids **8**, 229-268.

Iroshnikov, R. S. (1980), "On the turbulent shear instability of accretion disks," Sov. Astron. **24**, 565-7.

Ivanauskas, A., T. Ness, G. Seifert, and K. Graichen (1987), "Modelling of the flocculation process from a physical point of view. IV. Study of the floc stability in turbulent Couette flow," Chem. Technol. **39**, 60-3.

Ivanilov, I. P. and G. N. Iakovlev (1966), "The bifurcation of fluid flow between rotating cylinders," J. Appl. Math. Mech. (USSR) **30**, 910-916.

Jackson, P. A. and F. R. Mobbs (1975), "Visualization of secondary flow phenomena between rotating cylinders," in 3rd Symp. on Flow Visualization, pp. 125-130.

Jackson, P. A., B. Robati, and F. R. Mobbs (1975), "Secondary flows between eccentric rotating cylinders at subcritical Taylor numbers," in Superlaminar Flow in Bearings. Proc. of the second Leeds-Lyon Symposium on Tribology (Inst. of Mechanical Engineers, London), pp. 9-14.

Jacobs, D. A., C. W. Jacobs, and C. D. Andereck (1988), "Biological scattering of particles for laser Doppler velocimetry," Phys. Fluids **31**, 3457-3461.

Jae Min Hyun (1985), "Flow in an open tank with a free surface driven by the spinning bottom," J. Fluids Eng. **107**, 495-9.

Jain, M. K. and P. B. B. Rao (1966), J. Phys. Soc. Jpn. **25**.

Jain, P. C. and C. V. S. Prakash (1977), "Stability of time-dependent Couette flow in presence of magnetic field," Proc. Indian Natl. Sci. Acad., Part A **43**, 272-84.

Jana, R. N., N. Datta, and B. S. Mazumder (1977), "Magnetohydrodynamic Couette flow and heat transfer in a rotating system," J. Phys. Soc. Jpn. **42**, 1034-9.

Jeffery, G. B. (1923), "The moiton of ellipsoidal particles immersed in a viscous fluid," Proc. R. Soc. London, Ser. A **102**, 161-179.

Jenkins, J. T. (1978), "Flows of nematic liquid crystals," Ann. Rev. Fluid Mech., 197-219.

Jeong, K. and K. Park (1987), "Observation of a very-low-frequency oscillation in a Taylor-Couette flow," Phys. Rev. A **35**, 4854-5.

Jianhua, L. and D. Lili (1990), "The aysmptotiv dolutions of the almost rigid rotation of viscous fluid between two concentric spheres," J. Eng. Math. (China) **7**, 91-103.

Joanicot, M. and P. Pieranski (1985), "Taylor instabilities in colloidal crystals," J. Phys. (Paris), Lett. **46**, L91-6.

Johansson, L. B. A., A. Davidsson, G. Lindblom, and B. Norden (1978), "Linear dichroism as a tool for studying molecular orientation in membrane systems. II. Order parameters of guest molecules from linear dichroism and nuclear magnetic resonance," J. Phys. Chem. **82**, 2604-9.

Johnson, M. and R. D. Kamm (1986), "Numerical study of steady flow dispersion at low Dean number in a gently curving tube," J. Fluid Mech. **172**, 329-345.

Johnson, S. J. and G. G. Fuller (1987), "The dynamics of colloidal particles suspended in a second-order fluid," Faraday Discuss. Chem. Soc., 271-85.

Joly, M., C. Lacombe, and D. Quemada (1981), "Application of the transient flow rheology to the study of abnormal human bloods," Biorheology **18**, 445-52.

Jones, C. A. (1981), "Nonlinear Taylor vortices and their stability," J. Fluid Mech. **102**, 249-61.

Jones, C. A. (1982), "On flow between counter-rotating cylinders," J. Fluid Mech. **120**, 433-350.

Jones, C. A. (1985), "The transition to wavy Taylor vortices," J. Fluid Mech. **157**, 135-62.

Jones, W. M. (1979), "The flow of dilute aqueous solutions of macromolecules in various geometries: VI. Properties of the solutions," J. Phys. D **12**, 369-82.

Jones, W. M. (1988), "The effect of weak elasticity of Couette flow between rotating cylinders: spiral flow and eccentric cylinders," J. Non-Newtonian Fluid Mech. **28**, 255-63.

Jones, W. M., D. M. Davies, and M. C. Thomas (1973), "Taylor vortices and the evaluation of material constants: a critical assessment," J. Fluid Mech. **60**, 19-41.

Joseph, D. D. (1976), Stability of Fluid Motions I, (Springer-Verlag, Berlin; New York).

Joseph, D. D. (1986), "Historical perspectives on the elasticity of liquids," J. Non-Newtonian Fluid Mech. **19**, 237-249.

Joseph, D. D. (1990), Fluid Dynamics of Viscoelastic Liquids (Springer-Verlag, New York).

Joseph, D. D., G. S. Beavers, and R. L. Fosdick (1972), "The free surface on a liquid between cylinders roating at different speeds Part II," Arch. Ration. Mech. Anal. **49**, 381-401.

Joseph, D. D. and R. L. Fosdick (1972), "The free surface on a liquid between cylinders roating at different speeds Part I," Arch. Ration. Mech. Anal. **49**, 321-380.

Joseph, D. D. and W. Hung (1971), "Contributions to the nonlinear theory of stability of visous flow in pipes and rotating cylinders," Arch. Ration. Mech. Anal. **44**, 1-22.

Joseph, D. D. and B. R. Munson (1970), "Global stability of spiral flow," J. Fluid Mech. **43**, 545-575.

Joseph, D. D., A. Narain, and D. Liccius (1986), "Shear-wave speeds and elastic moduli for different liquids. Part 1. Theory," J. Fluid Mech. **171**, 289-308.

Joseph, D. D., O. Riccius, and M. Arney (1986), "Shear-wave speeds and elastic moduli for different liquids. Part 2. Experiments," J. Fluid Mech. **171**, 309-338.

Kai, Z. (1982), "Linear stability of flow of viscoelastic fluid between eccentric rotating cylinders," J. Non-Newtonian Fluid Mech. **11**, 201-7.

Kaloni, P. N. and P. O. Brunn (1986), "Circular Couette flow of concentrated polymer solutions: encapsulated finitely extendable nonlinear elastic (FENE) dumbbell model results," J. Chem. Phys. **84**, 6437-41.

Kamal, M. M. (1966), Trans. AMSE **88**, 717.

Kaneko, K. (1989), "Pattern dynamics in spatiotemporal chaos. Pattern selection, diffusion of defect and pattern competition intermittency," Physica D **34**, 1.

Karasudani, T. (1987), "Non-axis-symmetric Taylor vortex flow in eccentric rotating cylinders," J. Phys. Soc. Jpn. **56**, 855-8.

Karlsson, S. K. F. and H. A. Snyder (1965), "Observations on a thermally induced instability between rotating cylinders," Ann. Phys. (N.Y.) **31**, 314-324.

Kashiwamura, S., K. Miyake, K. Yamada, and J. Yamauchi (1987), "Stability of rotating Couette flow of superfluid /sup 4/He: analysis based on Bekarevich-Khalatnikov's hydrodynamics," Jpn. J. Appl. Phys. Suppl. **26**, 97-8.

Kataoka, K. (1970), Ph.D. Thesis, Kyoto Univ., Kyoto.

Kataoka, K. (1975), J. Chem. Eng. Jpn **8**, 271.

Kataoka, K. (1986), "Taylor Vortices and Instabilities in Circular Couette Flows," in Encyclopedia of Fluid Mechanics, vol. 1, edited by N. P. Cheremisinoff (Gulf Publishing, Houston), pp. 236-274 .

Kataoka, K., Y. Bitou, K. Hashioka, T. Komai, and M. Doi (1984), "Mass transfer in the annulus between two coaxial cylinders," in Heat and Mass Transfer in Rotating Machinery, edited by D. E. Metzger and N. H. Afgan (Hemisphere), pp. 143.

Kataoka, K., H. Doi, and T. Komai (1977), "Heat/mass transfer in Taylor vortex flow with constant axial flow rates," Int. J. Heat Mass Transf. **20**, 57-63.

Kataoka, K., H. Doi, T. Hongo, and M. Futagawa (1975), J. Chem. Eng. Jpn. **8**, 472.

Kataoka, K. and T. Takigawa (1981), AIChE J. **27**, 504.

Kawase, Y. and J. J. Ulbrecht (1983), "Heat and mass transfer in non-Newtonian fluid flow with power function velocity profiles," Can. J. Chem. Eng. **61**, 791-800.

Kawase, Y. and J. J. Ulbrecht (1988), "Laminar mass transfer between concentric rotating cylinders in the presence of Taylor vortices," Electrochim. Acta **33**, 199-203.

Kaye, J. and E. G. Elgar (1958), "Modes of adiabatic and diabatic fluid flow in an annulus with an inner cylinder rotating," J. Heat Transfer (Trans. ASME) **80**, 753.

Keefe, L. R. (1986), "Integrability and structural stability of solutions to the Ginzburg-Landau equation," Phys. Fluids **29**, 3135-3141.

Kendall, W. M. (1976), "Correction to the retarding torque experienced by two rotating concentric spheres," Phys. Fluids **19**, 1420-1421.

Keunings, R. and M. J. Crochet (1984), "Numerical simulation of the flow of a viscoelastic fluid through an abrupt contraction," J. Non-Newtonian Fluid Mech. **14**, 279-99.

Khayat, R. E. and B. C. Eu (1988), "Nonlinear transport processes and fluid dynamics: cylindrical Couette flow of Lennard-Jones fluids," Phys. Rev. A **38**, 2492-507.

Khayat, R. E. and B. C. Eu (1989a), "Generalized hydrodynamics, normal-stress effects, and velocity slips in the cylindrical Couette flow of Lennard-Jones fluids," Phys. Rev. A **39**, 728-44.

Khayat, R. E. and B. C. Eu (1989b), "Generalized hydrodynamics and Reynolds-number dependence of steady-flow properties in the cylindrical Couette flow of Lennard-Jones fluids," Phys. Rev. A **40**, 946.

Khidr, M. A. and M. A. Abdel-gaid (1980), "Cylindrical Couette flow with heat transfer of rarefied gas and porous surface," Rev. Roum. Sci. Tech. Ser. Mec. Appl. **25**, 549-57.

Khlebutin, G. N. (1968), "Stability of fluid motion between a rotating and stationary concentric sphere," Izv. Akad. Nauk SSSR, Mekh. Zhidk. Gaza **3** [Fluid Dyn. (USSR) **3**, 31 (1968)].

Kida, S. (1985), "Three-dimensional periodic flows with high-symmetry," J. Phys. Soc. Jpn. **54**, 2132-2136.

Kida, S., M. Yamada, and K. Ohkitani (1989), "A route to chaos and turbulence," Physica D **37D**, 116-125.

Kilgenstein, P., D. Lhuillier, and K. G. Roesner (1986), "Influence of the deformability of polymer molecules on Taylor-Görtler-vortices," Z. Angew. Math. Mech. **66**, T233-4.

Kiljanski, T. (1989), "A method for correction of the wall-slip effect in a Couette rheometer," Rheol. Acta **28**, 61-4.

Kim, C. S., J. W. Dufty, and A. Santos (1989), "Analysis of nonlinear transport in Couette flow," Phys. Rev. A **40**, 7165.

Kimura, T. and M. Tsutahara (1977), "Analysis of compressible flows around a uniformly expanding circular cylinder and sphere," J. Fluid Mech. **79**, 625-30.

Kind, R. J., F. M. Yowakim, and S. A. Sjolander (1989), "The law of the wall for swirling flow in annular ducts," J. Fluids Eng. **111**, 160.

King, G. P. (1983), "Limits of stability and irregular flow patterns in wavy vortex flow," Ph.D. Thesis, Univ. of Texas, Austin, Tx..

King, G. P., Y. Li, W. Lee, H. L. Swinney, and S. Marcus (1984), "Wave speeds in wavy Taylor-vortex flow," J. Fluid Mech. **141**, 365-90.

King, G. P. and H. L. Swinney (1983), "Limits of stability and irregular flow patterns in wavy vortex flow," Phys. Rev. A **27**, 1240-1243.

Kini, U. D. (1976), "The effect of magnetic fields and boundary conditions on the Couette flow of nematics," Pramana **7**, 223-35.

Kini, U. D. (1988), "Generalized Freedericksz transition in nematics. Geometrical threshold and quantization in cylindrical geometry," J. Phys. (Paris) **49**, 527-39.

Kirchgässner, K. (1961), "Die Instabilitat der Strömung zwischen zwei rotierenden Zylindern gegenuber Taylor-Wirbeln fur beliebige Spaltbreiten," Z. Angew. Math. Phys. **12**, 14-30.

Kirchgässner, K. (1975), "Bifurcation in Nonlinear Hydrodynamic Stability," SIAM Rev. **17**, 652-683.

Kirchgässner, K. and H. Kielhofer (1972), "Stability and bifurcation in fluid dynamics," Rocky Mountain J. Math. **3**, 275-318.

Kirchgassner, K. and P. Sorger (1969a), "Stability analysis of branching solutions of the Navier-Stokes equations," in Proc. of the Twelfth Intl. Cong. of Applied Mechanics, edited by M. Hetenyi and W. G. Vicenti (Springer-Verlag, Berlin), pp. 257-268.

Kirchgassner, K. and P. Sorger (1969b), "Branching analysis for the Taylor problem," Quart. J. Mech. Appl. Math. **22**, 183-209.

Kirchner, R. P. and C. F. Chen (1970), "Stability of time-dependent rotational Couette flow. Part 1. Experimental investigation," J. Fluid Mech. **40**, 39.

Kishinevskii, M. Kh. and T. S. Kornienko (1989), "Turbulent Mass Transport at the Rotating Cylinder Electrode," Sov. Electrochem. **25**, 737.

Klimontovich, yu. L. (1984), "Entropy and entropy production in laminar and turbulent flows," Pis'ma Zh. Tekh. Fiz. **10**, 80-3 [Sov. Tech. Phys. Lett. **10**, 33-4 (1984)].

Klimontovich, yu. L. and kh. Engel'-kherbert (1984), "Average steady Couette and Poiseuille turbulent flows in an incompressible fluid," Zh. Tekh. Fiz. **54**, 440-9 [Sov. Phys. Tech. Phys. **29**, 263-8 (1984)].

Knight, D. D. and P. G. Saffman (1977), "Model equation calculations of the turbulent flow between rotating cylinders and the structure of a turbulent vortex," in Structure and Mechanisms of Turbulence I., edited by H. Fiedler, Lecture Notes in Physics, vol. 75 (Springer, Berlin), pp. 136-143.

Knobloch, E. (1986), "On the degenerate Hopf bifurcation with O(2) symmetry," Contemporary Math. **56**, 193-201.

Knobloch, E. and R. Pierce, "Spiral vortices in finite cylinders," in Ordered and Turbulent Patterns in Taylor-Couette Flow, edited by C. D. Andereck and F. Hayot, Proceedings of a NATO Advanced Research Workshop (Plenum).

Ko, L. and E. G. D. Cohen (1987), "Propagating viscous modes in a Taylor-Couette system," Phys. Lett. A **125**, 231-4.

Kobayashi, M., H. Maekawa, and T. Takano (1990), "An Experimental Study on Turbulent Taylor Vortex Flow between Concentric Cylinders," JSME International J., Series II, Fluids **33**, 436.

Koga, J. K. and E. L. Koschmieder (1989), "Taylor vortices in short fluid columns," Phys. Fluids A **1**, 1475-1478.

Kogelman, S. and R. C. DiPrima (1970), "Stability of spatially periodic supercritical flows in hydrodynamics," Phys. Fluids **13**, 1-11.

Kohama, Y. and R. Kobayashi (1983), "Boundary layer transition and the behavior of spiral vortices on rotating spheres," J. Fluid Mech. **137**, 153-164.

Kohuth, K. R. and G. P. Neitzel (1988), "Experiments on the stability of an impulsively-initiated circular Couette flow," Exp. Fluids **6**, 199-208.

Kojima, Y., S. M. Miyama, and H. Kubotani (1989), "Effects of entropy distributions on non-axisymmetric unstable modes in differentially rotating tori and cylinders," Mon. Not. Roy. Astron. Soc. **238**, 753.

Kolesov, V. V. (1980a), "On stability conditions for the nonisothermic Couette flow," Appl. Math. Mech. **44**, 311-15.

Kolesov, V. V. (1980b), "Stability of nonisothermal Couette flow," Izv. Akad. Nauk SSSR, Mekh. Zhidk. Gaza **15**, 167-70 [Fluid Dyn. (USSR) **15**, 137-40 (1980)].

Kolesov, V. V. (1981a), "Calculation of auto-oscillations resulting from the loss of stability of a nonisothermal Couette flow," Izv. Akad. Nauk SSSR, Mekh. Zhidk. Gaza **16**, 25-32 [Fluid Dyn. (USSR) **16**, 344-50 (1981)].

Kolesov, V. V. (1981b), "Formation of Taylor vortices between heated rotating cylinders," Zh. Prikl. Mekh. Tekh. Fiz. **22**, 87-93 [J. Appl. Mech. Tech. Phys. **22**, 811-16 (1981)].

Kolesov, V. V. (1984), "Oscillatory rotationally symmetric loss of stability of nonisothermal Couette flow," Izv. Akad. Nauk SSSR, Mekh. Zhidk. Gaza **19**, 76-80 [Fluid Dyn. (USSR) **19**, 63-7 (1984)].

Koschmieder, E. L. (1975a), "Effect of finite disturbances on axisymmetric Taylor vortex flow," Phys. Fluids **18**, 499-503.

Koschmieder, E. L. (1975b), "Stability of supercritical Benard convection and Taylor vortex flow," Adv. Chem. Phys. **32**, 109-133.

Koschmieder, E. L. (1976), "Taylor vortices between eccentric cylinders," Phys. Fluids **19**, 1-4.

Koschmieder, E. L. (1979a), "Turbulent Taylor vortex flow," J. Fluid Mech. **93**, 515-527.

Koschmieder, E. L. (1979b), "Addendum to: Turbulent Taylor vortex flow," J. Fluid Mech. **93**, 801.

Koschmieder, E. L. (1980), "Transition from laminar to turbulent Taylor vortex flow," in Laminar-Turbulent Transition (IUTAM Symposium), edited by R. Eppler and H. Fasel (Springer-Verlag, Berlin), pp. 396-404.

Koschmieder, E. L. (1981), "Experimental aspects of hydrodynamic instabilities," in Order and Fluctuations in Equilibrium and Nonequilibrium Statistical Mechanics, edited by Nicolis, Dewel, and Turner (Wiley), pp. 159-207.

Koschmieder, L. (1992), Benard Cells and Taylor Vortices (Cambridge Univ. Press).

Kostelich, E. J. and J. A. Yorke (1988), "Noise reduction in dynamical systems," Phys. Rev. A **38**, 1649-52.

Kouitat, R., G. Maurice, and M. Lucius (1990), "Etude du regime de mise en mouvement dans un viscosimetre de Couette a cylindres coaxiaux," Rheol. Acta **29**, 332.

Kramer, L., E. Ben-Jacob, H. Brand, and M. C. Cross, "Wavelength selection in systems far from equilibrium," Phys. Rev. Lett. **49**, 1891-1894.

Kramer, L. and W. Zimmermann (1985), "On the Eckhaus instability for spatially periodic patterns," Physica D **16D**, 221-232.

Krause, E. (1982), Ed., Proceedings of the Eighth International Conference on Numerical Methods in Fluid Dynamics (Springer-Verlag, Berlin).

Krieger, I. M. (1976), "Some comments on a paper by A. Apelblat, J.C. Healy and M. Joly in rheol. Acta 14, 976-978 (1975)," Rheol. Acta **15**, 665-6.

Krishnam Raju, G. and V. V. Ramana rao (1978), "Hall effects on magnetohydrodynamic Couette flow and heat transfer in a rotating system," Acta Phys. Acad. Sci. Hung. **44**, 363-70.

Krueger, E. R. and R. C. DiPrima (1964), "The stability of viscous flow between rotating cylinders with an axial flow," J. Fluid Mech. **19**, 528-538.

Krueger, E. R., A. Gross, and R. C. DiPrima (1966), "On the relative importance of Taylor-vortex and non-axisymmetric modes in flow between rotating cylinders," J. Fluid Mech. **24**, 521-538.

Krugljak, Z. B., E. A. Kuznetsov, V. S. Lvov, Yu. E. Nesterikhin, A. A. Predtechensky, V. S. Sobolev, E. N. Utkin, and F. A. Zhuravel (1980), "Spectrum evolution at the transition to turbulence in a Couette flow," Phys. Lett. A **78A**, 269-72.

Krugljak, Z. B., E. A. Kuznetsov, V. S. L'vov, Yu. E. Nesterikhin, and et. al. (1980), "Laminar-turbulent transition in circular Couette flow," in Laminar-Turbulent Transition, edited by R. Eppler and H. Fasel (Springer-Verlag, Berlin), pp. 378-387.

Kubotani, H., M. Miyama, and M. Sekiya (1989), "The surface wave instability of stratified incompressible cylinders with differential rotation," Prog. Theor. Phys. **82**, 523.

Kuehn, T. H. and R. J. Goldstein (1976), "An experimental and theoretical study of natural convection in the annulus between horizontal concentric cylinders," J. Fluid Mech. **74**, 695-719.

Kuhlmann, H. (1985), "Model for Taylor-Couette flow," Phys. Rev. A **32**, 1703-7.

Kuhlmann, H., D. Roth, and M. Lücke (1989), "Taylor vortex flow under harmonic modulation of the driving force," Phys. Rev. A **39**, 745-62.

Kumar, K., J. K. Bhattacharjee, and K. Banerjee (1986), "Onset of the first instability in hydrodynamic flows: effect of parametric modulation," Phys. Rev. A **34**, 5000-6.

Kundu, P. K. (1990), "Instability," in Fluid Mechanics (Academic, San Diego).

Kuryachii, A. P. (1989), "The stability of Couette-Taylor electrohydrodynamic flow," J. Appl. Math. Mech. (USSR) **53**, 345.

Kurzweg, U. H. (1963a), Z. Angew. Math. Phys. **14**, 380-383.

Kurzweg, U. H. (1963b), "The stability of Couette flow in the presence of an axial magnetic field," J. Fluid Mech. **17**, 52-60.

Kurzweg, U. H. and A. H. Khalfaoui (1982), "Current induced instabilities in rotating hydromagnetic flows between concentric cylinders," Phys. Fluids **25**, 440-445.

Kusnetsov, E. A., V. S. L'vov, Yu. E. Nesterikhin, Y. F. Shmojlov, V. S. Sobolev, M. D. Spector, S. A. Timokhin, E. N. Utkin, and Yu. G. Vasilenko (1977), "About turbulence arising in Couette flow," Report No. 58 (Inst. of Automation and Electrometry, Siberian Branch, USSR Ac. Sci., Nowosibirsk).

Kutateladze, S. S., V. E. Nakoryakov, and M. S. Iskakov (1982), "Electrochemical measurements of mass transfer between a sphere and liquid in motion at high Peclet numbers," J. Fluid Mech. **125**, 453-62.

Kuzay, T. M. and C. J. Scott (1977), J. Heat Transfer **99**, 12.

Kuznetsov, E. A., V. S. L'vov, A. A. Predtechenskii, V. S. Sobolev, and E. N. Utkin (1979), "The problem of transition to turbulence in Couette flow," Pis'ma Zh. Eksp. Teor. Fiz. **30**, 207-210 [JETP Lett. **30**, 207-210 (1979)].

L'vov, V. S. (1981), "Statistical description of a chain of interacting Taylor vortexes in the direct interaction approximation," Zh. Eksp. Teor. Fiz. **80**, 1969-80 [Sov. Phys. JETP **53**, 1024-1029 (1981)].

L'vov, V. S. and A. A. Predtechensky (1981), "On Landau and stochastic attractor pictures in the problem of transition to turbulence," Physica D **2D**, 38-51.

L'vov, V. S., A. A. Predtechenskii, and A. I. Chernykh (1981), "Bifurcation and chaos in a Taylor vortex system: a natural and numerical experiment," Zh. Eksp. Teor. Fiz. **80**, 1099-121 [Sov. Phys. JETP **53**, 562 (1981)].

L'vov, V. S., A. A. Predtechenskii, and A. I. Chernykh (1983), "Bifurcationa nd chaos in the system of Taylor vortices - laboratory and numerical experiment," in Nonlinear Dynamics and Turbulence, edited by G. I. Barenblatt, G. Iooss, and D. D. Joseph (Pitman, Boston, London, Melbourne), pp. 238-280.

Ladeinde, F. and K. E. Torrance (1990), "Transient plumes from convective flow instability in horizontal cylinders," J. Thermophys. Heat Transfer **4**, 350.

Lambert, R. B., H. A. Snyder, and S. K. F. Karlsson (1965), "Hot thermistor anemometer for finite amplitude stability measurements," Rev. Sci. Instrum. **36**, 924-928.

Lambert, R. B. and H. A. Snyder (1966), "Experiments on the effect of horizontal shear and change of aspect ratio on convective flow in a rotating annulus," J. Geophys. Res. **71**, 5225-5234.

Landau, L. D. (1944), "On the problem of turbulence," C. R. (Dokl.) Acad. Sc. URSS **44**, 311-314.

Landau, L. D. and E. M. Lifshitz (1987), Fluid Mechanics, 2nd ed., Course in Theoretical Physics, Vol. 6 (Pergammon).

Langford, W. F. (1979), "Periodic and steady state mode interactions lead to tori," SIAM J. Appl. Math. **37**, 22-48.

Langford, W. F. (1980), "Interactions of Hopf and pitchfork bifurcations," in Bifurcation Problems and Their Numerical Solution (Birkhäuser, Boston), pp. 103-134.

Langford, W. F. (1983), "A review of interactions of Hopf and steady state bifurcations," in Nonlinear Dynamics and Turbulence, edited by G. I. Barenblatt, G. Iooss, and D. D. Joseph (Pitman, Boston), pp. 215-237.

Langford, W. F., R. Tagg, E. Kostelich, H. L. Swinney, and M. Golubitsky (1988), "Primary instabilites and bicriticality in flow between counter-rotating cylinders," Phys. Fluids **31**, 776-785.

Larson, R. G. (1989), "Taylor-Couette stability analysis for a Doi-Edwards fluid," Rheol. Acta **28**, 504.

Larson, R. G., E. S. G. Shaqfeh, and S. J. Muller (1990), "A purely elastic instability in Taylor-Couette flow," J. Fluid Mech. **218**, 573.

Lathrop, D. P., J. Fineberg, and H. L. Swinney (1992), "Turbulent flow between concnetric rotating cylinders at large Reynolds numbers," (preprint).

Laure, P. (1987), "Secondary bifurcations of quasi-periodic solutions in the Couette-Taylor problem. Computation of the normal form," C. R. Acad. Sci., Ser. A **305**, 493-6.

Laure, P. and Y. Demay (1988), "Symbolic computation and equation on the center manifold: application to the Couette-Taylor problem," Comput. Fluids **16**, 229-38.

Laurien, E. and H. Fasel (1988), "Numerical investigation of the onset of chaos in the flow between rotating cylinders," Z. Angew. Math. Mech. **68**, 311-13.

Ledovskaya, N. N. (1986), "Experimental investigation of the three-dimensional structure of detached flow in an axisymmetric annular diffusor," Inzh.-Fiz. Zh. **51**, 321-8 [J. Eng. Phys. **51**, 995-1000 (1986)].

Lee, C. M., H. B. Kim, M. C. Sung, W. K. Choi, C. N. Whang, and K. Jeong (1990), "Turbator region in wavy vortex flow," J. Korean Phys. Soc. **23**, 381-386.

Lee, J. S. and G. G. Fuller (1987), "The spatial development of transient Couette flow and shear wave propagation in polymeric liquids by flow birefringence," J. Non-Newtonian Fluid Mech. **26**, 57-76.

Lee, Y. N. and W. J. Minkowycz (1989), "Heat transfer characteristics of the annulus of two-coaxial cylinders with one cylinder rotating," Int. J. Heat Mass Transf. **32**, 711-22.

Lee, Y. N. (1989), "Heat transfer characteristics of the annulus of two-coaxial cylinders with one cylinder rotating," Int. J. Heat Mass Transf. **32**, 711.

Lega, J. (1991), "Secondary Hopf bifurcation of a one-dimensional periodic pattern," Eur. J. Mech. B/Fluids **10 (suppl.)**, 145-150.

Legrand, J. and F. Coeuret (1982), "Overall wall to liquid mass transfer for flows combining Taylor vortices and axial forced convection," Int. J. Heat Mass Transf. **25**, 345-51.

Legrand, J. and F. Coeuret (1987), "Wall to liquid mass transfer and hydrodynamics of two-phase Couette-Taylor-Poiseuille flow," Can. J. Chem. Eng. **65**, 237-43.

Legrand, J., F. Coeuret, and M. Billon (1983), "Dynamic structure and wall to liquid mass transfer in the laminar vortex regime of a Couette-Poiseuille flow," Int. J. Heat Mass Transf. **26**, 1075-85.

Leibovich, S., S. N. Brown, and Y. Patel (1986), "Bending waves on inviscid columnar vortices," J. Fluid Mech. **173**, 595-624.

Leighton, D. and A. Acrivos (1987a), "Measurement of shear-induced self-diffusion in concentrated suspensions of spheres," J. Fluid Mech. **177**, 109-131.

Leighton, D. and A. Acrivos (1987b), "The shear-induced migration of particles in concentrated suspensions," J. Fluid Mech. **181**, 415-439.

Leiman, V. G. (1987), "Rayleigh-Taylor instability in unneutralized electron beams," Fiz. Plazmy **13**, 1216-20 [Sov. J. Plasma Phys. **13**, 700-3 (1987)].

Leonard, R. A., G. J. Bernstein, R. H. Pelto, and A. A. Ziegler (1981), "Liquid-liquid dispersion in turbulent Couette flow," AICHE J. **27**, 495-503.

Leonov, A. and A. Srinivasan (1991), "On the modeling of fluidity loss phenomena in Couette and Poiseuille flows of elastic liquids," Rheol. Acta **30**, 14.

Lewis, E. (1979), "Steady flow between a rotating circular cylinder and fixed square cylinder," J. Fluid Mech. **95**, 497-513.

Lewis, J. W. (1928), "An experimental study of the motion of a viscous liquid contained between coaxial cylinders," Proc. R. Soc. London, Ser. A **117**, 388-407.

Lhuillier, D. (1981), "Molecular models and the Taylor stability of dilute polymer solutions," J. Non-Newtonian Fluid Mech. **9**, 329-37.

Lhuillier, D. and K. G. Roesner (1985), "Cylindrical Couette-flow for non-Newtonian fluids," in Flow of Real Fluids, edited by G. E. A. Meier and F. Obermeier, Lecture Notes in Physics, v. 235 (Springer-Verlag, Berlin), pp. 317-325.

Li, Y. S. and S. C. Kot (1981), "One-dimensional finite element method in hydrodynamic stability," Int. J. Numer. Methods Eng. **17**, 853-70.

Likashchuk, S. N. and A. A. Predtechenskii (1984), "Observation of resonant tori in the phase space of a Couette flow," Dok. Akad. Nauk SSSR **274**, 1317-20 [Sov. Phys. Dokl. **29**, 100-2 (1984)].

Lillie, H. R. (1930), Phys. Rev. **36**, 347.

Lin, C. C. (1955), The Theory of Hydrodynamic Stability (Cambridge Univ. Press, Cambridge, England).

Lin, C. C., "Hydrodynamics of Helium II.," in Rendicondti della Scuola Internazionale di Fisica "E. Fermi", XXI Corso, pp. 93-146.

Lin, S. H. and D. M. Hsieh (1980), "Heat transfer to generalized Couette flow of a non-Newtonian fluid in annuli with moving inner cylinder," J. Heat Transfer **102**, 786-9.

Lindner, P. and R. C. Oberthur (1984), "Apparatus for the investigation of liquid systems in a shear gradient by small angle neutron scattering (SANS)," Rev. Phys. Appl. **19**, 759-63.

Liseikina, T. A. (1978), "Analysis of the branching of a spiral flow between coaxial cylinders," Zh. Prikl. Mekh. Tekh. Fiz. **19**, 71-8 [J. Appl. Mech. Tech. Phys. **19**, 766-771 (1978)].

Liseikina, T. A. (1981), "Stability of a hollow vortex," Zh. Prikl. Mekh. Tekh. Fiz. **22**, 54-60 [J. Appl. Mech. Tech. Phys. **22**, 333-338 (1981)].

Liu, D. C. S. and C. F. Chen (1973), "Numerical experiments on time-dependent rotational Couette flow," J. Fluid Mech. **59**, 77-95.

Liu Shusheng, Sun Yongda, and Chai Lu (1988), "Bifurcations and chaos in a circular Couette system," Chin. Phys. Lett. **5**, 485-8.

Lorenzen, A., G. Pfister, and T. Mullin (1983), "End effects on the transition to time-dependent motion in the Taylor experiment," Phys. Fluids **26**, 10-13.

Lorenzen, A. and T. Mullin (1985), "Anomalous modes and finite-length effects in Taylor-Couette flow," Phys. Rev. A **31**, 3463-5.

Lortz, D., R. Meyer-Spasche, and P. Petroff, "A global analysis of secondary bifurcations in the Benard problem and the relationship between the Benard and Taylor problems," in <u>Methoden und Verfahren der mathematischen Physik</u>, (to appear), edited by B. Brosowski and E. Martensen, eds. (Verlag Peter D. Lang, Frankfurt a. Main).

Lucius, M. and J. C. Roth (1976), "Determination of the velocity profile in a non-Newtonian fluid for certain types of steady viscometric flows," J. Non-Newtonian Fluid Mech. **1**, 383-90.

Lücke, M., M. Mihelcic, B. Kowalski, and K. Wingerath (1987), "Structure formation by propogating fronts," in <u>The Physics of Structure Formation: Theory and Simulation</u>, edited by W. Guttinger and G. Dangelmayr (Springer-Verlag), pp. 97-116.

Lücke, M., M. Mihelcic, and K. Wingerath (1984), "Propagation of Taylor vortex fronts into unstable circular Couette flow," Phys. Rev. Lett. **52**, 625-8.

Lücke, M., M. Mihelcic, and K. Wingerath (1985), "Front propagation and pattern formation of Taylor vortices growing into unstable circular Couette flow," Phys. Rev. A **31**, 396-409.

Lücke, M., M. Mihelcic, K. Wingerath, and G. Pfister (1984), "Flow in a small annulus between concentric cylinders," J. Fluid Mech. **140**, 343-353.

Lücke, M. and D. Roth (1990), "Structure and dynamics of Taylor vortex flow and the effect of subcritical driving ramps," Z. Phys. B **78**, 147-158.

Luke, G. E. (1923), Trans. ASME **42**, 646.

Luyten, P. J. (1990), "The dynamical stability of differentially rotating, self-gravitating cylinders," Mon. Not. Roy. Astron. Soc. **242**, 447.

Lyon, D. J., J. R. Melcher, and M. Zahn (1988), "Couette charger for measurement of equilibrium and energization flow electrification parameters: application to transformer insulation," IEEE Trans. Electr. Insul. **23**, 159-76.

MacPhail, D. C. (1941), "Turbulence in a distorted passage and between rotating cylinders," Ph.D. Thesis, Cambridge University.

MacPhail, D. C. (1946), "Turbulence in a distorted passage and between rotating cylinders," in <u>Proc. 6th Int. Cong. Appl. Mech., Paris</u>.

Majumdar, A. K. and D. B. Spalding (1977), "Numerical computation of Taylor vortices," J. Fluid Mech. **81**, 295-304.

Makashev, N. K. and V. I. Nosik (1980), "Steady-state Couette flow (with heat transfer) of a gas of Maxwellian molecules," Dok. Akad. Nauk SSSR **253**, 1077-81 [Sov. Phys. Dokl. **25**, 589-91 (1980)].

Mallock, A. (1888), "Determination of the viscosity of water," Proc. R. Soc. London, Ser. A **45**, 126-132.

Mallock, A. (1896), "Experiments on fluid viscosity," Proc. R. Soc. London, Ser. A **187**, 41-56.

Malomed, B. A., I. E. Staroselsky, and A. B. Konstantinov (1989), "Corrections to the Eckhaus' stability criterion for one-dimensional stationary structures," Physica D **34D**, 270-6.

Manneville, P. (1990), <u>Dissipative Structures and Weak Turbulence</u> (Academic, Boston).

Manneville, P. and L. S. Tuckerman (1987), "Phenomenological modelling of the first bifurcations of spherical Couette flow," J. Phys. (Paris) **48**, 1461-9.

Marcus, P. S. (1984a), "Simulation of Taylor-Couette flow. I. Numerical methods and comparison with experiment," J. Fluid Mech. **146**, 45-64.

Marcus, P. S. (1984b), "Simulation of Taylor-Couette flow. II. Numerical results for wavy-vortex flow with one travelling wave," J. Fluid Mech. **146**, 65-113.

Marcus, P. S. and L. Tuckerman (1987), "Simulation of flow between concentric rotating spheres. I. Steady states," J. Fluid Mech. **185**, 1-30.

Marcus, P. S. and L. S. Tuckerman (1987), "Simulation of flow between concentric rotating spheres. 2. Transitions," J. Fluid Mech. **185**, 31-65.

Margules, M. (1881), "Uber die bestimmung des reibungs- und gleitungs-coefficient," Wiener Berichte (second series) **83**, 588-602.

Markho, P. H., C. D. Jones, and F. R. Mobbs (1977), "Wavy modes of instability in the flow between eccentric rotating cylinders," J. Mech. Eng. Sci. **19**, 76-80.

Marlin, P. C. (1978), "How is turbulence created?," Fiz. Sz. **28**, 121-32.

Marqués, F. (1990), "On boundary conditions for velocity potentials in confined flows: application to Couette flow," Phys. Fluids A **2**, 729-737.

Martin, B. W. and M. A. Hasoon (1976), "The stability of viscous axial flow in the entry region of an annulus with a rotating inner cylinder," J. Mech. Eng. Sci. **18**, 221-8.

Martin, B. W. and A. Payne (1972), "Tangential flow development for laminar axial flow in an annulus with a rotating inner cylinder," Proc. R. Soc. London, Ser. A **328**, 123-141.

Martin, P. C. (1982), "The onset of chaos in convecting fluids," in <u>Melting, Localization, and Chaos. Proceedings of the Ninth Midwest Solid State Theory Symposium</u>, edited by R. K. Kalia and P. Vashishta (North-Holland, New York), pp. 179-94.

Marx, K. and H. Haken (1988), "The generalized Ginzburg-Landau equations of wavy vortex flow," Europhys. Lett. **5**, 315-20.

Marx, K. and H. Haken (1989), "Numerical derivation of the generalized Ginzburg-Landau equations of wavy vortex flow," Z. Phys. B **75**, 393-411.

Massoudi, M. and I. Christie (1990), "Natural convection flow of a non-Newtonian fluid between two concentric vertical cylinders," Acta Mech. **82**, 11.

Mathis, D. M. and G. P. Neitzel (1985), "Experiments on impulsive spin-down to rest," Phys. Fluids **28**, 449-54.

Matic, D., B. Lovrecek, and D. Skansi (1978), "The rotating cylinder electrode," J. Appl. Electrochem. **8**, 391-8.

Matisse, P. and M. Gorman (1984), "Neutrally buoyant anisotropic particles for flow visualization," Phys. Fluids **27**, 759-760.

Matsui, T. (1981), "Flow visualization studies of vortices," Proc. Indian Acad. Sci. Eng. Sci. **4**, 239-57.

Matsumoto, N., S. Shirayama, K. Kuwahara, and F. Hussain (1989), "Three-dimensional simulation of Taylor-Couette flow," in Advances in Turbulence 2, edited by H. H. Fernholz and H. E. Fielder (Springer-Verlag, Berlin), pp. 366-371.

Matsuura, T. (1977), "Flowability of resin mortar," J. Soc. Mater. Sci. Jpn. **26**, 791-5.

Matusevich, N. P., V. K. Rakhuba, and V. A. Chernobai (1983), "Experimental investigation of hydrodynamic and thermal processes in magnetofluid seals," Magn. Gidrodin. **19**, 125-9 [Magnetohydrodynamics **19**, 102-6 (1983)].

Mau-ling wang and chon-fon chen (1984), "Optimization of batch crystallization processes via legendre polynomials," J. Chin. Inst. Eng. **7**, 27-35.

Maurice, G. and M. Lucius (1985), "Modelization and computation of a Couette flow. Determination of an approximate model and its application to the constitutive law of some fluids," C. R. Acad. Sci., Ser. B **301**, 217-20.

Mauss, J. and J. Pradere (1981), "On a viscoelastic law with relaxation time for blood," J. Biophys. Med. Nucl. **5**, 185-9.

Mazumder, B. S. (1991), "An exact solution of oscillatory Couette flow in a rotating system," J. Appl. Mech. **58**, 1104.

McCaughan, F. E. (1990), "Bifurcation Analysis of Axial Flow Compressor Stability," SIAM J. Appl. Math. **50**, 1232.

McFadden, G. B., S. R. Coriell, R. F. Boisvert, and M. E. Glicksman (1984), "Asymmetric instabilities in buoyancy-driven flow in a tall vertical annulus," Phys. Fluids **27**, 1359-1361.

McFadden, G. B., S. R. Coriell, and M. E. Glicksman (1989), "Instability of a Taylor-Couette flow interacting with a crystal-melt interface," Physicochemical hydrodynamics **11**, 387.

McFadden, G. B., S. R. Coriell, B. T. Murray, M. E. Glicksman, and M. E. Selleck (1990), "Effect of a crystal-melt interface on Taylor vortex flow," Phys. Fluids A **2**, 700-705.

McMillan, D. E., J. Strigberger, and N. G. Utterback (1987), "Blood's critical Taylor number and its flow behavior at supercritical Taylor numbers," Biorheology **24**, 401-10.

Mcmillan, D. E. and N. Utterback (1980), "Maxwell fluid behavior of blood at low shear rate," Biorheology **17**, 343-54.

Mcmillan, D. E., N. G. Utterback, and J. Stocki (1980), "Low shear rate blood viscosity in diabetes," Biorheology **17**, 355-62.

Mcmillan, D. E., N. G. Utterback, and J. B. Baldridge (1980), "Thixotropy of blood and red blood cell suspensions," Biorheology **17**, 445-56.

McMillan, D. E., N. G. Utterback, M. Nasrinasrabadi, and M. M. Lee (1986), "An instrument to evaluate the time dependent flow properties of blood at moderate shear rates," Biorheology **23**, 63-74.

Meiselman, H. J. (1978), "Rheology of shape-transformed human red cells," Biorheology **15**, 225-37.

Meister, B. and W. Muenzner (1966), "Das Talorsche Stabilitätsproblem mit Modulation," Z. Angew. Math. Phys. **17**, 537-540.

Meitlis, V. P. and N. N. Filonenko (1983), "Overheating instability in Couette flow," Teplofiz. Vys. Temp. **21**, 321-5 [High Temp. **21**, 249-53 (1983)].

Meksyn, D. (1946a), "Stability of viscous flow between rotating cylinders. I.," Proc. R. Soc. London, Ser. A **187**, 115-128.

Meksyn, D. (1946b), "Stability of viscous flow between rotating cylinders. II.," Proc. R. Soc. London, Ser. A **187**, 480-491.

Meksyn, D. (1946c), "Stability of viscous flow between rotating cylinders. III.," Proc. R. Soc. London, Ser. A **187**, 492-504.

Meksyn, D. (1950), "Stability of viscous flow over concave cylindrical surfaces," Proc. R. Soc. London, Ser. A **203**, 253-65.

Mello, T. M., P. H. Diamond, and H. Levine (1991), "Hydrodynamic modes of a granular shear flow," Phys. Fluids A **3**, 2067.

Merrill, E. W., H. S. Mickley, and A. Ram (1962), "Instability in Couette flow of solutions of macromolecules," J. Fluid Mech. **13**, 86-90.

Meyer, J. A. and H. Fasel (1988), "The numerical study of stability of Taylor-vortex flow in a wide slot," Z. Angew. Math. Mech. **68**, 317-19.

Meyer, K. A. (1966), Report No. LA-3497 (Los Alamos National Laboratory).

Meyer, K. A. (1967), Phys. Fluids **10**, 1874.

Meyer, K. A. (1969), "Three-dimensional study of flow between concentric rotating cylinders," Phys. Fluids Suppl. II (High Speed Computing in Fluid Dynamics), 165-170.

Meyer, W. H. (1981), "Properties of well-aligned polyacetylene obtained with a shear procedure," Mol. Cryst. Liq. Cryst. **77**, 137-46.

Meyer-Spasche, R., "Influence of double eigenvalues on the formation of flow patterns.," in Ordered and Turbulent Patterns in Taylor-Couette Flow, edited by C. D. Andereck and F. Hayot, Proceedings of a NATO Advanced Research Workshop (Plenum).

Meyer-Spasche, R. and H. B. Keller (1978), "Numerical study of Taylor-vortex flows between rotating cylinders, I.," Appl. Math Reports, Caltech.

Meyer-Spasche, R. and H. B. Keller (1980), "Computations of the axisymmetric flow between rotating cylinders," J. Comput. Phys. **35**, 100-9.

Meyer-Spasche, R. and H. B. Keller (1984), "Numerical study of Taylor-vortex flows between rotating cylinders, II.," Appl. Math Reports, Caltech.

Meyer-Spasche, R. and H. B. Keller (1985), "Some bifurcation diagrams for Taylor vortex flows," Phys. Fluids **28**, 1248-52.

Meyer-Spasche, R. and M. Wagner (1987a), "The basic (n,2n)-fold of steady axisymmetric Taylor vortex flows," in The Physics of Structure Formation: Theory and Simulation, edited by W. Guttinger and G. Dangelmayr (Springer-Verlag, Berlin), pp. 166-178.

Meyer-Spasche, R. and M. Wagner (1987b), "Steady axisymmetric Taylor vortex flows with free stagnation points of the poloidal flow," in Proceedings: Bifurcation: Analysis, Algorithms, Applications, edited by Kupper, Seydel, and e. Troger (Birkhauser Verlag), pp. 213 -- 221.

Michael, D. H. (1954), "The stability of an incompressible electrically-conducting fluid rotating about an axis when the current flops parallel to the axis," Mathematika **1**, 45-50.

Migun, N. P., P. P. Prokhorenko, and G. E. Konovalov (1983), "Viscous energy dissipation in a micropolar liquid between two coaxial cylinders," Vestsi Akad. Navuk Bssr Ser. Fiz. Energ. Navuk, 89-92.

Mikati, N., J. Nordh, and B. Norden (1987), "Scattering anisotropy of partially oriented samples. Turbidity flow linear dichroism (conservative dichroism) of rod-shaped macromolecules," J. Phys. Chem. **91**, 6048-55.

Missimer, J. R. and L. C. Thomas (1983), "Analysis of transitional and fully turbulent plane Couette flow," J. Lubr. Technol. **105**, 364-8.

Miyatake, O. and H. Iwashita (1990), "Laminar-flow heat transfer to a fluid flowing axially between cylinders with a uniform surface temperature," Int. J. Heat Mass Transf. **33**, 417.

Mizushina, T., R. Ito, K. Kataoka, S. Yokoyama, Y. Nakashima, and A. Fukuda (1968), Kagaku-Kogaku (Chem. Eng. Japan) **32**, 795.

Mizushina, T., R. Ito, K. Kataoka, Y. Nakashima, and A. Fukuda (1971), Kagaku-Kogaku (Chem. Eng. Japan) **35**, 1116.

Mizushina, T., R. Ito, T. Nakagawa, K. Kataoka, and S. Yokoyama (1976), Kagaku-Kogaku (Chem. Eng. Japan) **31**, 974.

Mobbs, F. R. and M. S. Ozogan (1984), "Study of sub-critical Taylor vortex flow between eccentric rotating cylinders by torque measurements and visual observations," Int. J. Heat Fluid FLow **5**, 251-3.

Mobbs, F. R., S. Preston, and M. S. Ozogan (1979), "An experimental investigation of Taylor vortex waves.," in Taylor Vortex Working Party.

Mobbs, F. R. and M. A. M. A. Younes (1974), "The Taylor vortex regime in the flow between eccentric rotating cylinders," J. Lubr. Technol. **?**, 127-134.

Molki, M., K. N. Astill, and E. Leal (1990), "Convective heat-mass transfer in the entrance region of a concentric annulus having a rotating inner cylinder," Int. J. Heat Fluid Flow **11**, 120.

Monin, A. S. (1986), "Hydrodynamic instability," Usp. Fiz. Nauk **150**, 61-105 [Sov. Phys. Usp. **29**, 843-68 (1986)].

Montgomery, D. (1982), "Thresholds for the onset of fluid and magnetofluid turbulence," Phys. Scr. **T2**, 506-10.

Morel, J., Z. Lavan, and B. Bernstein (1977), "Flow through rotating porous annuli," Phys. Fluids **20**, 726-733.

Morin, A. J., II, M. Zahn, and J. R. Melcher (1988), "Equilibrium electrification parameters inferred from Couette charger terminal measurements," in 1988 Annual Report, Conference on Electrical Insulation and Dielectric Phenomena (IEEE, New York), pp. 286-92.

Morin, A. J., M. Zahn, and J. R. Melcher (1991), "Fluid electrification measurements of transformer pressboard/oil insulation in a Couette charger," IEEE Trans. Electrical Insulation **26**, 870.

Moser, R. D. and P. Moin (1987), "The effects of curvature in wall-bounded turbulent flows," J. Fluid Mech. **175**, 497-510.

Moser, R. D., P. Moin, and A. Leonard (1983), "A spectral numerical method for the Navier-stokes equations with applications to Taylor-Couette flow," J. Comput. Phys. **52**, 524-44.

Mott, J. E. and D. D. Joseph (1968), "Stability of parallel flow between concentric cylinders," Phys. Fluids **11**, 2065-2073.

Mottaghy, K. and N. Luks (1978), "Investigation of suspended droplets in a Couette flow from the point of view of the rheology of blood. I," Biomed. Tech. **23**, 141/1-2.

Muller, S. J., R. G. Larson, and E. S. G. Shaqfeh (1989), "A purely elastic transition in Taylor-Couette flow," Rheol. Acta **28**, 499.

Mullin, T. (1982), "Mutations of steady cellular flows in the Taylor experiment," J. Fluid Mech. **121**, 207-18.

Mullin, T. (1984a), "Transition to turbulence in non-standard rotating flows," in Turbulence And Chaotic Phenomena In Fluids. Proceedings Of The International Symposium, edited by T. Tatsumi (North-Holland, A,sterdam, Netherlands), pp. 85-6.

Mullin, T. (1984b), "Cell number selection in Taylor-Couette flow," in Cellular Structures in Instabilities, edited by J. E. Wesfreid and S. Zaleski, Lecture Notes in Physics No. 210 (Springer-Verlag, Berlin), pp. 75-83.

Mullin, T. (1985), "Onset of time dependence in Taylor-Couette flow," Phys. Rev. A **31**, 1216-18.

Mullin, T. (1991), "Finite-dimensional dynamics in Taylor-Couette flow," IMA J. Appl. Math. **46**, 109.

Mullin, T. and T. B. Benjamin (1980), "Transition to oscillatory motion in the Taylor experiment," Nature (London) **288**, 567-9.

Mullin, T., T. B. Benjamin, K. Schatzel, and E. R. Pike (1981), "New aspects of unsteady Couette flow," Phys. Lett. A **83A**, 333-6.

Mullin, T. and K. A. Cliffe (1986), "Symmetry breaking and the onset of time dependence in fluid mechanical systems," in Nonlinear Phenomena and Chaos, edited by S. Sarkar (Adam Hilger, Bristol, England), pp. 96-112.

Mullin, T. and K. A. Cliffe (1985), "Symmetry breaking and the onset of time dependence," Preprint.

Mullin, T., K. A. Cliffe, and G. Pfister (1987), "Unusual time-dependent phenomena in Taylor-Couette flow at moderately low Reynolds numbers," Phys. Rev. Lett. **58**, 2212-15.

Mullin, T. and A. G. Darbyshire (1989), "Intermittency in a rotating annular flow," Europhys. Lett. **9**, 669-673.

Mullin, T. and A. Lorenzen (1985), "Bifurcation phenomena in flows between a rotating circular cylinder and a stationary square outer cylinder," J. Fluid Mech. **157**, 289-303.

Mullin, T., A. Lorenzen, and G. Pfister (1983), "Transition to turbulence in a non-standard rotating flow," Phys. Lett. A **96A**, 236-8.

Mullin, T., G. Pfister, and A. Lorenzen (1982), "New observations on hysteresis effects in Taylor-Couette flow," Phys. Fluids **25**, 1134-6.

Mullin, T. and T. J. Price (1989), "An experimental observation of chaos arising from the interaction of steady and time-dependent flows," Nature (London) **340**, 294-6.

Mullin, T., S. J. Tavener, and K. A. Cliffe (1989), "An experimental and numerical study of a codimension-2 bifurcation in a rotating annulus," Europhys. Lett. **8**, 251-256.

Munson, B. R. and R. W. Douglas (1979), "Viscous flow in oscillatory spherical annuli," Phys. Fluids **22**, 205-208.

Munson, B. R. and D. D. Joseph (1971a), "Viscous incompressible flow between concentric rotating spheres. Part 1. Basic flow," J. Fluid Mech. **49**, 289-303.

Munson, B. R. and D. D. Joseph (1971b), "Viscous incompressible flow between concentric rotating spheres. Part 2. Hydrodynamic stability," J. Fluid Mech. **49**, 305-318.

Munson, B. R. and M. Menguturk (1975), "Viscous incompressible flow between concentric rotating spheres. Part 3. Linear stability and experiments," J. Fluid Mech. **69**, 705.

Munson, B. R., A. A. Rangwalla, and J. A. Mann, III (1985), "Low Reynolds number circular Couette flow past a wavy wall," Phys. Fluids **28**, 2679-86.

Muralidhar, K. and S. I. Guceri (1986), "Comparative study of two numerical procedures for free-convection problems," Numer. Heat Transfer **9**, 631-8.

Murayama, T. (1987), "Theory of vorticity-quantum in turbulent flow. IV. Determination of turbulence quantum tau," Mem. Natl. Def. Acad. **27**, 43-51.

Murray, B. T., G. B. McFadden, and S. R. Coriell (1990), "Stabilization of Taylor-Couette flow due to time periodic outer cylinder oscillation," Phys. Fluids A **2**, 2147-2156.

Murthy, J. (1988), "Study of heat transfer from a finned rotating cylinder," J. Thermophys. Heat Transfer **2**, 250.

Murthy, S. N., M. R. K. Murthy, and P. C. Sridharan (1981), "Basic flows in MHD porous media. II," Proc. Natl. Acad. Sci. India, Sect. A **51**, 369-75.

Mutabazi, I. (1990), "Secondary instability of traveling inclined rolls in the Taylor-Dean system," in <u>Patterns, Defects and Instabilities in Materials</u>, edited by D. Walgraef and M. Ghoniem (Kluwer Academic Publishers), pp. 98.

Mutabazi, I. and C. D. Andereck (1991), "The transition from time-dependent to stationary flow patterns in the Taylor-Dean system," Preprint.

Mutabazi, I., J. J. Hegseth, C. D. Andereck, and J. E. Wesfreid (1988), "Pattern formation in the flow between two horizontal coaxial cylinders with a partially filled gap," Phys. Rev. A **38**, 4752-60.

Mutabazi, I., J. J. Hegseth, C. D. Andereck, and J. E. Wesfreid (1990), "Spatio-temporal modulation in the Taylor-Dean system," Phys. Rev. Lett. **64**, 1729-1732.

Mutabazi, I., C. Normand, H. Peerhossaini, and J. E. Wesfreid (1989), "Oscillatory modes in the flow between two horizontal corotating cylinders with a partially filled gap," Phys. Rev. A **39**, 763-71.

Mutabazi, I., H. Peerhossaini, and J. Wesfreid (1988), "New patterns in the flow between two horizontal coaxial rotating cylinders with partially filled gap," in <u>Propagation in Systems Far from Equilibrium</u>, edited by J. Wesfreid, H. Brand, P. Manneville, G. Albinet, and N. Boccara (Springer-Verlag, Berlin), pp. 331-337.

Mutabazi, I., H. Peerhossaini, J. Wesfreid, and C. D. Andereck (1988), "Non-axisymmetric traveling patterns in Taylor-Dean problem," Preprint.

Mutabazi, I. and J. E. Wesfreid (1991), "Spatio-temporal properties of hydrodynamic centrifugal instabilities," in <u>Instabilities and Nonequilibrium Structures</u>, edited by E. Tirapegui (Kluwer Academic Publishers).

Mutabazi, I., J. E. Wesfreid, J. J. J.J.Hegseth, and C. D. Andereck (1991), "Experimental results in the Taylor-Dean system," Eur. J. Mech. B/Fluids **10 (suppl.)**, 239-245.

Nadim, A. (1988), "The measurement of shear-induced diffusion in concentrated suspensions with a Couette device," Phys. Fluids **31**, 2781-5.

Nadim, A., R. G. Cox, and H. Brenner (1986), "Taylor dispersion in concentrated suspensions of rotating cylinders," J. Fluid Mech. **164**, 185-215.

Nagata, M. (1986), "Bifurcations in Couette flow between almost corotating cylinders," J. Fluid Mech. **169**, 229-50.

Nagata, M. (1988), "On wavy instabilities of the Taylor-vortex flow between corotating cylinders," J. Fluid Mech. **188**, 585-98.

Naimi, M., R. Deviene, and M. Lebouche (1990), "Etude dynamique et thermique de l'ecoulement de Couette-Taylor-Poiseuille; cas d'un fluide presentant un seuil d'ecoulement," Int. J. Mass Transfer **33**, 381-391.

Naitoh, T. and S. Ono (1978), "The shear viscosity of 500 hard spheres via non-equilibrium molecular dynamics," Phys. Lett. A **69A**, 125-6.

Naitoh, T. and S. Ono (1979), "The shear viscosity of a hard-sphere fluid via nonequilibrium molecular dynamics," J. Chem. Phys. **70**, 4515-23.

Nakabayashi, K. (1983), "Transition of Taylor-Görtler vortex flow in spherical Couette flow," J. Fluid Mech. **132**, 209-30.

Nakabayashi, K. and Y. Tsuchida (1988a), "Spectral study of the laminar-turbulent transition in spherical Couette flow," J. Fluid Mech. **194**, 101-32.

Nakabayashi, K. and Y. Tsuchida (1988b), "Modulated and unmodulated travelling azimuthal waves on the toroidal vortices in a spherical Couette system," J. Fluid Mech. **195**, 495-522.

Nakabayashi, K., Y. Yamada, and T. Kishimoto (1982), "Visous frictional torque in the flow between two concentric rotating rough cylinders," J. Fluid Mech. **119**, 409-422.

Nakabayashi, K., Y. Yamada, S. Mizuhara, and K. Hiraoka (1972), "Viscous frictional moment and pressure distribution between eccentric rotating cylinders, when the inner cylinder rotates.," Trans. Japan Soc. Mech. Engrs. **38-312**, 2075.

Nakabayashi K., Y. Yamada, and Y. Yamada (1977), "Flow between eccentric rotating cylinders, where the clearance is relatively large," Bull. Japan Soc. Mech. Engrs. **20-144**, 725.

Nakamura, I., Y. Toya, S. Yamashita, and Y. Ueki (1989), "An experiment on a Taylor vortex flow in a gap with a small aspect ratio (instability in Taylor vortex flows)," JSME International J., Series II, Fluids **32**, 388-394.

Nakamura, I., Y. Toya, and S. Yamashita (1990), "An experiment on a Taylor vortex flow in a gap with a small aspect ratio (bifurcation of flows in a symmetric system)," JSME International J., Series II, Fluids **33**, 685.

Nakaya, C. (1974), "Domain of stable periodic vortex flows in a viscous fluid between concentric circular cylinders," J. Phys. Soc. Jpn. **36**, 1164-1173.

Nakaya, C. (1975), "The second stability boundary for circular Couette flow," J. Phys. Soc. Jpn. **38**, 576-585.

Nanbu, K. (1984), "Analysis of cylindrical Couette flows by use of the direction simulation method," Phys. Fluids **27**, 2632-5.

Narasimhacharyulu, V. and N. C. Pattabhi Ramacharyulu (1978), "Steady flow through a porous region contained between two cylinders," J. Indian Inst. Sci. **60**, 37-42.

Narasimhan, M. N. L. and N. N. Ghandour (1982), "Taylor stability of a thermoviscoelastic fluid in Couette flow," Int. J. Eng. Sci. **20**, 303-9.

Neitzel, G. P. (1982a), "Marginal stability of impulsively initiated Couette flow and spin-decay," Phys. Fluids **25**, 226-32.

Neitzel, G. P. (1982b), "Stability of circular Couette flow with variable inner cylinder speed," J. Fluid Mech. **123**, 43-57.

Neitzel, G. P. (1984), "Numerical computation of time-dependent Taylor-vortex flows in finite-length geometries," J. Fluid Mech. **141**, 51-66.

Neitzel, G. P. and S. H. Davis (1981), "Centrifugal instabilities during spin-down to rest in finite cylinders. Numerical experiments," J. Fluid Mech. **102**, 329-52.

Newell, A. C. (1974), "Envelope equations," in Lectures in Applied Mechanics, vol. 15 (American Mathematical Society), pp. 157-163.

Newell, A. C. (1986), "Chaos and turbulence: is there a connection?," in Special Proceedings of the Conference on Mathematics Applied to Fluid Mechanics and Stability dedicated in Memory of Richard C. DiPrima (SIAM).

Newell, A. C. (1988), "The dynamics of patterns: a survey," in Propagation in Systems Far from Equilibrium, edited by J. Wesfreid, H. Brand, P. Manneville, G. Albinet, and N. Boccara (Springer-Verlag, Berlin), pp. 122-155.

Newell, A. C. (1989), "The dynamics of patterns," in Lectures in the Sciences of Complexity, edited by D. L. Stein (Addison-Wesley, Redwood City, CA), pp. 107-174.

Newell, A. C. and J. A. Whitehead (1969), "Finite bandwidth, finite amplitude convection," J. Fluid Mech. **38**, 279.

Newton, I. (1946), in Mathematical Principles, edited by F. Cajori (Univ. of California Press, Berkeley), pp. 385.

Ng, B. S. (1986), "On the nonlinear stability of spiral flow between rotating cylinders: results for a small-gap geometry," in Special Proceedings of the Conference on Mathematics Applied to Fluid Mechanics and Stability dedicated in Memory of Richard C. DiPrima (SIAM), pp. 218-228.

Ng, B. S. and E. R. Turner (1982), "On the linear stability of spiral flow between rotating cylinders," Proc. R. Soc. London, Ser. A **382**, 83-102.

Nguyen, Q. D. and D. V. Boger (1987), "Characterization of yield stress fluids with concentric cylinder viscometers," Rheol. Acta **26**, 508-15.

Niblett, E. R. (1958), "THe stability of Couette flow in an axial magnetic field," Canad. J. Phys. **36**, 1509-25.

Nickerson, E. C. (1969), "Upper bounds on the torque in cylindrical Couette flow," J. Fluid Mech. **38**, 807-815.

Nikishova, O. D. and N. A. Gorbatyuk (1991), "Development of Taylor-Goertler vortices in the boundary layer of a surface moving curvilinearly in the presence of polymer additives," J. Eng. Phys. **60**, 327.

Niklas, M. (1987), "Influence of magnetic fields on Taylor vortex formation in magnetic fluids," Z. Phys. B **68**, 493-501.

Niklas, M., M. Lücke, and H. Muller-Krumbhaar (1989a), "Propagating front of a propagating pattern: influence of group velocity," Europhys. Lett. **9**, 2337-42.

Niklas, M., M. Lücke, and H. Muller-Krumbhaar (1989b), "Velocity of a propagating Taylor-vortex front," Phys. Rev. A **40**, 493-6.

Niklas, M., H. Mueller-Krumbhaar, and M. Lücke (1989), "Taylor-vortex flow of ferrofluids in the presence of general magnetic fields," J. Magn. Magn. Mat. **81**, 29-38.

Niller, P. P. (1965), "Performance of a thermistor anemometer in constant density shear flow," Rev. Sci. Instrum. **36**, 921-924.

Ning, L. and G. Ahlers, Cannell, D. S. (1990), "Wave-number selection and traveling vortex waves in spatially ramped Taylor-Couette flow," Phys. Rev. Lett. **64**, 1235-1238.

Ning, L., G. Ahlers, and D. S. Cannell (1991), "Novel states in Taylor-Couette flow subjected to a Coriolis force," J. Stat. Phys. **64**, 927.

Ning, L., G. Ahlers, Cannell, D. S., and M. Tveitereid (1991), "Experimental and theoretical results for Taylor-Couette flow subjected to a Coriolis force," Phys. Rev. Lett. **66**, 1575-1578.

Ning, L., M. Tveitereid, G. Ahlers, and D. S. Cannell (1991), "Taylor-Couette flow subjected to external rotation," Phys. Rev. A **44**, 2505.

Nishimura, T., K. Yano, and T. Yoshino (1990), "Occurrence and structure of Taylor-Görtler vortices induced in two-dimensional wavy channels for steady flow," J. Chem. Eng. Jpn. **23**, 697.

Nordh, J., J. Deinum, and B. Norden (1986), "Flow orientation of brain microtubules studied by linear dichroism," Eur. Biophys. J. **14**, 113-22.

Normand, C., I. Mutabazi, and J. E. Wesfreid (1991a), "Recirculation eddies in the flow between two coaxial rotating cylinders," Eur. J. Mech. B/Fluids **10**, 1-14.

Normand, C., I. Mutabazi, and J. E. Wesfreid (1991b), "Recirculation eddies in the flow between two horizontal coaxial cylinders with a partially filled gap," Eur. J. Mech. B/Fluids **10**, 335.

Normand, C., I. Mutabazi, and J. E. Wesfreid (1992), " End circulation in non-axisymmetrical flows," in Ordered and Turbulent Patterns in Taylor-Couette Flow, edited by C. D. Andereck and F. H. F.Hayot (Plenum).

Nouar, C., R. Devienne, and M. Lebouche (1987), "Thermal convection for Couette flow with axial flow rate. Case of a pseudo-plastic fluid," Int. J. Heat Mass Transf. **30**, 639-47.

Nozaki, K. and N. Bekki (1983), "Pattern selection and spatiotemporal transition to chaos in the Ginzburg-Landau equation," Phys. Rev. Lett. **51**, 2171-2174.

O'brien, V. (1983), "Flows in pressure holes," J. Non-Newtonian Fluid Mech. **12**, 383-6.

Ohji, M., S. Shionoya, and K. Amagai (1986), "A note on modulated wavy disturbances to circular Couette flow," J. Phys. Soc. Jpn. **55**, 1032-3.

Oikawa, M., T. Karasudani, and M. Funakoshi (1989a), "Stability of flow between eccentric rotating cylinders with a wide gap," J. Phys. Soc. Jpn. **58**, 2209-10.

Oikawa, M., T. Karasudani, and M. Funakoshi (1989b), "Stability of flow between eccentric rotating cylinders," J. Phys. Soc. Jpn. **58**, 2355.

Onuki, A. (1985), "Conjectures on elongation of a polymer in shear flow," J. Phys. Soc. Jpn. **54**, 3656-9.

Onuki, A. and K. Kawasaki (1978), "Fluctuations in nonequilibrium steady states with laminar shear flow: classical fluids near the critical point," Prog. Theor. Phys. Suppl., 436-41.

Oolman, T., E. Walitza, and H. Chmiel (1986), "On the rheology of biosuspensions," Rheol. Acta **25**, 433-9.

Orszag, S. A., R. B. Pelz, and B. J. Bayly (1986), "Secondary instabilities, coherent structures and turbulence," in Supercomputers and Fluid Dynamics. Proceedings of the First Nobeyama Workshop, edited by K. Kuwahara, R. Mendez, and S. A. Orszag (Springer-Verlag, Berlin), pp. 1-14.

Ovchinnikov, A. I. and A. G. Perevozchikiv (1987), "On calculating turbulent Couette flow in a round annular slot," Izv. Akad. Nauk SSSR, Energ. Transp. **25**, 134-8 [Power Eng. (USSR) **25**, 121-5 (1987)].

Ovchinnikova, S. N. and V. I. Yudovich (1968), "Analysis of secondary steady flow between rotating cylinders," Prikl. Mat. Mek. **32**, 884-894 [J. Appl. Math. Mech. **32**].

Ovchinnikova, S. N. and V. I. Yudovich (1974a), "Stability and bifurcation of Couette flow in the case of a narrow gap between rotating cylinders," J. Appl. Math. Mech. (USSR) **34**, 1025-1030.

Ovchinnikova, S. N. and V. I. Yudovich (1974b), Prikl. Mat. Mek. **38**, 972-977 [J. Appl. Math. Mech.].

Ozogan, M. S. and F. R. Mobbs, "Superlaminar flow between eccentric rotating cylinders at small clearance ratios," in Proceedings?, pp. 181-189.

Paap, H. -G. and H. Rieke (1990), "Wave-number restriction and mode interaction in Taylor vortex flow: Appearance of a short-wavelength instability," Phys. Rev. A **41**, 1943-1951.

Paap, H. and H. Riecke (1991), "Drifting vortices in ramped Taylor vortex flow. Quantitative results from phase equation," Phys. Fluids A **3**, 1519-1532.

Pai, S. I. (1939), "Turbulent flow between rotating cylinders," Ph.D. Thesis, California Institute of Technology.

Pai, S. I. (1943), "Turbulent flow between rotating cylinders.," Report No. Technical Note No. 892 (National Advisory Committee for Aeronautics).

Panton, R. L. (1984), Incompressible Flow (Wiley, New York).\par

Papageorgiou, D. (1987), "Stability of the unsteady viscous flow in a curved pipe," J. Fluid Mech. **182**, 209-33.

Park, K. (1984), "Unusual transition sequence in Taylor wavy vortex flow," Phys. Rev. A **29**, 3458-60.

Park, K., C. Barenghi, and R. J. Donnelly (1980), "Subharmonic destabilization of Taylor vortices near an oscillating cylinder," Phys. Lett. A **78A**, 152-4.

Park, K. and G. L. Crawford (1983), "Deterministic transitions in Taylor wavy-vortex flow," Phys. Rev. Lett. **50**, 343-346.

Park, K., G. L. Crawford, and R. J. Donnelly (1981), "Determination of transition in Couette flow in finite geometries," Phys. Rev. Lett. **47**, 1448-50.

Park, K., G. L. Crawford, and R. J. Donnelly (1983), "Characteristic lengths in the wavy vortex state of Taylor-Couette flow," Phys. Rev. Lett. **51**, 1352-4.

Park, K. and R. J. Donnelly (1981), "Study of the transition of Taylor vortex flow," Phys. Rev. A **24**, 2277-9.

Park, K. and K. Jeong (1984), "Stability boundary of the Taylor vortex flow," Phys. Fluids **27**, 2201-3.

Park, K. and K. Jeong (1985), "Transition measurements near a multiple bifurcation point in a Taylor-Couette flow," Phys. Rev. A **31**, 3457-9.

Parry, M. A. and H. H. Billon (1990), "Flow behaviour of molten 2,4,6-trinitrotoluene (TNT) between concentric cylinders," Rheol. Acta **29**, 462.

Pavlovskii, V. A. (1981), "Systematization of experimental data on resistance for turbulent flow between rotating cylinders (Couette flow)," Dok. Akad. Nauk SSSR **261**, 305-9 [Sov. Phys. Dokl. **26**, 1033-6 (1981)].

Payne, A. and B. W. Martin (1974), <u>Proc. 5th Int. Heat Transfer Conf.</u>, vol. 2:FC2.7, pp. 80.

Pearson, C. E. (1967), "A numerical study of the time-dependent viscous flow between two rotating spheres," J. Fluid Mech. **28**, 323-336.

Pearson, D. S., R. R. Chance, A. D. Kiss, K. M. Morgan, and D. G. Peiffer (1989), "Flow induced birefringence of conjugated polymer solutions," Synth. Met. **28**, D689-97.

Pearson, J. B. and F. L. Curzon (1988), "Aspects of energy transport in a vortex-stabilized arc," J. Appl. Phys. **64**, 77-88.

Peev, G. (1985), "Mass transfer in cylindrical Couette flow of power law liquid," Chem. Eng. Sci. **40**, 1985-8.

Pelissier, R. (1979), "Stability of a pseudo-periodic flow," Z. Angew. Math. Phys. **30**, 577-85.

Pellew, A. and R. V. Southwell (1940), "On maintained convective motion in a fluid heated from below," Proc. R. Soc. London, Ser. A **176**, 312-43.

Pennings, A. J. and A. Zwijnenburg (1979), "Longitudinal growth of polymer crystals from flowing solutions. Vi. Melting behavior of continuous fibrillar polyethylene crystals," J. Polym. Sci., Polym. Phys. Ed. **17**, 1011-32.

Perrin, B. (1982), "Emergence of a periodic mode in the so-called turbulent region in a circular Couette flow," J. Phys. (Paris), Lett. **43**, L5-10.

Persen, L. N. (1983), "The break-up mechanism of a streamwise directed vortex," in <u>Agard Conference Proceedings No.342. Aerodynamics Of Vortical Type Flows In Three Dimensions (Agard-CP-342)</u> (AGARD, Neuilly-sur-Seine, France), pp. 24/1-5.

Petitjeans, P., J. E. Wesfreid, and H. Peerhossaini (1991), "Non linear investigation of the Taylor-Görtler vortices," Eur. J. Mech. B/Fluids **10 (suppl.)**, 327.

Pfandl, W., G. Link, and F. R. Schwarzl (1984), "Dynamic shear properties of a technical polystyrene melt," Rheol. Acta **23**, 277-90.

Pfister, G., "Period-doubling in rotational Taylor-Couette flow," Preprint.

Pfister, G. (1985), "Deterministic chaos in rotational Taylor-Couette flow," in <u>Flow of Real Fluids</u>, edited by G. E. A. G.E.A.Meier and F. Obermeier, Lecture Notes in Physics, vol.235 (Springer-Verlag, Berlin), pp. 199-209.

Pfister, G. and U. Gerdts (1981), "Dynamics of Taylor wavy vortex flow," Phys. Lett. **83A**, 23-25.

Pfister, G. and I. Rehberg (1981), "Space-dependent order parameter in circular Couette flow transitions," Phys. Lett. A **83A**, 19-22.

Pfister, G., K. Schatzel, and U. Gerdts (1984), "Velocity-correlation measurements of oscillating flow and turbulence in rotational Couette flow," in <u>Laser Anemometry in Fluid Mechanics</u> (Ladoan-Inst.SuperiorTecnico, Lisbon,Portugal), pp. 37-43.

Pfister, G., H. Schmidt, K. A. Cliffe, and T. Mullin (1988), "Bifurcation phenomena in Taylor-Couette flow in a very short annulus," J. Fluid Mech. **191**, 1-18.

Pfister, G., A. Schulz, and B. Lensch (1991), "Bifurcations and a route to chaos of an one-vortex state in Taylor-Couette flow," Eur. J. Mech. B/Fluids **10 (suppl.)**, 247-252.

Pfister, G., F. Schulz, and G. Geister (1985), "Period doubling and deterministic chaos in rotational Taylor-Couette flow," in <u>Optical Measurements in Fluid Mechanics 1985</u>, edited by P. H. Richards (Adam Hilger, Bristol), pp. 57-62.

Philander, S. G. H. (1971), "On the flow properties of a fluid between concentric spheres," J. Fluid Mech. **47**, 799-809.

Pilipenko, V. N. (1981), "Stability of the flow of a fiber suspension in the gap between coaxial cylinders," Dok. Akad. Nauk SSSR **259**, 554-8 [Sov. Phys. Dokl. **26**, 646-8 (1981)].

Polkowski, J. W. (1984), "Turbulent flow between coaxial cylinders with the inner cylinder rotating," J. Eng. Gas Turbines Power **106**, 128-35.

Pomeau, Y. (1986), "Front motion, metastability and subcritical bifurcations in hydrodynamics," Physica D **23**, 3-11.

Pomeau, Y. and P. Manneville (1979), J. Phys. (Paris), Lett. **40**, 610.

Pomeau, Y., A. Pumir, and W. R. Young (1988), "Transient effects in advection-diffusion of impurities," C. R. Acad. Sci., Ser. B **306**, 741-6.

Pomeau, Y. and S. Zaleski (1981a), "Wavelength selection in cellular structures: application to thermoconvection in porous materials," in <u>Symmetries And Broken Symmetries In Condensed Matter Physics. Proceedings Of The Colloque Pierre Curie</u>, edited by N. Boccara (IDSET, Paris), pp. 495-502.

Pomeau, Y. and S. Zaleski (1981b), "Wavelength selection in one-dimensional cellular structures," J. Phys. (Paris), Lett. **42**, 515-528.

Pomeau, Y. and S. Zaleski (1983), "Pattern selection in a slowly varying enviroment," J. Phys. (Paris), Lett. **44**, L135-41.

Proudman, I. (1956), "The almost-rigid rotation of viscous fluid between concentric spheres," J. Fluid Mech. **1**, 505-516.

Puchelle, E., J. M. Zahm, C. Duvivier, J. Didelon, J. Jacquot, and D. Quemada (1985), "Elasto-thixotropic properties of bronchial mucus and polymer analogs. I. Experimental results," Biorheology **22**, 415-23.

Rabinovich, M. I. (1978), "Stochastic self-oscillations and turbulence," Usp. Fiz. Nauk **125**, 123-68 [Sov. Phys. Usp. **21**, 443-69 (1978)].

Raffai, R. and P. Laure (1991), "Effects of the eccentricity on the primary instabilities in the Couette-Taylor problem," Eur. J. Mech. B/Fluids **10 (suppl.)**, 315.

Ramaswamy, M. and H. B. Keller (1991), "The calculation of the coefficients in the reduced bifurcation equation for the Taylor problem," Eur. J. Mech. B/Fluids **10 (suppl.)**, 316.

Ramesh, P. S. and M. H. Lean (1991), "A boundary integral equation method for Navier Stokes equations-application to flow in annulus of eccentric cylinders," Int. J. Numer. Methods Fluids **13**, 355.

Rand, D. (1980), "The preturbulent transitions and flows of a viscous fluid between concentric rotating cylinders" (Mathematics Institute, Univ. of Warwick, Coventry, England).

Rand, D. (1982), "Dynamics and symmetry. Predictions for modulated waves in rotating fluids," Arch. Ration. Mech. Anal. **79**, 1-37.

Raney, D. C. and T. S. Chang (1971), Z. Angew. Math. Phys. **22**, 680.

Rao, S. J., S. K. Bhatia, and D. V. Khakhar (1991), "Axial transport of granular solids in horizontal rotating cylinders. Part 2: Experiments in a non-flow system," Powder technology **67**, 153.

Rappl, P. H. O., L. F. Matteo Ferraz, H. J. Scheel, M. R. X. Barros, and D. Schiel (1984), "Hydrodynamic simulation of forced convection in Czochralski melts," J. Cryst. Growth **70**, 49-55.

Ravey, J. C., M. Dognon, and M. Lucius (1980), "Transient rheology in a new type of Couette apparatus. I. Theoretical study with Newtonian fluids," Rheol. Acta **19**, 51-9.

Rayey, J. C., S. Ikemoto, and J. F. Stoltz (1989), "Transient rheology in a new type of Couette apparatus. Application to blood," Rheol. Acta **28**, 423.

Rayleigh, L. (1914), Philos. Mag. **28**, 609.

Rayleigh, L. (1916), "On the dynamics of revolving fluids," Proc. R. Soc. London, Ser. A **93**, 148-154.

Reddy, P. G., Y. B. Reddy, and A. G. S. Reddy (1978), "The stability of non-Newtonian fluid between the two rotating porous cylinders (wide gap case)," Def. Sci. J. **28**, 145-52.

Rehberg, I. (1980), Diplom Thesis, Universitat Kiel, Kiel, Germany.

Reid, W. H. (1958), "On the stability of viscous flow in a curved-channel," Proc. R. Soc. London, Ser. A **244**, 186-198.

Renardy, M. (1982), "Bifurcation from rotating waves," Arch. Ration. Mech. Anal. **79**, 49-84.

Renardy, Y. (1987), "The thin layer effect and interfacial stability in a two-layer Couette flow with similar liquids," Phys. Fluids **30**, 1627-1637.

Renardy, Y. and D. D. Joseph (1985), "Couette flow of two fluids between concentric cylinders," J. Fluid Mech. **150**, 381-94.

Reynolds, W. C. and M. C. Potter (1967), "A finite amplitude state-selection theory for Taylor-vortex flow" (Dept. Mech. Eng. Stanford Univ.).

Richtmyer, R. D. (1981), "Ch. 29 Bifurcations in hydrodynamic stability problems; Ch. 30 Invariant manifolds in the Taylor problem; Ch. 31 The early onset of turbulence," in <u>Principles of Advanced Mathematical Physics</u>, vol. II (Springer, New York), pp. 244-311.

Riecke, H. (1986a), "Pattern selection by weakly pinning ramps," Europhys. Lett. **2**, 1-8.

Riecke, H. (1986b), "Wave-number restriction in quasi-one-dimensional pattern-forming systems," Ph.D. Thesis, Universitat Bayreuth, Bayreuth, Germany.

Riecke, H. (1988), "Imperfect wave-number selection by ramps in a model for Taylor vortex flow," Phys. Rev. A **37**, 636-8.

Riecke, H. (1990a), "Stable wave-number kinks in parametrically excited standing waves," Europhys. Lett. **11**, 213-218.

Riecke, H. (1990b), "On the stability of parametrically excited standing waves," in <u>Nonlinear Evolution of Spatio-temporal Structures in Dissipative Continuous Systems</u>, edited by F. H. Busse and L. Kramer (Plenum).

Riecke, H., J. D. Crawford, and E. Knobloch (1988), "Temporal modulation of a subcritical bifurcation to travelling waves," in <u>New Trends in Nonlinear Dynamics and Pattern Forming Phenomena: the Geometry of Nonequilibrium</u>, edited by P. Coullet and P. Huerre, NATO Advanced Study Institute (Plenum).

Riecke, H. and H. Paap (1986), "Stability and wave-vector restriction of axisymmetric Taylor vortex flow," Phys. Rev. A **33**, 547-553.

Riecke, H. and H. -G. Paap (1987), "Perfect wave-number selection and drifting patterns in ramped Taylor vortex flow," Phys. Rev. Lett. **59**, 2570-3.

Riecke, H. and H. Paap (1991), "Spatio-temporal chaos through ramp-induced Eckhaus instability," Europhys. Lett. **14**, 433.

Riecke, H. and H. -G. Paap (1991), "Parity Breaking and Hopf Bifurcation in Axisymmetric Taylor Vortex Flow," (preprint).

Riecke, H. and H. -G. Paap (1992), "Chaotic Phase Diffusion through the Interaction of Phase Slip Processes," in <u>Ordered and Turbulent Patterns in Taylor-Couette Flow</u>, edited by C. D. Andereck and F. Hayot, Proceedings of a NATO Advanced Research Workshop (Plenum).

Rietveld, J. and A. J. Mchugh (1983), "Flow-induced crystallization by surface growth of polyethylene fibers," J. Polym. Sci., Polym. Phys. Ed. **21**, 1513-26.

Rigby, D. and P. M. Eagles (1992), "Taylor-Goertler instability for thin-film flows over moving curved beds," Phys. Fluids A **4**, 86.

Riley, P. J. and R. L. Laurence (1976), "Linear stability of modulated circular Couette flow," J. Fluid Mech. **75**, 625-646.

Riley, P. J. and R. L. Laurence (1977), "Energy stability of modulated circular Couette flow," J. Fluid Mech. **79**, 535-52.

Rios, P. P. and U. van Kolck (1988), "Intermittency in a Couette-Taylor system," <u>Phys 380N student report</u>.

Riste, T. (1982), Ed., <u>Nonlinear Phenomena at Phase Transitions and Instabilities. Proceedings of a NATO Advanced Study Institute</u> (Plenum, New York).

Ritchie, G. S. (1968), "On the stability of viscous flow between eccentric rotating cylinders," J. Fluid Mech. **32**, 131-144.

Rivlin, R. S. (1983), "Spin-up in Couette flow," Appl. Anal. **15**, 227-53.

Roberts, G. W., A. R. Davies, and T. N. Phillips (1991), "Three-dimensional spectral approximations to Stokes flow between eccentrically rotating cylinders," Int. J. Numer. Methods Fluids **13**, 217.

Roberts, P. H. (1960), "Characteristic value problems posed by differential equations arising in hydrodynamics and hydromagnetics," J. Math. Anal. Appl. **1**, 195-214.

Roberts, P. H. (1965), "The solution to the characteristic value problems (appendix to R. J. Donnelly and K. W. Schwarz)," Proc. R. Soc. London, Ser. A **283**, 550-556.

Roberts, G. W., A. R. Davies, and T. N. Phillips (1991), "Three-dimensional spectral approximations to Stokes flow between eccentrically rotating cylinders," Int. J. Numer. Methods Fluids **13**, 217.

Roesner, K. G. (1978), "Hydrodynamic stability of cylindrical Couette-flow," Arch. Mech. **30**, 619-27.

Roesner, K. G. (1986), "Sur une solution exacte des equations de Navier-Stokes pour un fluide non-Newtoniien," C. R. Acad. Sci..

Rogers, C. and W. F. Ames (1989), "Group analysis of the Navier-Stokes equations. Application to a rotating cylinder problem," in <u>Nonlinear Boundary Value Problems in Science and Engineering</u> (Academic, San Diego), pp. 308-314.

Rohan, K. and G. Lefebvre (1991), "Hydrodynamic aspects in the rotating cylinder erosivity test," Geotechnical Testing Journal **14**, 166.

Romashko, E. A. (1977), "Nonlinear stability of motion of viscous liquid between concentric rotating cylinders," Inzh.-Fiz. Zh. **33**, 719-27 [J. Eng. Phys. **33**, 1231-8 (1977)].

Rosenblat, S. (1968), "Centrifugal instability of time-dependent flows. Part 1. Inviscid, periodic flows," J. Fluid Mech. **33**, 321-336.

Rosenblatt, J. S., D. S. Soane, and M. C. Williams (1987), "A Couette rheometer design for minimizing sedimentation and red-cell-aggregation artifacts in low-shear blood rheometry," Biorheology **24**, 811-16.

Ross, M. P. and A. K. M. Fazle Hussain (1987), "Effects of cylinder length on transition to doubly periodic Taylor-Couette flow," Phys. Fluids **30**, 607-9.

Rosso, F. (1984), "On the pointwise stability of a viscous liquid between rotating coaxial cylinders," J. Mec. Theor. Appl. **98**, 251-69.

Roth, D. and M. Lücke (1991), "Phase diffusion for spatially nonumniform systems," Eur. J. Mech. B/Fluids **10 (suppl.)**, 333.

Roth, D., M. Lücke, M. Kamps, and R. Schmitz, "Structure of Taylor vortex flow and the influence of spatial amplitude variations on phase dynamics," in <u>Ordered and Turbulent Patterns in Taylor-Couette Flow</u>, edited by C. D. Andereck and F. Hayot, Proceedings of a NATO Advanced Research Workshop (Plenum).

Ruelle, D. (1979), "Sensitive dependence on initial condition and turbulent behavior of dynamical systems," Ann. N. Y. Acad. Sci. **316**, 408-416.

Ruelle, D. and F. Takens (1971), Commun. Math. Phys. **20**, 167.

Rustad, J. R., D. A. Yuen, and F. J. Spera (1989), "Nonequilibrium molecular dynamics of liquid sulfur in Couette flow," J. Chem. Phys. **91**, 3662.

Ruzmaikin, A. A., D. D. Sokolov, and A. A. Solov'ev (1989), "Couette-Poiseuille Flow as a Screw Dynamo," Magn. Gidrodin. **25** [Magnetohydrodynamics **25**, 6 (1989)].

Sagues, F. and W. Horsthemke (1986), "Diffusive transport in spatially periodic hydrodynamic flows," (preprint).

Salinas, A. and J. F. T. Pittman (1981), "Bending and breaking fibers in sheared suspensions," Polym. Eng. Sci. **21**, 23-31.

Saric, W. S. and Z. Lavan (1971), "Stability of circular Couette flow of binary mixtures," J. Fluid Mech. **47**, 65-80.

Sastry, V. U. K. and T. Das (1985), "Stability of Couette flow and Dean flow in micropolar fluids," Int. J. Eng. Sci. **23**, 1163-77.

Sattinger, D. H. (1980), "Bifurcation and symmetry breaking in Applied Mathematics," Bull. Am. Math. Soc. **3**, 779-817.

Savage, S. B. and S. Mckeown (1983), "Shear stresses developed during rapid shear of concentrated suspensions of large spherical particles between concentric cylinders," J. Fluid Mech. **127**, 453-72.

Savage, S. B. and M. Sayed (1984), "Stresses developed by dry cohesionless granular materials in an annular shear cell," J. Fluid Mech. **142**, 391-430.

Savas, O. (1985), "On flow visualization using reflective flakes," J. Fluid Mech. **152**, 235-248.

Sawatzki, O. and J. Zierep (1970), "Das Stromfeld im Splat zwischen zwei konzentrischen Kugelflachen, von denen die innere rotiert," Acta Mech. **9**, 13.

Schaeffer, D. G. (1980), "Qualitative analysis of a model for boundary effects in the Taylor problem," Math. Proc. Camb. Phil. Soc. **87**, 307-337.

Schlegel, D. (1980), "Determination of the wall slip function of blood with a Couette rheometer under consideration of the wall effect," Rheol. Acta **19**, 375-80.

Schlegel, D. and H. -J. Welter (1986), "Wall effects in disperse plastic materials," Rheol. Acta **25**, 618-31.

Schmidt, H. (1983), Diplom Thesis, Universitat Kiel, Kiel, Germany.

Schneider, W. (1989), "On Reynolds stress transport in turbulent Couette flow," Z. Flugwiss. Weltraumforsch. **13**, 315.

Schneyer, G. P. and S. A. Berger (1971), "Linear stability of the dissipative, two-fluid, cylindrical Couette problem. Part 1. The stably-stratified hydrodynamic problem," J. Fluid Mech. **45**, 91-110.

Schramm, G. (1984), "An introduction to rheology and viscometry: rotary viscosimeters," Regul. And Mando Autom. **18**, 99-105.

Schrauf, G. (1986), "The first instability in spherical Taylor-Couette flow," J. Fluid Mech. **166**, 287-303.

Schrauf, G. and E. Krause (1984), "Symmetric and asymmetric Taylor vortices in a spherical gap," in Second IUTAM Symposium on Laminar-Turbulent Transition (Springer).

Schroder, W. and H. B. Keller (1990), "Wavy Taylor-Vortex flows via multigrid-continuation methods," J. Comput. Phys. **91**, 197.

Schultz-Grunow, F. (1959), "Zur stabilitat der Couette-Stromung," Z. Angew. Math. Mech. **39**, 101-110.

Schultz-Grunow, F., "On the stability of Couette flow," Report No. 265 (Nato Advisory Group for Aeronautical Research and Development).

Schultz-Grunow, F. (1963), "Stabilitat einer rotierenden Flussigkeit," Z. Angew. Math. Mech. **43**, 411-415.

Schultz-Grunow, F. and H. Hein (1956), "Beitrag zur Couetteströmung," Z. Flugwiss. **4**, 28-30.

Schulz-Dubois, E. O. (1984), "Super-critical bifurcation in the Taylor vibration experiment, and illustration by the Landau approach," Z. Angew. Math. Mech. **64**, T89-90.

Schummer, P. and W. Zang (1982), "Measurement of dynamic properties of dilute polymer solutions with a laser-doppler velocimeter," Rheol. Acta **21**, 517-20.

Schwarz, K. W., B. E. Springett, and R. J. Donnelly (1964), "Modes of instability in spiral flow between rotating cylinders," J. Fluid Mech. **20**, 281-289.

Segel, L. A. (1969), "Distant sidewalls cause slow slow amplitude modulation of cellular convection," J. Fluid Mech. **38**, 203-224.

Seminara, G. (1979), "The centrifugal instability of a Stokes layer: asymmetric disturbances," Z. Angew. Math. Phys. **30**, 615-25.

Seminara, G. and P. Hall (1976), "Centrifugal instability of a Stokes layer: linear theory," Proc. R. Soc. London, Ser. A **350**, 299-316.

Seminara, G. and P. Hall (1977), "The centrifugal instability of a Stokes layer: nonlinear theory," Proc. R. Soc. London, Ser. A **354**, 119-126.

Serrin, J. (1959), "On the stability of viscous fluid motions," Arch. Ration. Mech. Anal. **3**, 1-13.

Shadrina, N. Kh. (1978), "Shear flows of a thixotropic fluid," Izv. Akad. Nauk SSSR, Mekh. Zhidk. Gaza **13**, 3-12 [Fluid Dyn. (USSR) **13**, 341-9 (1978)].

Shahinpoor, M. and S. P. Lin (1982), "Rapid Couette flow of cohesionless granular materials," Acta Mech. **42**, 183-96.

Shaw, R. S., C. D. Andereck, L. A. Reith, and H. L. Swinney (1982), "Superposition of traveling waves in the circular Couette system," Phys. Rev. Lett. **48**, 1172-5.

Shearer, M. and I. C. Walton (1981), "On bifurcation and symmetry in Benard convection and Taylor vortices," Stud. Appl. Math. **65**, 85-93.

Shen Ching and Chao Quo-Ying (1980), "Couette flow between cylinders in transition regime," Acta Mech. Sin., 100-4.

Sherman, J. and J. Mclaughlin (1978), "Power spectra of nonlinearly coupled waves," Commun. Math. Phys. **58**, 9-17.

Sherwood, J. D. (1990), "The hydrodynamic forces on a cylinder touching a permeable wellbore," Phys. Fluids A **2**, 1754-1759.

Shivamoggi, B. K. (1986), <u>Theory of Hydromagnetic Stability</u> (Gordon and Breach, New York).

Shoji, Y., S. Uchida, and T. Ariga (1982), "Dissolution of solid copper cylinder in molten tin-lead alloys under dynamic conditions," Metall. Trans. B **13B**, 439-45.

Shonhiwa, T. and M. B. Zaturska (1987), "Disappearance of criticality in thermal explosion for reactive viscous flows," Combust. Flame **67**, 175-7.

Shul'man, Z. P., B. M. Khusid, and A. D. Matsepuro (1977), "Structure formation in electrorheological suspensions in an electric field. Ii. Quantitative," Vestsi Akad. Navuk Bssr Ser. Fiz. Energ. Navuk, 122-7.

Shul'man, Z. P. and V. F. Volchenok (1977), "Generalized Couette flow of a nonlinearly viscoplastic fluid," Inzh.-Fiz. Zh. **33**, 880-8 [J. Eng. Phys. **33**, 1340-6 (1977)].

Shusheng, L., S. Ernan, and Y. M. Lan, W. (1990), "Phase portraits and flow regimes in a circular Couette system with counter-rotating cylinders," Chin. Phys. Lett. **8**, 9-12.

Signoret, F. and G. Iooss (1988), "A codimension 3 singularity in Taylor-Couette problem," J. Mec. Theor. Appl. **7**, 545-72.

Simmers, D. A. and J. E. R. Coney (1979a), "The effect of Taylor vortex flow on the development length in concentric annuli," J. Mech. Eng. Sci. **21**, 59-64.

Simmers, D. A. and J. E. R. Coney (1979b), "A Reynolds analogy solution for the heat transfer characteristics of combined Taylor vortex and axial flows," Int. J. Heat Mass Transf. **22**, 679-89.

Simmers, D. A. and J. E. R. Coney (1980), "Velocity distributions in Taylor vortex flow with imposed laminar axial flow and isothermal surface heat transfer," Int. J. Heat Fluid Flow **2**, 85-91.

Simon, S. A., M. F. Czysz, III., K. Everett, and C. Field (1981), "Polytropic, differentially rotating cylinders," Am. J. Phys. **49**, 662-665.

Simon, S. A. and D. Hurley (1981), "Simplified models of completely degenerate, differentially rotating stars," Am. J. Phys. **49**, 20-24.

Sinevic, V., R. Kuboi, and A. W. Nienow (1986), "Power numbers, Taylor numbers and Taylor vortices in viscous Newtonian and non-Newtonian fluids," Chem. Eng. Sci. **41**, 2915-23.

Smith, G. P. and A. A. Townsend (1982), "Turbulent Couette flow between concentric cylinders at large Taylor numbers," J. Fluid Mech. **123**, 187-217.

Smith, A. M. O. (1955), "On the growth of Taylor-Görtler vortices along highly concave walls," Quart. Appl. Math. **13**, 233-62.

Smith, L. M. (1991), "Turbulent Couette flow profiles that maximize the efficiency function," J. Fluid Mech. **227**, 509.

Smook, J., J. C. Torfs, P. F. Van Hutten, and A. J. Pennings (1980), "Ultra-high strength polyethylene by hot drawing of surface growth fibers," Polym. Bull. **2**, 293-300.

Snyder, H. A. (1962), "Experiments on the stability of spiral flow at low axial Reynolds numbers," Proc. R. Soc. London, Ser. A **265**, 198-214.

Snyder, H. A. (1965), "Experiments on the stability of two types of spiral flow," Ann. Phys. (N.Y.) **31**, 292-313.

Snyder, H. A. (1968a), "Stability of rotating Couette flow. I. Asymmetric waveforms," Phys. Fluids **11**, 728-734.

Snyder, H. A. (1968b), "Stability of rotating Couette flow. II. Comparison with numerical results," Phys. Fluids **11**, 1599-1605.

Snyder, H. A. (1968c), "Experiments on rotating flows between noncircular cylinders," Phys. Fluids **11**, 1606-1611.

Snyder, H. A. (1969a), "Wavenumber selection at finite amplitude in rotating Couette flow," J. Fluid Mech. **35**, 273-298.

Snyder, H. A. (1969b), "Change in wave-form and mean flow associated with wavelength variations in rotating Couette flow. Part I," J. Fluid Mech. **35**, 337-352.

Snyder, H. A. (1969c), "Rotating cylinder viscometer," Rev. Sci. Instrum. **40**, 992-997.

Snyder, H. A. (1969d), "End effects and length effects in rotating Couette flow," Unpublished.

Snyder, H. A. (1970), "Waveforms in rotating Couette flow," Int. J. Non-Linear Mech. **5**, 659-685.

Snyder, H. A. (1972), "Rotating Couette flow of superfluid helium," in Proceedings of LT13 (Plenum), pp. 283-287.

Snyder, H. A. and S. K. F. Karlsson (1964), "Experiments on the stability of Couette motion with a radial thermal gradient," Phys. Fluids **7**, 1696-1706.

Snyder, H. A. and R. B. Lambert (1966), "Harmonic generation in Taylor vortices between rotating cylinders," J. Fluid Mech. **26**, 545-562.

Sohn, C. W. and M. M. Chen (1981), "Microconvective thermal conductivity in disperse two-phase mixtures as observed in a low velocity Couette flow experiment," J. Heat Transfer **103**, 47-51.

Solomon, T. H. and J. P. Gollub (1988), "Chaotic particle transport in time-dependent Rayleigh-Benard convection," Phys. Rev. A **38**, 6280.

Soloski, S. C. and T. H. K. Frederking (1979), "Axial heat transport through an annular duct with rotating inner cylinder (He II)," in Proceedings Of The Seventh International Cryogenic Engineering Conference (IPC Sci. and Technol. Press, Guildford, England), pp. 222-7.

Solov'yev, A. A. (1985), "A description of the value range of parameters of the spiral Couette current of a conducting liquid at which the excitation of a magnetic field is possible," Izv. Akad. Nauk SSSR, Fiz. Zemli **21** [Izv. Acad. Sci. USSR, Phys. Solid Earth **21**, 927-32 (1985)].

Sorour, M. M. and J. E. R. Coney (1979a), "The characteristics of spiral vortex flow at high Taylor numbers," J. Mech. Eng. Sci. **21**, 65-71.

Sorour, M. M. and J. E. R. Coney (1979b), "An experimental investigation of the stability of spiral vortex flow," J. Mech. Eng. Sci. **21**, 397-402.

Sorour, M. M. and J. E. R. Coney (1979c), "The effect of temperature gradient on the stability of flow between vertical, concentric, rotating cylinders," J. Mech. Eng. Sci. **21**, 403-9.

Soundalgekar, V. M. and P. R. L. Sarma (1987), "Finite-difference solution of heat transfer in developing flow in an annulus of two rotating cylinders," Int. J. Energy Res. **11**, 599-608.

Soundalgekar, V. M., M. A. Ali, and H. S. Takhar (1990), "Effects of a radial temperature gradient on the stability of a wide-gap annulus," Int. J. Energy Res. **14**, 597-603.

Soundalgekar, V. M., H. S. Takhar, and T. J. Smith (1981), "Effects of radial temperature gradient on the stability of viscous flow in an annulus with a rotating inner cylinder," Waerme- & Stoffuebertrag. **15**, 233-8.

Soundalgekar, V. M. and P. R. L. Sarma (1986), "Finite-difference solution of laminar developing flow in an annulus between two rotating cylinders," Appl. Energy **23**, 47-60.

Soward, A. M. and C. A. Jones (1983), "The linear stability of the flow in the narrow gap between two concentric rotating spheres," Quart. J. Mech. Appl. Math. **36**, 19-41.

Sparrow, E. M. and S. H. Lin (1964), Phys. Fluids **8**, 229.

Sparrow, E. M., W. D. Munro, and V. K. Jonsson (1964), "Instability of the flow between rotating cylinder: the wide-gap problem," J. Fluid Mech. **20**, 35-46.

Specht, H., M. Wagner, and R. Meyer-Spasche (1989), "Interactions of secondary branches of Taylor vortex solutions," Z. Angew. Math. Mech. **69**, 339 - 352.

Speziale, C. G. (1983), "On the nonlinear stability of rotating Newtonian and non-Newtonian fluids," Acta Mech. **49**, 263-73.

Squire, T. H., D. F. Jankowski, and G. P. Neitzel (1986), "Experiments with deceleration from a Taylor-vortex flow," Phys. Fluids **29**, 2742-3.

Sran, K. S. (1989), "Unsteady viscous flow between two coaxial porous cylinders subjected to a series of random pulses," Ind. Math. **39**, 1.

Sritharan, S. S. (1990), Invariant Manifold Theory for Hydrodynamic Transition (Longman, Harlow, England).

Stein, H. N., E. H. Logtenberg, A. J. G. Can Diemen, and P. J. Peters (1986), "Coagulation of suspensions in shear fields of different characters," Colloids Surf. **18**, 223-40.

Steinberg, V., J. Fineberg, and E. Moses (1989), "Pattern selection and transition to turbulence in propagating waves," Physica D **37**, 359.

Steinman, H. (1956), "The stability of viscous flow between rotating cylinders," Quart. J. Appl. Math. **14**, 27-33.

Stern, C. E. (1988), "Azimuthal mode interaction in counter-rotating Couette flow," Ph.D. Thesis, University of Houston.

Stern, C. and P. Chossat (1991), "Experimental and numerical study of mode interaction and chaos in counter-rotating Taylor-Couette flow," Eur. J. Mech. B/Fluids **10 (suppl.)**, 261-266.

Stern, C., P. Chossat, and F. Hussain (1990), "Azimuthal mode interaction in counter-rotating Taylor-Couette flow," Eur. J. Mech. B/Fluids **9**, 93.

Stewart, I. (1984), "Applications of nonelementary catastrophe theory," IEEE Trans. Circuits Syst. **CAS-31**, 165-74.

Stiles, P. J., M. Kagan, and J. B. Hubbard (1987), "On the Couette-Taylor instability in ferrohydrodynamics," J. Colloid Interface Sci. **120**, 430-8.

Stoff, H. (1980), "Incompressible flow in a labyrinth seal," J. Fluid Mech. **100**, 817-29.

Stokes, G. G. (1880), Mathematical and Physical Papers, vol. 1 (Cambridge Univ. Press, Cambridge, England), pp. 102.

Stokes, G. G. (1905), Mathematical and Physical Papers, vol. 5 (Cambridge Univ. Press, Cambridge, England), pp. 102.

Streett, C. L. and M. Y. Hussaini (1987), "Finite length Taylor Couette flow," in Stability of Time Dependent and Spatially Varying Flows, edited by D. L. Dwoyerand and M. Y. Hussaini (Springer, Berlin), pp. 312-334.

Streett, C. L. and M. Y. Hussaini (1991), "A numerical simulation of the appearance of chaos in finite-length Taylor-Couette flow," Appl. Num. Math. **7**, 41.

Stuart, J. T. (1958), "On the non-linear mechanics of hydrodynamic stability," J. Fluid Mech. **4**, 1-21.

Stuart, J. T. (1960), "On the nonlinear mechanics of wave disturbances in stable and unstable parallel flows Part 1. The basic behaviour in plane Poiseuille flow," J. Fluid Mech. **9**, 353-370.

Stuart, J. T. (1963), "Hydrodynamic Stability," in Laminar Boundary Layers, edited by L. Rosenhead (Oxford), pp. 492-579.

Stuart, J. T. (1971), "Nonlinear Stability Theory," Ann. Rev. Fluid Mech. **3**, 347-370.

Stuart, J. T. (1977), "Bifurcation theory in nonlinear hydrodyamical stability," in Applications of Bifurcation Theory, edited by P. H. Rabinowitz, Applications of Bifurcation Theory (Academic Press, New York), pp. 127-147.

Stuart, J. T. (1986), "Taylor-Vortex flow: a dynamical system," SIAM Rev. **28**, 315-42.

Stuart, J. T. (1987), "Instability and transition," in Advances in Turbulence. Proceedings of the First European Turbulence Conference , edited by G. Comte-Bellot and J. Mathieu (Springer-Verlag, Berlin), pp. 2-6.

Stuart, J. T. and R. C. DiPrima (1978), "The Eckhaus and Benjamin-Feir resonance mechanisms," Proc. R. Soc. London, Ser. A **362**, 27-41.

Stuart, J. T. and R. C. Diprima (1980), "On the mathematics of Taylor-vortex flows in cylinders of finite length," Proc. R. Soc. London, Ser. A **372**, 357-65.

Su, T. C. (1985), "Obtaining the exact solutions of the Navier-Stokes equations," Int. J. Non-linear Mech. **20**, 9-19.

Sugata, S. and S. Yoden (1991), "The effects of centrifugal force on the stability of axisymmetric viscous flow in a rotating annulus," J. Fluid Mech. **229**, 471.

Swanson, C. E. and R. J. Donnelly (1987), "The appearance of vortices in the flow of helium II between rotating cylinders," J. Low Temp. Phys. **67**, 185-93.

Swanson, C. J. and R. J. Donnelly (1991), "Instability of Taylor-Couette flow of helium II," Phys. Rev. Lett. **67**, 1578.

Swift, J. B., M. Gorman, and H. L. Swinney (1982), "Modulated wavy vortex flow in laboratory and rotating reference frames," Phys. Lett. **87A**, 457-460.

Swinney, H. L. (1979), "Transition to turbulence in Couette-Taylor flow (observed by laser velocimetry)," in Light Scattering In Solids, edited by J. L. Birman, H. Z. Cummins, and K. K. Rebane (Plenum, New York), pp. 15-22.

Swinney, H. L. (1982), "Experiments in Nonlinear Dynamics I. Transition to turbulence in circular Couette flow," Unpublished lecture notes: Morris Loeb Lectures in Physics,Harvard Univ..

Swinney, H. L. (1983), "Observations of order and chaos in nonlinear systems," Physica D **7D**, 3-15.

Swinney, H. L. (1985), "Observations of complex dynamics and chaos," in Fundamental Problems in Statistical Mechanics VI. Proceedings of the Sixth International Summer School, edited by E. G. D. Cohen (North-Holland, Amsterdam, Netherlands), pp. 253-89.

Swinney, H. L. (1988), "Instabilities and chaos in rotating fluids," in Nonlinear Evolution and Chaotic Phenomena. Proceedings of a NATO Advanced Study Institute, edited by G. Gallavotti and P. F. Zweifel (Plenum, New York), pp. 319-26.

Swinney, H. L., P. R. Fenstermacher, and J. P. Gollub (1977a), "Transition to turbulence in circular Couette flow," in Turbulent Shear Flow Symposium, pp. 17.1-17.6.

Swinney, H. L., P. R. Fenstermacher, and J. P. Gollub (1977b), "Transition to turbulence in a fluid flow," in Synergetics, a Workshop, edited by H. Haken (Springer-Verlag, Berlin), pp. 60-69.

Swinney, H. L. and J. P. Gollub (1978), "The transition to turbulence," Phys. Today **31**, 41-3,46-9.

Swinney, H. L. and J. P. Gollub (1985), Eds., Hydrodynamic Instabilities and the Transition to Turbulence, 2nd ed. (Springer, Berlin).

Synge, J. L. (1933), "The stability of heterogeneous liquids," Trans. R. Soc. Canada **27**, 1-18.

Synge, J. L. (1938), "On the stability of a viscous liquid between two rotating coaxial cylinders," Proc. R. Soc. London, Ser. A **167**, 250-256.

Szilas, A. P. (1984), "Grid-shell theory, a new concept to explain thixotropy," Rheol. Acta **23**, 70-4.

Tabeling, P. (1981), "Magnetohydrodynamic Taylor vortex flows," J. Fluid Mech. **112**, 329-45.

Tabeling, P. (1982), "Sequence of instabilities in electromagnetically driven flows between conducting cylinders," Phys. Rev. Lett. **49**, 460-463.

Tabeling, P. (1983), "Dynamics of the phase variable in the Taylor vortex system," J. Phys. (Paris), Lett. **44**, L665-72.

Tabeling, P. (1984), "Stability of cellular systems in Taylor-Couette instability," in Cellular Structures in Instabilities, edited by J. E. Wesfreid and S. Zaleski, Lect. Notes in Physics, vol. 210 (Springer, Berlin), pp. 172-176.

Tabeling, P. (1985), "Sudden increase of the fractal dimension in a hydrodynamic system," Phys. Rev. A **31**, 3460-3462.

Tabeling, P. and J. P. Chabrerie (1981a), "Magnetohydrodynamic Taylor vortex flow under a transverse pressure gradient," Phys. Fluids **24**, 406-12.

Tabeling, P. and J. P. Chabrerie (1981b), "Magnetohydrodynamic secondary flows at high Hartmann numbers," J. Fluid Mech. **103**, 225-239.

Tabeling, P. and I. Marsan (1980), "Magnetohydrodynamic supercritical Taylor-Couette flows," Phys. Lett. A **75A**, 217-19.

Tabeling, P. and C. Trakas (1984a), "Spiral vortices in a Taylor instability subjected to an external magnetic field," J. Phys. (Paris), Lett. **45**, L159-67.

Tabeling, P. and C. Trakas (1984b), "Structures spiralees dans une instabilite de Taylor en presence de champ magnetique," J. Phys. (Paris), Lett. .**45**, L159-L167.

Tabeling, P. and C. Trakas (1984c), "Temporal and spatial aspects of the onset of chaos in a Taylor instability subjected to a magnetic field," in Cellular Structures in Instabilities, edited by J. E. Wesfried and S. Zaleski, Lect. Notes in Physics, vol.210 (Springer, Berlin), pp. 285-293.

Tachibana, F. and S. Fukui (1964), Bull. JSME **7**.

Tachibana, F., S. Fukui, and H. Mitsumura (1959), Trans. JSME **25**, 788.

Tachibana, F., S. Fukui, and H. Mitsumura (1963), Trans. JSME **29**, 1366.

Tagg, R. (1986), "Multicritical points in flow between independently rotating cylinders," in Multiparameter Bifurcation Theory, edited by M. Golubitsky and J. Guckenheimer, Contemporary Mathematics 56 (American Math. Society, Providence, RI), pp. 383-387.

Tagg, R., W. S. Edwards, and H. L. Swinney (1990), "Convective versus absolute instability in flow between counterrotating cylinders," Phys. Rev. A **42**, 831-837.

Tagg, R., W. S. Edwards, H. L. Swinney, and P. S. Marcus (1989), "Nonlinear standing waves in Couette-Taylor flow," Phys. Rev. A **39**, 3734-7.

Takeda, Y. (1986a), "Velocity profile measurement by ultrasound Doppler shift method," Int. J. Heat Fluid FLow **7**, 313-18.

Takeda, Y. (1986b), "Velocity profile measurement by ultrasound Doppler shift method," Fluid Control and Measurement **2**, 851-6.

Takeda, Y. (1987), "Measurement of velocity profile of mercury flow by ultrasound Doppler shift method," Nuclear Technology **79**.

Takeda, Y. (1989), "Development of ultrasound velocity profile monitor and its experience," in Fourth International Topical Meeting on Nuclear Reactor Thermal-Hydraulics: Nureth-4, pp. 418-423.

Takeda, Y. (1991), "Development of an ultrasound velocity profile monitor," Nuclear Engineering & Design **126**, 277-284.

Takeda, Y. and W. E. Fischer (1992), "Spatial characteristics of dynamic properties of modulated wavy vortex flow in a rotating Couette system," Exp. Fluids (to appear).

Takeda, Y. and M. Haefeli (1992), "Evaluation of shape reproducibility in an instantaneous velocity profile measurement," in Proceedings of the Second World Conference on Experimental Heat Transfer, Fluid Mechanics and Thermodynamics dynamics (Elsevier).

Takeda, Y., K. Kobashi, and W. E. Fischer (1990), "Observation of the transient behaviour of Taylor vortex flow between rotating concentric cylinders after sudden start," Exp. Fluids 9, 317.

Takens, F. (1983), "Distinguishing deterministic and random systems," in Nonlinear Dynamics and Turbulence, edited by G. I. Barenblatt, G. Iooss, and D. D. Joseph (Pitman, Boston, London, Melbourne), pp. 314-333.

Takeuchi, D. I. and D. F. Jankowski (1981), "A numerical and experimental investigation of the stability of spiral Poiseuille flow," J. Fluid Mech. **102**, 101-26.

Takhar, H. S., M. A. Ali, and V. M. Soundalgekar (1988a), "Effect of radial temperature gradient on the stability of flow in an annulus with constant heat flux at the inner cylinder: wide-gap problem," J. Franklin Inst. **325**, 609-19.

Takhar, H. S., M. A. Ali, and V. M. Soundalgekar (1988b), "Effects of radial temperature gradient on the stability of flow in a narrow-gap annulus with constant heat flux at the inner rotating cylinder," Waerme- & Stoffuebertrag. **22**, 23-8.

Takhar, H. S., M. A. Ali, and V. M. Soundalgekar (1989a), "Stability of MHD Couette flow in a narrow gap annulus," Appl. Sci. Res. **46**, 1-24.

Takhar, H. S., M. A. Ali, and V. M. Soundalgekar (1989b), "Stability of the flow between rotating cylinders-wide-gap problem," J. Fluids Eng. **111**, 97-9.

Takhar, H. S., T. J. Smith, and V. M. Soundalgekar (1985), "Effects of radial temperature gradient on the stability of flow of a viscous incompressible fluid between two rotating cylinders," J. Math. Anal. Appl. **111**, 349-52.

Takhar, H. S., V. M. Soundalgekar, and M. A. Ali (1989), "Some cell patterns in a wide-gap Taylor stability problem," Phys. Lett. A **137**, 389-92.

Tal thau, R. and B. Gal-or (1979), "Heat and mass transfer in reacting flow systems: an approximate theory," Int. J. Heat Mass Transf. **22**, 557-63.

Tam, W. Y. and H. L. Swinney (1987), "Mass transport in turbulent Couette-Taylor flow," Phys. Rev. A **36**, 1374-81.

Tavener, S. and A. Cliffe (1987), "Primary flow exchange mechanisms in Taylor-Couette flow applying non-flux boundary conditions," (preprint).

Tavener, S., T. Mullin, and A. Cliffe (1987), "Bifurcation in Taylor-Couette flow with rotating ends," (preprint).

Tavener, S. J., T. Mullin, and K. A. Cliffe (1991), "Novel bifurcation phenomena in a rotating annulus," J. Fluid Mech. **229**, 483-497.

Taylor, G. I. (1923), "Stability of a viscous liquid contained between two rotating cylinders," Phil. Trans. Roy. Soc. London, Ser. A **223**, 289-343.

Taylor, G. I. (1935), "Distribution of velocity and temperature between concentric rotating cylinders," Proc. R. Soc. London, Ser. A, 494-512.

Taylor, G. I. (1936a), "Fluid friction between rotating cylinders I. - torque measurements," Proc. R. Soc. London, Ser. A **157**, 546-564.

Taylor, G. I. (1936b), "Fluid friction between rotating cylinders II. - distribution of velocity between concentric cylinders when outer one is rotating and inner one is at rest," Proc. R. Soc. London, Ser. A **157**, 565-578.

Tieu, H. A., D. D. Joseph, and G. S. Beavers (1984), "Interfacial shapes between two superimposed rotating simple fluids," J. Fluid Mech. **145**, 11-70.

Thomas, R. H. and K. Walters (1964a), "The stability of elastico-viscous fluid between rotating cylinders. Part 1," J. Fluid Mech. **18**, 33-43.

Thomas, R. H. and K. Walters (1964b), "The stability of elastico-viscous fluid between rotating cylinders. Part 2," J. Fluid Mech. **19**, 557-560.

Thompson, J. M. T. and H. B. Stewart (1986), <u>Nonlinear Dynamics and Chaos</u> (Wiley).

Thual, O. and S. Fauve (1988), "Localized structures generated by subcritical instabilities," J. Phys. (Paris) **49**, 1829-1833.

Tillmann, W. (1961a), "Zur Turbulenzentstehung bei der Stromung zwischen rotierenden Zylindern," Z. Angew. Phys. **13.**, 468-475.

Tillmann, W. (1961b), "Zum Reibungsmoment der turbulenten Stromung zwischen rotierenden Zylindern," Forsch. Ingenieurwes. **27**, 189-194.

Tillmann, W. (1962), "Zum erloschen der Turbulenz unter dem Einfluss einer stabilen Zentrifugalkraftschichtung," in <u>Miszellaneen der Angewandten Mechanik: Festschrift Walter Tollmien</u>, edited by M. Schafer (Academie-Verlag, Berlin), pp. 316-319.

Tillmann, W. and H. Schlieper (1979), "A device for the calibration of hot-film wall shear probes in liquids," J. Phys. E **12**, 373-80.

Tillmann, W., M. Waschmann, M. Herold, and G. Haussinger (1981), "Hot-film wall shear probe for measurements at flexible walls," J. Phys. E **14**, 692-4.

Tomita, Y. and T. Jotaki (1978), "Elongational viscosity and stability of rotating Couette flow," Bull. JSME **21**, 1264-7.

Tong-Kun Lim, Un-Yob Shim, and Seong-Oon Choi (1987), "The wavenumber spectrum of Couette flow at laminar to turbulent transition point," New Phys. (Korean Phys. Soc.) **27**, 251-6.

Torrest, R. S. (1982), "Rheological properties of aqueous solutions of the polymer natrosol 250 hhr," J. Rheol. **26**, 143-51.

Townsend, A. A. (1984), "Axisymmetric Couette flow at large Taylor numbers," J. Fluid Mech. **144**, 329-62.

Toy, M. L., L. E. Scriven, and C. W. Macosko (1991), "Nonhomogeneities in Couette flow of ferrite suspensions," J. Rheol. **35**, 887.

Tritton, D. J. (1988), <u>Physical Fluid Dynamics</u>, 2nd ed. (Oxford University Press, Oxford).

Trouilhet, Y. and F. Widmer (1978), "Measurement error correction in a Couette viscosimeter caused by dissipated heat and non-newtonian effects," Tech. Mess. - Atm **45**, 167-71.

Tsameret, A. and V. Steinberg (1991), "Convective vs absolute instability in Couette-Taylor flow with an axial flow," Europhys. Lett. **14**, 331-36.

Tsiveriotis, K. and R. A. Brown (1989), "Bifurcation structure and the Eckhaus instability," Phys. Rev. Lett. **63**, 2048-2051.

Tuckerman, L. and D. Barkley (1990), "Bifurcation analysis for the Eckhaus instability," Physica D **46**, 57-86.

Tuckerman, L. S. and P. S. Marcus (1985), "Formation of Taylor vortices in spherical Couette flow," in Proc. of the Ninth International Conference on Numerical Mehtods in Fluid Dynamics, edited by Soubbarameyer and J. P. Boukot, Lecture Notes in Physics, vol. 218 (Springer-Verlag, Berlin).

Turchi, P. J., D. L. Book, and R. L. Burton (1979), "Optimization of stabilized imploding liner fusion reactors," in Fusion Technology 1978 (Pergamon, Oxford, England), pp. 121-5.

Tustaniwskyj, J. I. and S. Carmi (1980), "Nonlinear stability of modulated finite gap Taylor flow," Phys. Fluids **23**, 1732-9.

Tuszynski, J. A. and M. Otwinowski (1990), "Pattern formation and pattern selection in the Landau-Ginsburg model of critical phenomena," Can. J. Phys. **68**, 760.

Twineham, M., M. Hoare, and D. J. Bell (1984), "The effects of protein concentration on the break-up of protein precipitate by exposure to shear," Chem. Eng. Sci. **39**, 509-13.

Ueno, S. (1991), "Stochastic approach to the Cauchy problem of linearized Couette flow," Applied mathematics and computation **45**, 207.

Umegaki, K. and S. Uchikawa (1985), "Numerical simulation of two-dimensional incompressible viscous flow driven by rotating boundaries using boundary-fitted coordinate system," in Numerical Methods in Laminar and Turbulent Flow. Proceedings of the Fourth International Conference, vol.2, edited by C. Taylor, M. D. Olson, P. M. Gresho, and W. G. Habashi (Pineridge Press, Swansea, Wales), pp. 1735-46.

Urintsev, A. L. (1976), "Calculation of autooscillations of the type of azimuthal waves occurring at loss of stability of flow of a viscous fluid between concentric cylinders rotating in opposite directions," Zh. Prikl. Mekh. Tekh. Fiz. **17**, 68-75 [J. Appl. Mech. Tech. Phys. **17**, 198-203 (1976)].

Usui, T. (1983), "Hydrodynamic stability of superfluid helium between coaxially rotating cylinders," Prog. Theor. Phys. **70**, 1454-6.

Valyaev, N. I., E. P. Olofinskii, B. S. Stepin, and I. A. Chinenkov (1978), "Application of the Reichardt dependence to calculation of pressurized turbulent Couette flow in a circular annular slot," Izv. Akad. Nauk SSSR, Energ. Transp. **16**, 167-70 [Power Eng. (USSR) **16**, 151-4 (1978)].

Van Atta, C. (1966a), "Exploratory measurements in spiral turbulence," J. Fluid Mech. **25**, 495-512.

Van Atta, C. (1966b), "Measured distortion of a laminar circular Couette flow by end effects," J. Fluid Mech. **25**, 513-521.

Van Buskirk, R. (1986), "Non-hysteretic non-intermittent transitions directly to chaos," Preprint.

van Houten, H. and J. J. M. Beenakker (1985), "Flow birefringence in gases at room temperature: new absolute values," Physica A **130A**, 23.

van Saarloos, W. (1987), "Dynamical velocity selection: marginal stability," Phys. Rev. lett. **58**, 2571-2574.

van Saarloos, W. (1988), Phys. Rev. A **37**, 211.

van Saarloos, W. (1989), "Front propagation into unstable states. II. Linear versus nonlinear marginal stability and rate of convergence," Phys. Rev. A **39**, 6367.

van Saarloos, W. and P. C. Hohenberg (1990), "Pulses and fronts in the complexf Ginzburg-Landau equation near a subcritical bifurcation," Phys. Rev. Lett. **64**, 749.

Vasilenko, Yu. G., E. A. Kuznetsov, V. S. L'vov, Yu. E. Nesterikhin, V. S. Sobolev, M. D. Spektor, S. A. Timokhin, E. N. Utkin, and N. F. Shmoilov (1980), "Generation of Taylor vortices in Couette flow," Zh. Prikl. Mekh. Tekh. Fiz. **21** [J. Appl. Mech. Tech. Phys. **21**, 206-211 (1980)].

Vastano, J. A. and R. D. Moser (1991), "Short-time Lyapunov exponent analysis and the transition to chaos in Taylor-Couette flow," J. Fluid Mech. **233**, 83.

Vastano, J. A. and H. L. Swinney (1988), "Information transport in spatiotemporal systems," Phys. Rev. Lett. **60**, 1773-1776.

Vehrenkamp, R., K. Schatzel, G. Pfister, B. S. Fedders, and E. O. Schulz-Dubois (1979), "A comparison between analog LDA, photon correlation LDA, and rate correlation techniques," Phys. Scr. **19**, 379-382.

Vehrenkamp, R., K. Schatzel, G. Pfister, and E. O. Schulz-Dubois (1979), "Direct measurement of velocity correlation functions using the Erdmann-Gellert rate correlation technique," J. Phys. E **12**, 119-25.

Velte, W. (1964), "Stabilitatsverhalten und Verzweigung stationarer Losungene der Navier-Stokes-schen Gleichungen," Arch. Ration. Mech. Anal. **16**, 97-125.

Velte, W. (1966), "Stabilitat und Verzweigung stationarer Losungen der Navier-Stokesschen Glecihungen beim Taylor-Problem," Arch. Ration. Mech. Anal. **22**, 1-14.

Vera, M. and J. B. Grutzner (1986), "The Taylor vortex: the measurement of viscosity in NMR samples," J. Am. Chem. Soc. **108**, 1304-1306.

Verma, P. D., R. C. Chaudhary, and S. C. Rajvanshi (1978), "On the flow of a visco-elastic fluid between two rotating non-concentric cylinders for small eccentricity," Def. Sci. J. **28**, 109-16.

Verma, P. D. S., D. V. Singh, and K. Singh (1980), "Couette flow of micro-thermopolar fluids between concentric rotating cylinders," Acta Tech. Csav **25**, 369-83.

Versteegen, P. L. and D. F. Jankowski (1969), Phys. Fluids **12**, 1138.

Vid'machenko, A. P. (1986), "Some dynamical parameters of Jupiter's atmosphere," Kinematika Fiz. Nebesnykh Tel **2**, 48-51.

Vidyanidhi, V., G. Krishnam raju, and V. V. Ramana rao (1980), "Magnetohydrodynamic Couette flow and heat transfer in a rotating system," Def. Sci. J. **30**, 143-8.

Vilgelmi, T. A. and V. N. Shtern (1974), "Stability of spiral flow in an annulus," Izv. Akad. Nauk SSSR, Mekh. Zhidk. Gaza **3**, 35-44 [Fluid Dyn. (USSR) **3** (1974)].

Vislovich, A. N., V. A. Novikov, and A. K. Sinitsyn (1986), "Influence of a magnetic field on the Taylor instability in magnetic fluids," Zh. Prikl. Mekh. Tekh. Fiz. **27**, 79-86 [J. Appl. Mech. Tech. Phys. **27**, 72-8 (1986)].

Vives, C. (1988), "Effects of a forced Couette flow during the controlled solidification of a pure metal," Int. J. Heat Mass Transf. **31**, 2047-62.

Vohr, J. H. (1968), "An experimental study of Taylor vortices and turbulence in flow between eccentric rotating cylinders," J. Lubr. Technol. **90**, 285-296.

Voisin, P., C. Guimont, and J. F. Stoltz (1985), "Experimental investigation of the rheological activation of blood platelets," Biorheology **22**, 425-35.

von Karman, T. (1934), "Some aspects of the Turbulence problem," in Proc. 4th Int. Cong. Appl. Mech., pp. 54-91.

Vysotskii, V. V. and S. P. Bakanov (1990), "Mechanodiffusion of a moderately low-density binary gas mixture under Couette flow conditions," Sov. Phys. JETP **71**, 697.

Wada, K. and H. Hayafuji (1986a), "Some characteristics of axial laminar Couette flow contained between circular double tubes. III. Results of sample calculation," Rep. Fac. Sci. Technol. Meijo Univ., 59-64.

Wada, K. and H. Hayafuji (1986b), "Some characteristics of axial laminar Couette flow contained between circular double tubes. I. Theoretical analysis for eccentric circular double tubes model," Rep. Fac. Sci. Technol. Meijo Univ., 45-53.

Wada, K. and H. Hayafuji (1987), "Some characteristics of axial Couette flow contained between circular crown section double tubes. I. Theoretical analysis," Rep. Fac. Sci. Technol. Meijo Univ., 39-45.

Wagner, E. M. (1932), Ph.D. Thesis, Stanford Univ..

Wagner, N. J., G. G. Fuller, and W. B. Russel (1988), "The dichroism and birefringence of a hard-sphere suspension under shear," J. Chem. Phys. **89**, 1580-7.

Walden, R. W. (1978), "Transition to turbulence in Couette flow between concentric cylinders," Ph.D. Thesis, Univ. of Oregon.

Walden, R. W. ?., "Experiments in non-linear circular Couette flow," Preprint.

Walden, R. W. and R. J. Donnelly (1979), "Reemergent order of chaotic circular Couette flow," Phys. Rev. Lett. **42**, 301-4.

Walgraef, D. (1986), "End effects and phase instabilities in a model for Taylor-Couette systems," Phys. Rev. A **34**, 3270-8.

Walgraef, D., "Drift flow and phase instabilities in a model for Taylor-Couette systems," (preprint).

Walgraef, D., P. Borckmans, and G. Dewel (1982), "Fluctuation effects in the transition to Taylor vortex flow in finite geometries," Phys. Rev. A **25**, 2860-2.

Walgraef, D., P. Borckmans, and G. Dewel (1984), "Onset of wavy Taylor vortex flow in finite geometries," Phys. Rev. A **29**, 1514-1519.

Wallace, D. J., C. Moreland, and J. J. C. Picot (1985), "Shear dependent of thermal conductivity in polyethylene melts," Polym. Eng. Sci. **25**, 70-4.

Walowitt, J., S. Tsao, and R. C. DiPrima (1964), "Stability of flow between arbitrarily spaced concentric cylindrical surfaces including the effect of a radial temperature gradient," J. Appl. Mech. **31**, 585-593.

Walsh, T. J. and R. J. Donnelly (1988), "Taylor-Couette flow with periodically corotated and counterrotated cylinders," Phys. Rev. Lett. **60**, 700-3.

Walsh, T. J., W. T. Wagner, and R. J. Donnelly (1987), "Stability of modulated Couette flow," Phys. Rev. Lett. **58**, 2543-6.

Walters, K. (1970), "Rheometrical flow systems. Part 1. Flow between concentric spheres rotating about different axes," J.Fluid Mech. **40**, 191-203.

Walton, I. C. (1978), "The linear stability of the flow in a narrow spherical annulus," J. Fluid Mech. **86**, 673-93.

Walton, I. C. (1980), "The transition to Taylor vortices in a closed rapidly rotating cylindrical annulus," Proc. R. Soc. London, Ser. A **372**, 201-218.

Wan, C. C. and J. E. R. Coney (1980), "Transition modes in adiabatic spiral vortex flow in narrow and wide annular gaps," Int. J. Heat Fluid FLow **2**, 131-8.

Wang, P. K. C. (1985), "Feedback control of onset of turbulence in hydrodynamic systems," in Proceedings of the 24th IEEE Conference on Decision and Control, vol. 2 (IEEE, New York), pp. 1163-7.

Watson, J. (1960), "On the nonlinear mechanics of wave disturbances in stable and unstable parallel flows Part 2. The development of a solution for plane Poiseuille flow and for plane Couette flow," J. Fluid Mech. **9**, 371-389.

Weidman, P. D. (1989), "Measurement techniques in laboratory rotating flows," in Advances in Fluid Mechanics Measurements, edited by M. Gad-el-Hak (Springer-Verlag, Berlin), pp. 401-534.

Weidman, P. D. and G. Mehrdadtehranfar (1985), "Instability of natural convection in a tall vertical annulus," Phys. Fluids **28**, 776-787.

Weinstein, M. (1977a), "Wavy vortices in the flow between two long eccentric rotating cylinders. I. Linear theory," Proc. R. Soc. London, Ser. A **354**, 441-57.

Weinstein, M. (1977b), "Wavy vortices in the flow between two long eccentric rotating cylinders. II. Nonlinear theory," Proc. R. Soc. London, Ser. A **354**, 459-89.

Weinstein, M. (1990), "An analytical solution of the basic Couette flow affected by a three-dimensional magnetic field," J. Franklin Inst. **327**, 13.

Weisshaar, E., F. H. Busse, and M. Nagata (1991), "Twist vortices and their instabilities in the Taylor-Couette system," J. Fluid Mech. **226**, 549.

Weitz, D. A., W. D. Dozier, and P. M. Chaikin (1985), "Periodic structures in driven colloidal crystals," J. Phys. (Paris) Colloq. **46**, 257-68.

Wendt, F. (1933), "Turbulente Stromungen zwischen zwei rotierenden konaxialen Zylindern," Ing. Arch. **4**, 577.

Wesfreid, J. E., H. R. Brand, P. Manneville, G. Albinet, and N. Boccara (1988), Eds., Propagation in Systems Far from Equilibrium (Springer-Verlag, New York).

Wesfreid, J. E. and S. Zaleski (1984), "Cellular structures in instabilities: an introduction," in Cellular Structures in Instabilities, edited by J. E. Wesfreid and S. Zaleski, Lecture Notes in Physics, v. 210 (Springer-Verlag, Berlin), pp. 1-32.

White, F. M. (1990), Viscous Fluid Flow, 2nd ed. (McGraw-Hill, New York).

Whitham, G. B. (1974), Linear and nonlinear waves (Wiley, New York).

Wichterle, K. and P. Mitschka (1986), "A novel approach to high shear rate rheometry," Rheol. Acta **25**, 331-4.

Wiener, R. J. (1991), "Stability of Taylor-Couette flow subjected to a Coriolis force," Ph.D. Thesis, Univ. of Oregon.

Wiener, R. J., P. W. Hammer, C. E. Swanson, and R. J. Donnelly (1990), "Stability of Taylor-Couette flow subject to an external Coriolis force," Phys. Rev. Lett. **64**, 1115-1118.

Wiener, R. J., P. W. Hammer, C. E. Swanson, D. C. Samuels, and R. J. Donnelly (1991), "Effect of a Coriolis force on Taylor-Couette flow," J. Stat. Phys. **64**, 913.

Wiener, R. J., P. W. Hammer, and R. Tagg (1991), "Perturbation analysis of the primary instability in Taylor-Couette flow subjected to a Coriolis force," Phys. Rev. A **44**, 3653.

Wiener, R. J., P. W. Hammer, and R. Tagg, "Instability of Taylor-Couette flow subjected to a Coriolis force," in Ordered and Turbulent Patterns in Taylor-Couette Flow, edited by C. D. Andereck and F. Hayot, Proceedings of a NATO Advanced Research Workshop (Plenum).

Wilks, G. and D. M. Sloan (1976), "Invariant imbedding, Riccati transformations and eigenvalue problems," J. Inst. Math. Its Appl. **18**, 99-116.

Wilson, G. (1988), "A time-dependent nodal-integral method for the investigation of bifurcation and nonlinear phenomena in fluid flow and natural convection," Nucl. Sci. Eng. **100**, 414.

Wimmer, M., "Experimentelle Untersuchungen der Stromung im Splat zwischen zwei konzentrischen Kugeln, die beide um einen gemeinsamen Durchmesser rotieren," Ph.D. Thesis, Universitat (T.H.) Karlsruhe.

Wimmer, M. (1976), "Experiments on a viscous fluid flow between concentric rotating shperes," J. Fluid Mech. **78**, 317.

Wimmer, M. (1977), "Experimentelle Untersuchung der Stromung zwischen konzentrischen, rotierenden Kugeln," Z. Angew. Math. Mech. **57**, T218.

Wimmer, M. (1981), "Experiments on the stability of viscous flow between two concentric rotating spheres," J. Fluid Mech. **103**, 117-31.

Wimmer, M. (1983), Z. Angew. Math. Mech. **63**, T299.

Wimmer, M. (1985), "Effects of geometrical shape on a Taylor vortex," Z. Angew. Math. Mech. **65**, T255-6.

Wimmer, M. (1988), Prog. Aeronaut. Sci..

Winters, K. H. (1987), "A bifurcation study of laminar flow in a curved tube of rectangular cross-section," J. Fluid Mech. **180**, 343-369.

Withjack, E. M. and C. F. Chen (1974), "An experimental study of Couette instability of stratified fluids," J. Fluid Mech. **66**, 725-737.

Withjack, E. M. and C. F. Chen (1975), "Stability analysis of rotational Couette flow of stratified fluids," J. Fluid Mech. **68**, 157-175.

Wolf, A., J. B. Swift, H. L. Swinney, and J. A. Vastano (1985), "Determining Lyapunov exponents from a time series," Physica D **16D**, 285-317.

Wolf, P. E. and B. Perrin (1979), "Taylor vortex instability in Couette flow of normal helium," Phys. Lett. A **73A**, 324-5.

Wolf, P. E., B. Perrin, J. P. Hulin, and P. Elleaume (1981), "Rotating Couette flow of helium II," J. Low Temp. Phys. **44**, 569-93.

Wortmann, F. X. (1981), "Boundary-layer waves and transition," in Advances In Fluid Mechanics. Proceedings Of A Conference, edited by Krause, (Springer-Verlag, Berlin), pp. 268-79.

Wronski, S. and M. Jastrzebski (1990a), "The stability of the helical flow of pseudoplastic liquids in a narrow annular gap with a rotating inner cylinder," Rheol. Acta **29**, 442.

Wronski, S. and M. Jastrzebski (1990b), "Experimental investigations of the stability limit of the helical flow of pseudoplastic liquids," Rheologica acta **29**, 453.

Wu, J. C. and M. M. Wahbah (1976), "Numerical solution of viscous flow equations using integral representations," in Proceedings of the 5th International Conference On Numerical Methods In Fluid (Springer-Verlag, Berlin), pp. 448-53.

Wu, M. and C. D. Andereck (1991a), "Phase modulation of wavy vortex flow," Phys. Rev. A **43**, 2074.

Wu, M. and C. D. Andereck (1991b), "Phase dynamics of wavy vortex flow," Phys. Rev. Lett. **67**, 1258.

Wu, X. and J. B. Swift (1989), "Onset of secondary flow in the modulated Taylor-Couette system," Phys. Rev. A **40**, 7197-7201.

Wunderlich, A. M. and P. O. Brunn (1989), "The complex rheological behavior of an aqueous cationic surfactant solution investigated in a couette-type viscometer," Colloid Polym. Sci. **267**, 627.

Yahata, H. (1977a), "Slowly-varying amplitude of the Taylor vortices near the instability point," Prog. Theor. Phys. **57**, 347-60.

Yahata, H. (1977b), "Slowly-varying amplitude of the Taylor vortices near the instability point. II. Mode-coupling-theoretical approach," Prog. Theor. Phys. **57**, 1490-6.

Yahata, H. (1978a), "Dynamics of the Taylor vortices above higher instability points," Prog. Theor. Phys. **59**, 1755-6.

Yahata, H. (1978b), "Temporal development of the Taylor vortices in a rotating fluid," Prog. Theor. Phys. Suppl., 176-85.

Yahata, H. (1979), "Temporal development of the Taylor vortices in a rotating fluid. II," Prog. Theor. Phys. **61**, 791-800.

Yahata, H. (1980a), "A simple model for the Taylor vortex flow," Prog. Theor. Phys. Suppl., 200-11.

Yahata, H. (1980b), "Temporal development of the Taylor vortices in a rotating fluid. III," Prog. Theor. Phys. **64**, 782-93.

Yahata, H. (1981), "Temporal development of the Taylor vortices in a rotating fluid. IV," Prog. Theor. Phys. **66**, 879-91.

Yahata, H. (1982), "Transition to turbulence in the Taylor-Covette and Rayleigh-Benard system," in Us-Japan Joint Institute For Fusion Theory Workshop On 'Nonequilibrium Statistical Physics Problems In Fusion Plasmas - Stochasticity And Chaos -' (IPPJ 578) (Inst. Plasma Phys., Nagoya, Japan), pp. 125-30.

Yahata, H. (1983), "Temporal development of the Taylor vortices in a rotating fluid. V," Prog. Theor. Phys. **69**, 396-402.

Yahata, H. (1984), "Transition to turbulence in the Taylor-Couette flow and the Rayleigh-Benard convection," in Turbulence And Chaotic Phenomena In Fluids. Proceedings Of The International Symposium (IUTAM), edited by T. Tatsumi (North Holland, Amsterdam, Netherlands), pp. 209-14.

Yakhot, V. (1988), "Reynolds number scaling of turbulent diffusivity in wall flows," Phys. Fluids **31**, 709-10.

Yamada, T. and H. Jujisaka (1977), "A discrete model exhibiting successive bifurcations leading to the onset of turbulence," Z. Phys. B **28**, 239-46.

Yamada, Y. and S. Imao (1986), "Flow of a fluid contained between concentric cylinders, both rotating," Bull. JSME **29**, 1691-7.

Yamada, Y. and M. Ito (1976), "On the viscous frictional resistance of enclosed rotating cones (2nd report, effects of surface roughness)," Bull. JSME **19-134**, 943.

Yamada, Y., K. Nakabayashi, and Y. Suzuki (1969), "Viscous frictional moment between eccentric rotating cylinders when the outer cylinder rotates," Bull. JSME **12**, 1024.

Yamaguchi, H. (1988), "Spherical Couette flow of magnetic fluid," Sci. Eng. Rev. Doshisha Univ. **29**, 15-21.

Yamaguchi, H. and I. Kobori (1990), " Spherical Couette flow in magnetic fluid," Magnetohydrodynamics **26**, 284.

Yanase, S. (1991), "Time dependent analysis of bifurcating flows through a curved tube," Eur. J. Mech. B/Fluids **10 (suppl.)**, 321.

Yardin, G. and H. J. Meiselman (1989), "Effects of cellular morphology on the viscoelastic behavior of high hematocrit RBC suspensions," Biorheology **26**, 153-75.

Yang, R. -J. (1989), "Numerical study of spherical Taylor-Couette flow," in <u>Advances in Fluid Dynamics: Proceedings of a Symposium in Honor of Maurice Holt on his 75th Birthday</u>, edited by W. F. Ballhau and M. Y. Hussaini (Springer, New York).

Yavorskaya, I. M., N. M. Astaf'eva, and N. D. Vvedenskaya (1978), "Stability and nonuniqueness of liquid flow in rotating spherical layers," Dok. Akad. Nauk SSSR **241**, 52-5 [Sov. Phys. Dokl. **23**, 461-3 (1978)].

Yavorskaya, I. M. and Yu. N. Belyaev (1991), "Nonuniqueness and multiparametric study of transition to chaso on the spherical Couette flow," Eur. J. Mech. B/Fluids **10**, 267-274.

Yavorskaya, I. M., Yu. N. Belyaev, and A. A. Monakhov (1977), "Stability investigation and secondary flows in rotating spherical layers at arbitrary Rossby numbers," Dok. Akad. Nauk SSSR **237**, 804-7 [Sov. Phys. Dokl. **22**, 717-19 (1977)].

Yavorskaya, I. M., yu. N. Belyayev, and A. A. Monakhov (1978), "Experimental study of loss of stability in spherical Couette flows," Fluid Mech. - Sov. Res. **7**, 56-66.

Yih, C. S. (1972a), "Spectral theory of Taylor vortices. Part I," Arch. Ration. Mech. Anal. **46**, 218-240.

Yih, C. S. (1972b), "Spectral theory of Taylor vortices. Part II: proof of nonoscillation," Arch. Ration. Mech. Anal. **47**, 288-300.

Yoshimura, A. and P. K. Prud'homme (1988), "Wall slip corrections for Couette and parallel disk viscometers," J. Rheol. **32**, 53-67.

Yowakim, F. M. (1988), "Mean flow and turbulence measurements of annular swirling flows," J. Fluids Eng. **110**, 257.

Yu, X. and J. T. C. Liu (1991), "The secondary instability in Görtler flow," Phys. Fluids A **3**, 1845.

Yudovich, V. I. (1966a), "Secondary flows and fluid instability between rotating cylinders," Prikl. Mat. Mek. **30**, 688-698 [J. appl. Math. Mech.].

Yudovich, V. I. (1966b), "Secondary flows and fluid instability between rotating cylinders," J. Appl. Math **22**, 822-833.

Yudovich, V. I. (1966c), "The bifurcation of a rotating flow of a liquid," Sov. Phys. Dokl. **11**, 566-568.

Zaleski, S. (1984), "Cellular patterns with boundary forcing," J. Fluid Mech. **149**, 101-125.

Tabeling, P. (1984), "Wavelength selection through boundaries in 1-D cellular structures," in <u>Cellular Structures in Instabilities: Lect. Notes in Physics, vol.210</u>, edited by J. E. Wesfreid and S. Zaleski (Springer, Berlin), pp. 84-103.

Zaleski, S., P. Tabeling, and P. Lallemand (1985), "Flow structures and wave-number selection in spiraling vortex flows," Phys. Rev. A **32**, 655-658.

Zarti, A. S. and F. R. Mobbs, "Wavy Taylor vortex flow between eccentric rotating cylinders," in <u>Proceedings of ASME Winter Annual Meeting ??</u>, pp. 103-116.

Zeldovich, Y. B. (1981), "On the friction of fluids between rotating cylinders," Proc. R. Soc. London, Ser. A **374**, 299-312.

Zhang, L. and H. L. Swinney (1985), "Nonpropagating oscillatory modes in couette-taylor flow," Phys. Rev. A **31**, 1006-9.

Zhang, L., L. A. Reith, and H. L. Swinney, "Waves on turbulent Taylor vortices," Preprint.

Zhang Tianling and Sun Yisui (1988), "Measure-preserving mapping in the Couette-Taylor system," Sci. Sin. A **31**, 87-97.

Zhdan, L. A. and V. A. Samsonov (1990), "Motion of two viscous immiscible incompressible liquids in a rotating cylinder," Izv. Akad. Nauk SSSR, Mekh. Zhidk. Gaza **25** [Fluid Dyn. (USSR) **25**, 77 (1990)].

Zheleva, I. M. and V. P. Stulov (1978), "Asymptotic investigation of two-phase Couette flow," Izv. Akad. Nauk SSSR, Mekh. Zhidk. Gaza **13**, 67-73 [Fluid Dyn. (USSR) **13**, 539-44 (1978)].

Zhukov, M. Yu., V. V. Kolesov, and O. A. Tsyvenkova (1988), "Numerical analysis of the stability of nonisothermal Couette flow," Izv. Akad. Nauk SSSR, Mekh. Zhidk. Gaza **23**, 70-6 [Fluid Dyn. (USSR) **23**, 701-6 (1988)].

Zhuravel', F. A., M. S. Iskakov, S. N. Lukashchuk, and A. A. Predtechenskii (1983), "Application of multichannel systems of data gathering and processing for studying amplitude waves in Couette flow using an electrodiffusion method," Avtometriya, 69-76 [Autom. Monit. Meas.].

Zielinska, B. J. A. and Y. Demay (1988), "Couette-Taylor instability in viscoelastic fluids," Phys. Rev. A **38**, 897-903.

Zierep, J. (1978), "Instabilities in flows of viscous, heat conducting media. 21st Ludwig Prandtl Memorial Lecture, Brussels, March 28, 1978," Z. Flugwiss. Weltraumforsch. **2**, 143-50.

Zierep, J. and K. Bühler, <u>Strömungsmechanik</u>, (to appear) (Springer, Berlin).

Zierep, J. and O. Sawatzki (1970), "Das Stromfeld im Spalt zwischen zwei konzentrischen Kugelflachen von denen diw innere rotiert," Acta Mech., 13.

Zierep, J. and O. Sawatski (1970), "Three-dimensional instabilities and vortices between two rotating spheres," in Proc. Eighth Symposium on Naval Hydrodynamics, pp. 275-288.

Zimmermann, G. (1985), "Reduction to finite dimension of continuous systems having only a few amplified modes," in Flow of Real Fluids, edited by G. E. A. Meier and F. Obermeier, Lecture Notes in Physics, vol.235 (Springer-Verlag, Berlin), pp. 181-187.

Zinenko, Zh. A., A. M. Stolin, and F. A. Khrisostomov (1977), "Thermal conditions in Couette flow of a viscous liquid," Zh. Prikl. Mekh. Tekh. Fiz. **18**, 103-9 [J. Appl. Mech. Tech. Phys. **18**, 363-367 (1977)].

Zuniga, I. (1990), "Orientational instabilities in Couette flow of non-flow-aligning nematic liquid crystals," Phys. Rev. A **41**, 2050.

INDEX

Absolute instability, 188-192
Acceleration, influence on flow states, 108, 209
Analysis of data, general techniques, 22
Angular momentum, 152
Anomalous modes, 209
Attractors
  low-dimensional, 44
  strange, 22-23, 71

Base flow, three-dimensional, 123-124, 141-142, 206
Bénard problem, 182-185, 187-196
Benjamin-Feir instability, 173-175
Benjamin-Feir-Newell instability, 167-170
Bifurcation
  codimension-two, 55-56
  harmonic, 181-182
  heteroclinic, 88
  Hopf, 56, 83-84
  Hopf, weakly inverted, 167, 170
  secondary, 180-181, 184-185
  stationary, forward, 167
  stationary, weakly inverted, 167, 170
  steady, 54
  subcritical, 54
  supercritical, 174
  transcritical, 181
  wavelength-halving, 179
Blasius mean flow, 255
Boundary conditions
  ends, 3-4, 20
  no-slip, 3
Boundary layer
  free convection, 289-295
  laminar, 102, 245
  turbulent, 253, 256
Boussinesq approximation, 182, 187, 289

Centrifugal instability, 245-246, 281
Chaos, 150
  Lagrangian, 264-265
  ramp-induced, 70

Chaotic advection, 264-265
Codimension-two point, 300
Complex Ginzburg-Landau equation; *see* Ginzburg-Landau equation
Conducting cylinders, 221-222
Control of parameters, experimental, 21-22
Convection, finite amplitude, 293
Convective instability, 72, 188-192
Coriolis force
  instability induced by, 297, 299
  bifurcations, 110-112
  Taylor-Couette flow, 107
  bifurcation diagram, 132
  broken symmetry, 108, 131
  stability analysis of, 122-127, 143
  time-dependent states, 135-139
  turbulence, 111-113, 138
Correlation length, 176
Counter-rotating cylinders, 83
Curved channels, 263-266, 271, 273-279, 281

Dean instability, 94, 301
Dean number, 264, 281
Dean vortices, 263, 281
Defects, space-time, 175

Eckhaus instability, 68-72, 167, 193-194, 217, 273-280
Eckhaus stable band, 168
Eckhaus valley, 278-280
Eigenvalue, double, 180
Ekman regions, 92-95, 97
Energy balance, turbulent, 260
Envelope equation, 168-170
Exchange of stabilities, 200

Finite cylinders, 83-85
Floquet modes, 46
Fluids; *see also* Helium II
  glycerine, 18
  mercury, 13
  solvents, 19
Frequency locking, 40

355

Fronts, 93, 159, 165, 190

Ginzburg-Landau equation, 63, 85
  complex, 56, 94-97, 146, 173, 189
Görtler instability, 245
Görtler vortices, 253
  secondary instability of, 258
Grashof number, 291

Hall-Vinen-Bekarevich-Khalatnikov equations, 213
Helical vortices, 210-211
Helium II, 7-8, 10-11, 19, 213, 219
  Taylor transition in, 213-219
Herring bone streaks, 157
Hysteresis, 22, 133, 138, 215

Inertial manifold, 48
Inflectional point, 256-257
Inflectional velocity profile, 256-257
  linear stability of, 257-258
Instability
  absolute, 188-192
  Benjamin-Feir, 173-175
  Benjamin-Feir-Newell, 167-170
  centrifugal, 245-246, 298
  convective, 72, 188-192
  Coriolis, 297, 299
  Dean, 301
  Eckhaus, 167, 193-194, 295
  Görtler, 245
  MHD, 221-222
  Rayleigh-Bénard, 289
  subharmonic, 293-295
Ising model, 174

Kuramoto-Shivashinsky equation, 174

Linear stability analysis, small gaps, 52-53

Magnetic field, 7, 222, 239
  effect on Taylor transition, 239, 242
Magnetic Prandtl number, 228
MHD instability, 221-222
Mode coexistence, 39-40
Mode selection, 30-32, 39
Mushroom, 249, 255, 285

Neutral stability curve, small gaps, 52-53
Noise-sustained structure, 190-192
Numerical methods
  Adams-Bashforth-Moulten predictor-corrector, 125
  finite difference, 60, 181, 189
  Galerkin, 274, 279
  pseudospectral, 254
  spectral, 149, 174, 274, 292-293

spectral, adaptive, 43, 48-49

Orr-Sommerfeld equation, 101

Period-doubling cascade, 70
Phase dynamics, 63, 76
Phase diffusion, 63, 76
Phase equations, 68, 174, 192
  coupled, 78
Phase pinning, 192-193
Phase-slip processes, 67
Phase turbulence, 174
Plane Couette flow, 159, 161
Poiseuille velocity profile, 187, 273, 279
Porous cylinders, 197-203
Prandtl number, 228, 293
  magnetic, 228

Quantized vortex lines, 214

Radial flow, superimposed, 197-203
Radial temperature gradient, 222
Ramps, spatial, 131, 134
  corners, 69
  parabolic, 70
Ramping rate, 22; *see also* Acceleration
Rayleigh circulation criterion, 300
Rayleigh discriminant, 300
Reaction-diffusion model, 68
Recirculation eddies, 104-106
Re-emergent order, 114
Relaminarization, 299-301
Reynolds stress, 160
Ribbons, 84
Roll coalescence, 290; *see also* Vortex merging
Rossby number, 298
Rotating channels, 273, 297
Rotating cones, 205-211
Rotating cylinder flows
  Newton, 2
  Stokes, 2

Second sound, 19, 215
Small gap approximation, 51
Spiral turbulence, 160
Spectral degrees of freedom, 109-110
Spectral number distribution, 109-110
Symmetry, broken, 108, 182-184
Symmetries, system, 84

Taylor number, local, 207-208
Taylor vortices, 3, 12, 15-16, 29-30, 45, 67-68, 71-72, 76, 198-203, 205, 273-274
  between cones and cylinders, 205
  linear stability of, 199-202

with throughflow, 198-203
Taylor-Dean system, 91-92
  free surfaces in, 100
  fronts in, 93
  hysteresis, 95-96
  mode interactions, 95-96
  recirculation eddies, 104-106
    oscillatory modes of, 105
  subcritical regime, 94
  traveling rolls in, 92-94
Throughflow, 187-196
Torque, 152-153
Torque measurements, 5, 7-13
  effect of turbulence on, 5
Transition, subcritical, 164
Transition to turbulence, 149
  direct, 115
  in Görtler problem, 253, 256, 261
Traveling rolls, 188, 190-191, 195
Traveling vortices, 208
Tricritical point, 167-168
Turbulence
  Coriolis force induced, 138
  in curved channels, 282
  intermittent, 159
  phase, 174
  small scale, 152
  spiral, 160
  weak, 155
Turbulent spots, 160, 163-164

Viscometer, rotating
  Couette, 5-6
  Mallock, 3-5
  Margules, 3
Viscometer, torsion
  Bearden, 9
  Donnelly, 10-13
  Hollis Hallett, 10
  Taylor, 8-9
Visualization
  experimental
    electrochemical, 15-16, 289
    Fluorescein salt, 245, 267-268
    ink, 14-15
    Iriodin, 162
    Kalliroscope, 75, 92, 108, 131
    particles, 17-18
    Pyroceram, 162
    smoke, 282, 287
  numerical, 153-157
Vortices
  Dean, 263
  Görtler, 253
  longitudinal, 246
  modulated wavy, 22, 45
  parameter dependences of behavior

  aspect ratio Γ, 31-32, 88
  radius ratio η, 31
  wavelength λ, 33
  spiral, 83-85, 87-88
  Taylor, 3, 12, 15-16, 29-30, 45, 67-68, 71-72, 76, 198-203, 205, 273-274
    structure of, 60
    tilted, 108, 132, 142
    tilted, chaotic, 132
  turbulent modulated wavy, 150
  turbulent Taylor, 80, 152
  undulating, 287
  wavy, 33, 36-38, 45, 78, 274
    instability of, 46
    tilted, 108, 135
Vortex coalescence, 265
Vortex merging, 134, 276-280, 282-287
Vortex pair twisting, 282, 287
Vortex splitting, 276-280, 282-287

Wavenumber selection, 278-279
Waves
  finite amplitude, 160
  travelling, 44, 56, 80, 84, 135, 173
    linear stability of, 173
Weakly nonlinear stability analysis, small gaps, 53-57
Winding number, 174